# Advances in Agricultural Systems Modeling

## Transdisciplinary Research, Synthesis, and Applications

Volume 6

Lajpat R. Ahuja, Series Editor

Advances in Agricultural Systems Modeling
Transdisciplinary Research, Synthesis, and Applications
Volume 6
Lajpat R. Ahuja, Series Editor

# Synthesis and Modeling of Greenhouse Gas Emissions and Carbon Storage in Agricultural and Forest Systems to Guide Mitigation and Adaptation

Stephen J. Del Grosso, Lajpat R. Ahuja, and William J. Parton, Editors

American Society of Agronomy, Inc.
Crop Science Society of America, Inc.
Soil Science Society of America, Inc.
5585 Guilford Road, Madison, WI 53711-5801 USA

agronomy.org | soils.org | crops.org
dl.sciencesocieties.org
SocietyStore.org

ISBN: 978-0-89118-345-7 (print)
ISBN: 978-0-89118-346-4 (digital)
doi:10.2134/advagricsystmodel6
Library of Congress Control Number: 2015960316

Advances in Agricultural Systems Modeling
ISSN: 2163-2790 (online)
ISSN: 2163-2774 (print)

Cover design: Patricia Scullion

Printed in the United States of America.

The manufacturer's authorized representative according to the EU General Product Safety Regulation is Wiley-VCH GmbH, Boschstr. 12, 69469 Weinheim, Germany, e-mail: Product_Safety@wiley.com.

# Contents

# Foreword

In the early 21st century, climate change has emerged as one of the great societal challenges. Taking effective actions to address this concern is complicated by many technological hurdles and socio-economic challenges. Agriculture is a critical player in this arena, because it disproportionately influences climate change as a fraction of its contribution to the global economy. Agriculture also disproportionately suffers from climate change through the effects of weather extremes that can result in crop and animal losses. But the agricultural sector also offers some of the most promising solutions to climate change through improved production practices and efficiencies that can reduce greenhouse gas emissions, as well as through opportunities around carbon sequestration in soils.

This book addresses an important aspect of the scientific needs around this issue: the quantification of agricultural greenhouse gas emissions through the use of numerical simulation models. Why are computer models critical to this effort? First, it is practically impossible to measure greenhouse gas emissions from agricultural practices on a continuing basis, which would be required if we want to document climate benefits from changes in agriculture. Second, we need to project outcomes from different management scenarios into the future to predict potential long-term benefits. Accurate simulation models have the potential to allow us to meet these needs at low cost. This book discusses the recent advances that have been made in this area, as well as scientific issues around the effective use of such models. This helps advance our ability to address the climate change challenges of the present and the future.

The Soil Science Society of America, American Society of Agronomy, and Crop Science Society of America believe that agriculture and science are important elements of the solution to the climate challenge, and this book is an important contribution to that effort.

**Harold van Es,** 2016 President, Soil Science Society of America
**Paul Fixen,** 2016 President, American Society of Agronomy
**Michael A. Grusak,** 2016 President, Crop Science Society of America

# Preface

In the past ~130 years (1883–2012), the global surface temperature has increased by an average of about 0.85°C (IPCC, 2014). In the last 50 years, the increase has been much faster. Some precipitation patterns have also changed. The year 2004 has been the warmest year on record since the 1880. These changes are already having effects on physical (e.g., glaciers cover) and biological systems in the Northern Hemisphere (IPCC, 2007, 2014). Lobell et al. (2011) have shown that in some major cropping regions of the world, global yields of corn and wheat have declined by 3.8 and 5.5%, respectively, from 1980 to 2008 due to climate change. There is evidence to suggest that the climatic changes are very likely due to increase in greenhouse gas (GHG) emissions due to human activity (IPCC, 2007, 2014). These emissions have been increasing since the start of the Industrial Revolution in 1750, with a 70% increase between 1970 and 2004 (IPCC, 2007). The $CO_2$ emissions account for 77% of the total GHG emissions, the rest being mainly $CH_4$ and $N_2O$ gases. Agriculture directly accounts for approximately 13% of the total anthropogenic emissions, and deforestation, mainly for agriculture, contributes another ~15 to 20% of human-caused emissions. In the 21st century, further average temperature increases are estimated to range from 1.8 to 4.0°C, depending on the different GHG emission scenarios. Precipitation is likely to increase at high latitudes and decrease in most subtropical land regions of the world, which are important agricultural regions, by as much as 20%, along with changes in snow, glacier water supply, and extreme events. Cereal crop productivity may increase slightly at mid to high latitudes with temperature increases between 1 to 3°C and decrease above 3°C. At low latitudes, the cereal crop production will likely decrease in the whole range of temperature increases, 1 to 5°C. In order to minimize the harmful effects on agricultural production, we need to reduce the GHG emissions that are causing the climatic changes.

The purpose of this Volume 6 of the ASA, CSSA, and SSSA Advances in Agricultural Systems Modeling series is to document recent advances in understanding, synthesis, and modeling of GHG emissions and changes in carbon storage in agricultural and forest systems, and how changes in soil and crop management affect these processes. The scope spans from simple empirical models based on how key driving variables impact emissions and carbon stock changes to mechanistic models that represent the processes that control emissions and their interactions. Validation of the ability of different models to quantify and synthesize the experimental data from different systems, and how management practices effect emissions and carbon storage are emphasized. The validated models and decision support tools developed from these will estimate GHG emission and C changes in agricultural systems and guide strategies for mitigation and adaptations to climatic changes.

Additional motivation for this volume is to bring together scientists from different disciplines, worldwide, to encourage further transdisciplinary research in the area of GHG emissions and C changes, and its synthesis to: (i) fill knowledge gaps in our scientific understanding of processes and management effects in the current models, (ii) improve existing concepts and approaches in the models, (iii) advance their applicability to larger spatial areas, and (iv) develop system-based strategies and tools for mitigation of the GHG effects.

Based on a review of the subject matter by the organizers, some of the specific areas of knowledge gaps needing better understanding and quantification for system models are listed below. Contributions for this volume listed under the next section have been selected around these areas.

- Effects of soil, environmental, crop and management factors on GHG emissions and C changes
- Effects of climate variables and global change on GHG emissions, crop yields, livestock production
- Quantification of model uncertainty
- Comparisons of model outputs with other models and different types of measurements.

**Stephen J. Del Grosso, Lajpat R. Ahuja, and William J. Parton,** Editors

## References

IPCC. 2007. Climate Change 2007: The physical science basis. Contribution of Working Group 1 to the Fourth Assessment Report of the Intergovernmental Panel on Climate Change. S. Solomon, D. Qin, M. Manning, Z. Chen, M. Marquis, K.B. Averyt, M. Tignor, and H.L. Miller, editors. Cambridge, UK. Cambridge University Press, Cambridge, UK. http://www.ipcc.ch/publications_and_data/publications_ipcc_fourth_assessment_report_wg1_report_the_physical_science_basis.htm (accessed 6 Oct. 2015).

IPCC. 2014. Climate Change 2014: Synthesis report. Contribution of Working Groups I, II and III to the Fifth Assessment Report of the Intergovernmental Panel on Climate Change. R.K. Pachauri and L.A. Meyer, editors. IPCC, Geneva, Switzerland. https://www.ipcc.ch/pdf/assessment-report/ar5/syr/AR5_SYR_FINAL_Front_matters.pdf (accessed 6 Oct. 2015).

Lobell, D.B., W. Schlenker, and J. Costa-Roberts. 2011. Climate trends and global crop production since 1980. Science 333:616–620.

# Contributors


| | |
|---|---|
| Abdalla, Mohammed | Univ. of Aberdeen, Inst. of Biological and Environmental Sciences, 23 St. Machar Drive, Aberdeen, AB24 3RY (mabdalla@abdn.ac.uk) |
| Ahuja, Lajpat R. | USDA-ARS, Agricultural Systems Research Unit, 2150 Centre Ave., Bldg. D, Fort Collins, CO 80526-8119 (Laj.Ahuja@ars.usda.gov) |
| Albanito, Fabrizio | Univ. of Aberdeen, Inst. of Biological and Environmental Sciences, 23 St Machar Drive, Aberdeen, AB24 3RY (f.albanif@abdn.ac.uk) |
| Bannink, Andre | Wageningen Univ. and Research Centre (UR) Livestock Research, 8219 PH Lelystad, The Netherlands (andre.bannink@wur.nl) |
| Bellarby, Jessica | Lancaster Environment Centre, Lancaster Univ., Bailrigg, Lancaster, LA1 4YQ, UK (jbellarby@gmail.com) |
| Buendia, Leandro | 2113-A Pula Street, College Ville, Los Banos, Laguna 4030, Philippines (leandro.buendia@gmail.com) |
| Butterbach-Bahl, Klaus | Karlsruhe Inst. of Technology, Inst. of Meteorology and Climate Research (IMK-IFU), Kreuzeckbahnstr. 19, 82467 Garmisch-Partenkirchen, Germany (klaus.butterbach-bahl@kit.edu) |
| Campbell, Con A. | Eastern Cereal and Oilseed Research Centre, Agriculture and Agri-Food Canada, 960 Carling Ave., Ottawa, ON K1A 0C6, Canada (con.campbell@agr.gc.ca) |
| Datta, Arindam | Centre for Environmental Studies, Earth Sciences and Climate Change Division, The Energy and Resources Inst., New Delhi, 110003, India (a.datta@abdn.ac.uk) |
| Del Grosso, Stephen J. | 2150 Centre Ave., Bldg. D, Suite 100, Fort Collins, CO 80526 (steve.delgrosso@ars.usda.gov) |
| Desjardins, Raymond L. | Eastern Cereal and Oilseed Research Centre, Agriculture and Agri-Food Canada, 960 Carling Ave., Ottawa, ON K1A 0C6, Canada (ray.desjardins@agr.gc.ca) |
| Dijkstra, Jan | Animal Nutrition Group, Wageningen Univ., De Elst 1, 6708WD, Wageningen, The Netherlands (jan.dijkstra@wur.nl) |
| Dondini, Marta | Univ. of Aberdeen, Inst. of Biological and Environmental Sciences, 23 St Machar Drive, Aberdeen, AB24 3RY (m.dondini@abdn.ac.uk) |
| du Toit, Linde | Dep. of Animal Science, Faculty of Science, Tshwane University of Technology, Building 5 Room 627, Private Bag x680, Pretoria 0001, South Africa (DuToitCJL@tut.ac.za) |
| Ellis, Jennifer L. | Animal Nutrition Group, Wageningen Univ., De Elst 1, 6708WD, Wageningen, The Netherlands (jen.ellis.stpierre@gmail.com); also Centre for Nutrition Modelling, Dep. of Animal and Poultry Science, Univ. of Guelph, Guelph, ON, N1G 2W1 Canada |
| Fitton, Nuala | Univ. of Aberdeen, Inst. of Biological and Environmental Sciences, 23 St Machar Drive, Aberdeen, AB24 3RY (n.fitton@abdn.ac.uk) |
| France, James | Centre for Nutrition Modelling, Dep. of Animal and Poultry Science, Univ. of Guelph, Guelph, ON, N1G 2W1 Canada (jfrance@uoguelph.ca) |
| Gollany, Hero | USDA-ARS-CPCRC, 48037 Tubbs Ranch Rd., Adams, OR 97810 (hero.gollany@ars.usda.gov) |
| Gollany, Hero | USDA-ARS-CPCRC, 48037 Tubbs Ranch Rd., Adams, OR 97810 (hero.gollany@ars.usda.gov) |
| Grant, Brian B. | Eastern Cereal and Oilseed Research Centre, Agriculture and Agri-Food Canada, 960 Carling Ave., Ottawa, ON K1A 0C6, Canada (brian.grant@agr.gc.ca) |
| Gurwick, Noel P. | Smithsonian Environmental Research Center, 647 Contees Wharf Rd., Edgewater, MD 21037 (noel.gurwick@gmail.com) |
| Haas, Edwin | Karlsruhe Inst. of Technology, Inst. of Meteorology and Climate Research (IMK-IFU), Kreuzeckbahnstr. 19, 82467 Garmisch-Partenkirchen, Germany (Edwin.haas@kit.edu) |
| Hallett, Paul | Rm. 2.14, Cruickshank Bldg., Inst. of Biological and Environmental Sciences, Univ. of Aberdeen, Aberdeen, AB24 3UU, UK (paul.hallett@abdn.ac.uk) |
| Hartman, Melannie | Natural Resource Ecology Lab., Campus Delivery 1499, Colorado State University, Fort Collins, CO 80523 (Melannie.Hartman@colostate.edu) |

| | |
|---|---|
| Hastings, Astley | Univ. of Aberdeen, Inst. of Biological and Environmental Sciences, 23 St Machar Drive, Aberdeen, AB24 3RY (astley.hastings@abdn.ac.uk) |
| Heath, Linda S. | USDA Forest Service, Northern Research Station, 271 Mast Road, Durham, NH 03824 (Lheath@fs.fed.us) |
| Hillier, Jon | Univ. of Aberdeen, Inst. of Biological and Environmental Sciences, 23 St Machar Drive, Aberdeen, AB24 3RY (j.hillier@abdn.ac.uk) |
| Hu, Chunsheng | Key Laboratory of Agricultural Water Resources, Center for Agricultural Resources Research, Inst. of Genetic and Developmental Biology, The Chinese Academy of Sciences, 286 Huaizhong Road, Shijiazhuang 050021, Hebei, China (cshu@sjziam.ac.cn) |
| Jones, Edward | 113 LSHTM, Keppel Street, London, WC1E 7HT (e.o.jones@abdn.ac.uk) |
| Kebreab, Ermias | Dep. of Animal Science, 2111 Meyer Hall, One Shields Ave., Univ. of California, Davis, CA 95616 (ekebreab@ucdavis.edu) |
| Kiese, Ralf | Karlsruhe Inst. of Technology, Inst. of Meteorology and Climate Research (IMK-IFU), Kreuzeckbahnstr. 19, 82467 Garmisch-Partenkirchen, Germany (ralf.kiese@kit.edu) |
| Klatt, Steffen | Karlsruhe Inst. of Technology, Inst. of Meteorology and Climate Research (IMK-IFU), Kreuzeckbahnstr. 19, 82467 Garmisch-Partenkirchen, Germany |
| Kraus, David | Karlsruhe Inst. of Technology, Inst. of Meteorology and Climate Research (IMK-IFU), Kreuzeckbahnstr. 19, 82467 Garmisch-Partenkirchen, Germany (david.kraus@kit.edu) |
| Kröbel, Roland | Lethbridge Research Centre, Agriculture and Agri-Food Canada, 5403 1st Ave. South, Lethbridge, AB T1J 4B1, Canada (Roland.Kroebel@agr.gc.ca) |
| Kuhnert, Matthias | Univ. of Aberdeen, Inst. of Biological and Environmental Sciences, 23 St Machar Drive, Aberdeen, AB24 3RY (matthias.kuhnert@abdn.ac.uk) |
| Lafond, Guy P. | deceased, formerly Indian Head Research Farm, Agriculture and Agri-Food Canada, P.O. Box 760, Indian Head, SK S0G 2K0, Canada |
| Lemke, Reynald L. | Agriculture and Agri-Food Canada, 107 Science Place, Saskatoon, SK S7N 0X2, Canada. (reynald.lemke@agr.gc.ca) |
| McConkey, Brian G. | Semiarid Prairie Agricultural Research Centre, Agriculture and Agri-Food Canada, 1 Airport Road, P.O. Box 1030, Swift Current, SK S9H 3X2, Canada (Brian.McConkey@agr.gc.ca) |
| Nayak, Dali | Univ. of Aberdeen, Inst. of Biological and Environmental Sciences, 23 St Machar Drive, Aberdeen, AB24 3RY (d.nayak@abdn.ac.uk) |
| Ogle, Stephen M. | Natural Resource Ecology Lab., Campus Delivery 1499, Colorado State University, Fort Collins, CO 80523; also Dep. of Ecosystem Science and Sustainability, Fort Collins, CO 80523 (stephen.ogle@colostate.edu) |
| Parton, William J. | Natural Resource Ecology Lab, Colorado State University, Campus Delivery 1499, Fort Collins, CO 80523 (William.parton@colostate.edu) |
| Pogson, Mark | Engineering, Sports and Sciences Academic Group, Univ. of Bolton, Deane Road, Bolton, BL3 5AB (m.pogson@abdn.ac.uk) |
| Qin, Shuping | Key Laboratory of Agricultural Water Resources, Center for Agricultural Resources Research, Inst. of Genetic and Developmental Biology, The Chinese Academy of Sciences, 286 Huaizhong Road, Shijiazhuang 050021, Hebei, China (qinshuping@sjziam.ac.cn) |
| Rahn, Karl-Heinz | Karlsruhe Inst. of Technology, Inst. of Meteorology and Climate Research (IMK-IFU), Kreuzeckbahnstr. 19, 82467 Garmisch-Partenkirchen, Germany (rahn.karlheinz@yahoo.de) |
| Reyes-Fox, Melissa | 2150 Centre Ave., Bldg. D, Suite 100, Fort Collins, CO 80526 (Melissa.Reyes@ars.usda.gov) |
| Richards, Mark | Univ. of Aberdeen, Inst. of Biological and Environmental Sciences, 23 St Machar Drive, Aberdeen, AB24 3RY (m.richards@abdn.ac.uk) |
| Smith, Elwin G. | Lethbridge Research Centre, Agriculture and Agri-Food Canada, 5403 1st Ave. South, Lethbridge, AB T1J 4B1, Canada (Elwin.Smith@agr.gc.ca) |
| Smith, James E. | USDA Forest Service, Northern Research Station, 271 Mast Road, Durham, NH 03824 (Jsmith11@fs.fed.us) |

| | |
|---|---|
| Smith, Jo | Univ. of Aberdeen, Inst. of Biological and Environmental Sciences, 23 St Machar Drive, Aberdeen, AB24 3RY (jo.smith@abdn.ac.uk) |
| Smith, Pete | Univ. of Aberdeen, Inst. of Biological and Environmental Sciences, 23 St Machar Drive, Aberdeen, AB24 3RY (pete.smith@abdn.ac.uk) |
| Smith, Ward N. | Eastern Cereal and Oilseed Research Centre, Agriculture and Agri-Food Canada, 960 Carling Ave., Ottawa, ON K1A 0C6, Canada (ward.smith@agr.gc.ca) |
| Spencer, Shannon | Natural Resource Ecology Lab., Campus Delivery 1499, Colorado State University, Fort Collins, CO 80523 (Shannon.Spencer@colostate.edu) |
| Stevens, Luanne | School of Geo and Spatial Science, Unit for Environmental Sciences and Management, Bldg. E4, Room G3, North-West University (NWU), Potchefstroom campus, Hoffman Street, Potchefstroom 2531, South Africa (luanne@jacali.net). |
| Tedeschi, Luis | Dep. of Animal Science, Room 133 Kleberg, Texas A&M Univ. College Station, TX 77843-2471 (luis.tedeschi@tamu.edu) |
| Tonitto, Christina | Cornell International Inst. for Food, Agriculture and Development, Cornell University, B75 Mann Library, Ithaca, NY 14853 (ctonitto@cornell.edu) |
| Vetter, Sylvia | Univ. of Aberdeen, Inst. of Biological and Environmental Sciences, 23 St Machar Drive, Aberdeen, AB24 3RY (sylvia.vetter@abdn.ac.uk) |
| Wang, Yuying | Key Laboratory of Agricultural Water Resources, Center for Agricultural Resources Research, Inst. of Genetic and Developmental Biology, The Chinese Academy of Sciences, 286 Huaizhong Road, Shijiazhuang 050021, Hebei, China (wangyuying@sjziam.ac.cn) |
| Werner, Christian | Karlsruhe Inst. of Technology, Inst. of Meteorology and Climate Research (IMK-IFU), Kreuzeckbahnstr. 19, 82467 Garmisch-Partenkirchen, Germany (christian.werner@senkenberg.de) |
| Witi, Jongikhaya | Dep. of Environmental Affairs, Environment House, 473 Steve Biko Street, Arcadia, Pretoria 0001, South Africa (Jwiti@environment.gov.za) |
| Woodbury, Peter B. | Section of Soil and Crop Sciences, School of Integrative Plant Science, 1017 Bradfield Hall, 306 Tower Rd., Cornell University, Ithaca, NY 14853 (peterwoodbury@cornell.edu) |
| Yeluripati, Jagadeesh | The James Hutton Inst., Craigiebuckler, Aberdeen AB15 8QH, UK (jagadeesh.yeluripati@hutton.ac.uk) |
| Zhang, Yuming | Key Laboratory of Agricultural Water Resources, Center for Agricultural Resources Research, Inst. of Genetic and Developmental Biology, The Chinese Academy of Sciences, 286 Huaizhong Road, Shijiazhuang 050021, Hebei, China (ymzhang@sjziam.ac.cn) |

# Nitrous Oxide Emission from Uplands in Northern China

**Shuping Qin, Chunsheng Hu,* Lajpat R. Ahuja, Yuming Zhang, and Yuying Wang**


## Abstract

Northern China, as one of the key agricultural regions, is an important nitrous oxide ($N_2O$) emission source because of increasing use inorganic nitrogen (N) fertilization to achieve high crop yields. Nitrous oxide is usually emitted through soil nitrification and denitrification processes regulated by microbial and environmental factors. Microbial nitrification may dominate $N_2O$ emission in this region. Influenced by meteorological and agronomical practice, soil $N_2O$ emission shows large spatial and temporal variation. In the typical winter wheat (*Triticum aestivum* L.)–summer corn (*Zea mays* L.) double-cropping system, $N_2O$ concentration in soil air increased with soil depth from 0 to 30 cm and approached stable values at soil depths deeper than 60 cm. Annual $N_2O$ emission increased with N fertilization rates. Nitrous oxide emission in the corn season is more than in the wheat season. Yield-scaled $N_2O$ emission ($N_2O$ emission per unit crop yield) is proposed in view of the challenges of increasing food production and simultaneously decreasing $N_2O$ emissions. According to this concept, N fertilizer rate of 310 kg N $ha^{-1}$ $yr^{-1}$ was reported to be suitable for both crop production and mitigation of $N_2O$ emissions under the double-cropping rotation in northern China. The $N_2O$ emission factor is simple to estimate $N_2O$ emission from agricultural soil. The reported conceptual first-order reaction modeling approach, with the effects of fertilizer source and plant uptake, is useful to enhance the simple emission factor approach. The process-based models are promising methods for accurate estimation in the future studies of soil $N_2O$ fluxes.


Abbreviations: NT-CB, no-till corn–soybean rotation; CT-CC, conventionally tilled continuous corn; HIP, "hole in the pipe"; IPCC, Intergovernmental Panel on Climate Change; NT-CC, no-till continuous corn; UAN, urea ammonium nitrate.

Shuping Qin (qinshuping@sjziam.ac.cn), Chunsheng Hu, Yuming Zhang (ymzhang@sjziam.ac.cn), Yuying Wang (wangyuying@sjziam.ac.cn), Key Laboratory of Agricultural Water Resources, Center for Agricultural Resources Research, Institute of Genetic and Developmental Biology, The Chinese Academy of Sciences, 286 Huaizhong Road, Shijiazhuang 050021, Hebei, China. *Corresponding author (cshu@sjziam.ac.cn). Lajpat R. Ahuja, USDA-ARS, Agricultural Systems Research Unit, 2150 Centre Avenue, Bldg D, Fort Collins, CO 80526-8119 (Laj.Ahuja@ars.usda.gov).

doi:10.2134/advagricsystmodel6.2013.0006.5


Synthesis and Modeling of Greenhouse Gas Emissions and Carbon Storage in Agricultural and Forest Systems to Guide Mitigation and Adaptation
S. Del Grosso, L. Ahuja, and W. Parton, Editors
Advances in Agricultural Systems Modeling, Volume 6. Lajpat R. Ahuja, Series Editor

Northern China is one of the dominant grain producing areas in China (Qin et al., 2012). The climate in this area is temperate semiarid with monsoons. The winter wheat (*Triticum aestivum* L.)–summer corn (*Zea mays* L.) double-cropping system is the dominant cropping system in northern China. Winter wheat is usually planted in early October and harvested in the following June. Summer corn is usually planted in early June, immediately after the winter wheat harvest, and harvested in October. This cropping system is called as intensive cropping system since two crops are planted within a single year without a fallow period. Other important crops in this area are soybean [*Glycine max* (L.) Merr.] and cotton (*Gossypium hirsutum* L.).

Farmlands in northern China may be a significant emitter of nitrous oxide ($N_2O$) because the local farmers usually apply excessive inorganic N fertilizer to achieve high crop yield. The N fertilization rate in this area is increasing year by year (Zhang et al., 2003; Hu et al., 2006). The average N fertilizer rate for winter wheat–summer corn double-cropping systems is over 500 kg N $ha^{-1}$ $yr^{-1}$ (Cui et al., 2012; Meng et al., 2012), which exceeds the crop requirements of 200 to 300 kg N $ha^{-1}$ $yr^{-1}$ in this area (Cui et al., 2010) and consequently results in serious environmental problems including $N_2O$ emission (Cui et al., 2012). Flood irrigation is usually applied immediately after the surface broadcasting of inorganic N fertilizer, which further promotes soil $N_2O$ emission as well as nitrate leaching. Furthermore, precipitation in northern China comes during the period from June to September (Jin et al., 1999). This period also has relative high air temperature. Summer is the major period for soil $N_2O$ emission in northern China as a result of the co-occurrence of rain and hot weather during this season (Pei et al., 2012; Wang et al., 2013).

When considering the increasing food demand in China, increasing grain yields is a condition for "greening" Chinese agriculture. Northern China, as the center of wheat and corn production, has experienced increasing serious environmental problems such as $N_2O$ emissions. The responses of $N_2O$ emission and crop yield to increased fertilizer rates have not been well investigated in northern China. Such investigations will be useful for modeling $N_2O$ emissions under different agricultural managements and for optimizing fertilizer rate to reduce $N_2O$ emissions. The aims of the study were to (i) investigate the responses of crop yield, $N_2O$ emissions, and yield-scaled $N_2O$ emissions to fertilizer rates; (ii) investigate the $N_2O$ emission factors under different fertilizer rates; and (iii) demonstrate the use of the first-order conceptual model to interpret the seasonal $N_2O$ emissions in future studies and enhance the emission factors approach.

## Factors Controlling Nitrous Oxide Emissions in Northern China

Nitrous oxide emissions from soil microbial nitrification and denitrification processes contribute substantially to global $N_2O$ emissions (~70%; Braker and Conrad 2011, Butterbach-Bahl and Dannenmann 2011; Syakila and Kroeze 2011; Butterbach-Bahl et al., 2013). Soil nitrification and denitrification processes are significantly affected by several factors such as soil water-filled pore space (Dobbie and Smith 2003), temperature (Keeney et al., 1979), and mineral N contents (nitrate and ammonia) (Ryden and Lund 1980; Granli and Boeckman 1994; Butterbach-Bahl et al., 2013). Among these factors, soil water-filled pore space is considered as one of the most dominant factors (Dobbie and Smith 2003), as it regulates the water and oxygen availability to soil microorganisms. Nitrification dominates soil $N_2O$ emission when the soil water-filled pore space is lower than 60%, while denitrification dominates when soil water-filled pore space is higher than 70% (Bateman and Baggs 2005). In northern China, especially in the North China Plain, the groundwater table is usually more than 10 m deep (Yuan and Shen 2013), as a result, most of time the soil water-filled pore space is below 70% unless rain and irrigation occurred (Cui et al., 2012). These conditions generally favor $N_2O$ emission from nitrification rather than from denitrification. Urea, the most commonly available fertilizer in northern China, is hydrolyzed into ammonium by soil urease, and then the ammonium is transformed into nitrite and nitrate through microbial nitrification process. The microbial transformation from urea to nitrate is rapid in soils in northern China (Qin et al., 2010). Furthermore, soil-dissolved organic C is limited under summer corn–winter wheat double-cropping systems in the North China Plain (generally 30–80 mg C $kg^{-1}$ dried soil; Dong et al., 2009; Cui et al., 2012), which also limits soil denitrification process. Thus, microbial nitrification may dominate $N_2O$ emission in northern China (Wan et al., 2009; Ju et al., 2011; Cui et al., 2012).

## Spatial and Temporal Variation of $N_2O$ Emission Rate

Soil $N_2O$ emission in northern China shows large spatial and temporal variation (Ding et al., 2007; Pei et al., 2012; Qin et al., 2012; Tian et al., 2012; Yan 2012; Shi et al., 2013, Wang et al., 2013). Soil water, nitrate, and dissolved organic C contents were reported to be the major factors accounting for spatial variation of soil $N_2O$ emission (Fu and Li, 2012). In northern China, nitrate accumulates in soil profiles as a result of over application of N fertilizer (Li et al., 2007), resulting in large spatial variation of $N_2O$ concentration in soil profiles (Wang et al., 2013). Nitrous oxide concentration in soil air was reported to increase with soil depth from 0 to 30 cm and approach stable values at soil depths deeper than 60 cm (Wang et al.,

2013). Agriculture managements, such as irrigation and N fertilization, significantly affected the $N_2O$ concentration in soil air as deep as 300 cm (Wang et al., 2013). Although the growing season of summer corn (~5 mo) is shorter than winter wheat (~7 mo), the $N_2O$ emission in the corn season is more than in the wheat season because the soil moisture and temperature are relatively higher in summer. Nitrous oxide emissions during the freezing and thawing periods (about 3–4 wk) were rarely reported in the literature (Wang et al., 2013), which could have led to underestimation in $N_2O$ emission in growing season of winter wheat (Qin et al., 2012). Soil $N_2O$ emission during the freezing and thawing period in the North China Plain was reported to account for as much as 9.7 and 5.1%, respectively, of the annual $N_2O$ emission under different agricultural managements (Shi et al., 2013). However, other studies have shown that emissions are very low from late December to early February in the North China Plain as a result of low soil moisture and low temperature (Ding et al., 2007). Nitrous oxide emissions increased significantly immediately after irrigation and N fertilization (Ye et al., 2005; Sun et al., 2008; Cui et al., 2012; Wang et al., 2013), especially when N fertilization followed by flooding irrigation or rainfall, which dominated annual $N_2O$ emissions (generally contributes more than 60%) (Ding et al., 2007; Cui et al., 2012; Pei et al., 2012). This trend is in line with $N_2O$ emission studies in other regions (Reay et al., 2012; Rees et al., 2013). The $N_2O$ emission pulse stimulated by the N application in July for summer corn is higher than that for winter wheat probably as a result of the higher temperature in July. The temporal variation of $N_2O$ emission suggests that higher frequency of $N_2O$ emission measurement (e.g., daily) is needed for precise budget of annual $N_2O$ emission (Cui et al., 2012).

## Nitrous Oxide Seasonal Average Emissions as Functions of Nitrogen Fertilization Rate

The "hole in the pipe" (HIP) theory was proposed to describe the relationship between N input and $N_2O$ emission (Firestone and Davidson, 1989; Davidson et al., 1991). Based on this theory, the cumulative $N_2O$ emission was determined by the sizes of the "pipe" (the amount of N being cycled) and the HIP (Smith et al., 1998; Davidson and Verchot 2000). To some degree, the slope of $N_2O$ emission vs. N fertilization rate is a reflection of the diameter of the "hole." As can been seen from Fig. 1 through 4, in northern China, annual $N_2O$ emission increased with N fertilization rates. For the whole year of winter wheat–summer corn double-cropping system, both linear and exponential models well describe the relationship between $N_2O$ emission and N fertilizer rates (Fig. 1). For the winter-wheat growing season, a linear model is superior to an exponential model (Fig. 2). For the summer-corn growing season, an exponential model is superior to a linear model

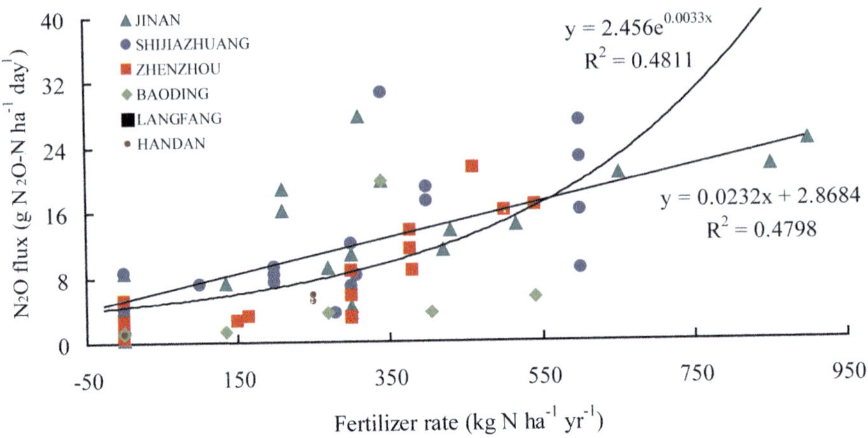


**Fig. 1. Nitrous oxide annual emission as a function of N fertilizer rates. The flux data was collected from the literatures (Firestone and Davidson, 1989; Davidson et al., 1991; Smith et al., 1998; Jin et al., 1999; Shen et al., 2002; Zhang et al., 2003; Ye et al., 2005; Hu et al., 2006; Ding et al., 2007; Li et al., 2007; Sun et al., 2008; Cui et al., 2010; Ding et al., 2011; Zhang et al., 2011; Cui et al., 2012; Fu and Li, 2012; Meng et al., 2012; Pei et al., 2012; Qin et al., 2012; Reay et al., 2012; Tian et al., 2012; Yan, 2012; Zhang et al., 2012; Hu et al., 2013; Rees et al., 2013; Shi et al., 2013; Wang et al., 2013; Yuan and Shen, 2013).**

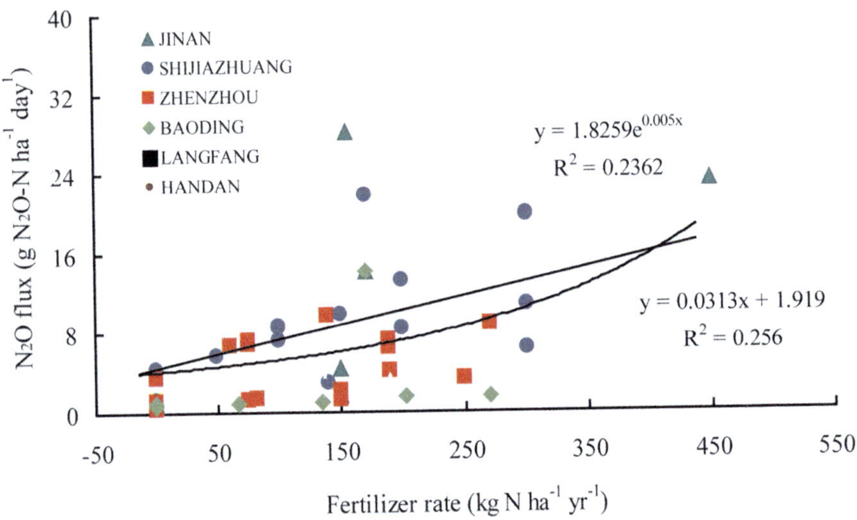


**Fig. 2. Nitrous oxide emission in winter wheat season as a function of N fertilization rates. The flux data was collected from the literatures (Firestone and Davidson, 1989; Davidson et al., 1991; Smith et al., 1998; Jin et al., 1999; Shen et al., 2002; Zhang et al., 2003; Ye et al., 2005; Hu et al., 2006; Ding et al., 2007; Li et al., 2007; Sun et al., 2008; Cui et al., 2010; Ding et al., 2011; Zhang et al., 2011; Cui et al., 2012; Fu and Li, 2012; Meng et al., 2012; Pei et al., 2012; Qin et al., 2012; Reay et al., 2012; Tian et al., 2012; Yan, 2012; Zhang et al., 2012; Hu et al., 2013; Rees et al., 2013; Shi et al., 2013;, Wang et al., 2013; Yuan and Shen, 2013).**

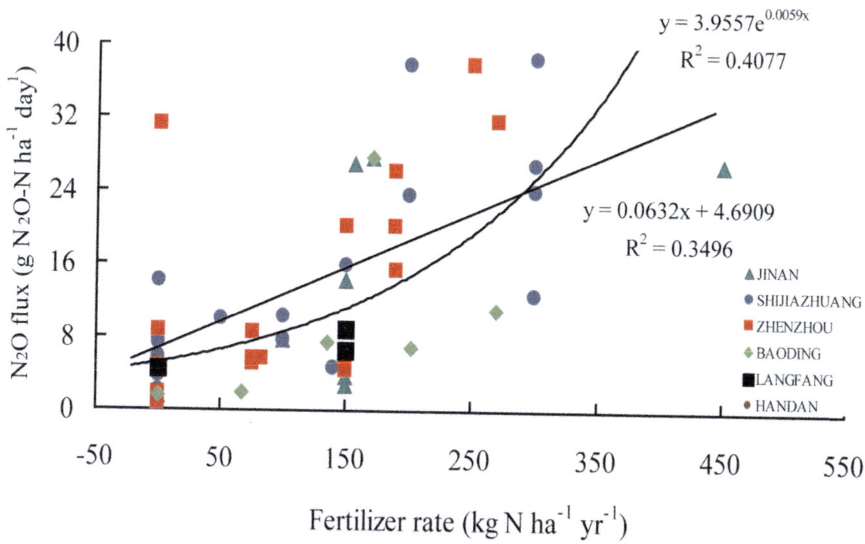


**Fig. 3. Nitrous oxide emission in summer corn season as a function of N fertilization rates. The flux data was collected from the literatures (Firestone and Davidson, 1989; Davidson et al., 1991; Smith et al., 1998; Jin et al., 1999; Shen et al., 2002; Zhang et al., 2003; Ye et al., 2005; Hu et al., 2006; Ding et al., 2007; Li et al., 2007; Sun et al., 2008; Cui et al., 2010; Ding et al., 2011; Zhang et al., 2011; Cui et al., 2012; Fu and Li, 2012; Meng et al., 2012; Pei et al., 2012; Qin et al., 2012; Reay et al., 2012; Tian et al., 2012; Yan, 2012; Zhang et al., 2012; Hu et al., 2013; Rees et al., 2013; Shi et al., 2013; Wang et al., 2013; Yuan and Shen, 2013).**

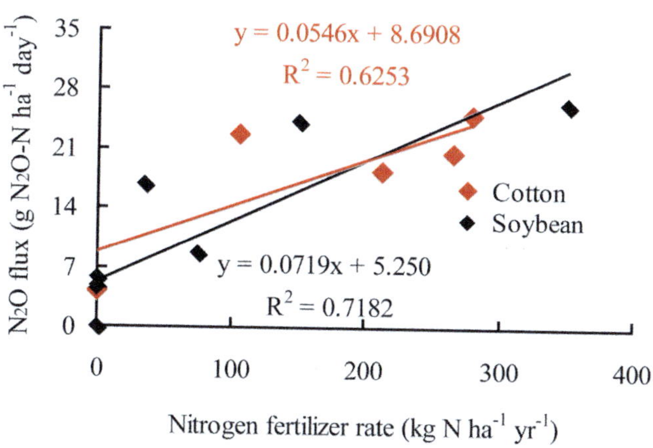


**Fig. 4. Nitrous oxide emission from cotton and soybean farmlands as a function of N fertilization rates. The flux data was collected from the literatures (Guohong et al., 1995; Huang et al., 1995; Yu et al., 1995; Xing, 1998; Xu et al., 2000; Ding et al., 2001; Xiong and Xing, 2002; Chen et al., 2005; Wang et al. 2006a,b, 2008; Liu et al., 2010).**

(Fig. 3). For the cotton and soybean farmlands, $N_2O$ emission generally linearly increased with N fertilizer rate (Fig. 4). The difference in the responses of $N_2O$ to N fertilization rates among different crops and rotations has been reported in other studies (Hoben et al., 2011). The slope of $N_2O$ annual emission vs. N fertilizer rates is 6.1 g $N_2O$-N $kg^{-1}$ fertilizer N. When taken separately, seasonal $N_2O$ emissions from both winter wheat and summer corn growing seasons are significantly and linearly increased with N fertilizer rates (Fig. 2 and 3). The slope of $N_2O$ seasonal emission vs. N fertilizer rate is higher in the summer-corn growing season (6.3 g $N_2O$-N $kg^{-1}$ fertilizer N) than that in the winter-wheat growing season (5.2 g $N_2O$-N $kg^{-1}$ fertilizer N). For cotton and soybean farmlands, the slope is 8.2 and 6.3 g $N_2O$-N $kg^{-1}$ fertilizer N, respectively. These results indicate that N fertilizer application causes difference in $N_2O$ emission between different crop systems. Soil water content and temperature are significantly higher in the growing season of summer corn than winter wheat (Wang et al., 2013), which probably accounts for the difference in $N_2O$ emission between the two seasons.

There are two general approaches to estimate $N_2O$ fluxes from soils. In the statistical models, $N_2O$ flux is proportional to easily quantifiable factors such as N fertilizer input, soil available N, and water contents. The details of the N cycle, for example, nitrification and denitrification processes, are not explicitly represented (Hoben et al., 2011). The widely used Intergovernmental Panel on Climate Change (IPCC) Tier 1 methodology (IPCC, 2006) and the HIP model are such models. Process-based models, for example, the denitrification–decomposition model and Daycent model (Giltrap et al., 2013), attempt to consider many or all of the processes involved in the N cycle. The process-based models require more details about the ecological processes than the statistical models. Quantifying the responses of $N_2O$ emission to N fertilization rates is helpful for modeling field- or region-scaled $NO_2$ flux (Butterbach-Bahl et al., 2013).

## Yield-Scaled Nitrous Oxide Emissions as a Function of Nitrogen Fertilization Rate

Nitrogen fertilization increases both $N_2O$ emission and crop yield (Mosier et al., 1998; Davidson 2009). The high fertilizer rate achieves high crop yield (Qin et al., 2012), but inevitably causes environmental issues such as $N_2O$ emissions (Ding et al., 2007). To simultaneously achieve relative high crop yield and minimal environmental cost, fertilizer rate should be optimized based on an integrated index involving both crop yield and environmental parameters. Considering that the food demand (staple grains) has been relatively stable for China (Gale and Huang 2007), minimizing $N_2O$ emission per unit production will lead to minimization of total $N_2O$ emissions from agricultural soils. Therefore, the yield-scaled $N_2O$ emissions are useful indicators for optimizing N fertilizer management for crop production and $N_2O$ emission mitigation. Recently, it has been suggested that the best management options for $N_2O$-efficient agronomic production are similar to those promoting agronomic efficiency in general (Van Groenigen et al., 2010). Based on a meta-analysis of 19 studies, it was found that yield-scaled $N_2O$ emissions (emissions expressed as grams $N_2O$-N per kilogram aboveground N uptake) are lowest at medium N application rates. The concept of yield-scaled $N_2O$ emissions is relevant in view of the global challenges of increasing food production and decreasing emissions concomitantly (Tilman et al., 2002; Foley et al., 2011). In the recent years, yield-scaled $N_2O$ emission has been studied in many agricultural ecosystems for sustainable evaluation (Venterea et al., 2011; Sanz-Cobena et al., 2012; Schellenberg et al., 2012; Abalos et al., 2013; Cui et al., 2013). The yield-scaled $N_2O$ emission ($N_2O$ emission per unit crop yield) was proposed and reported to decrease slightly with N fertilizer rate from 0 to 187 kg N $ha^{-1}$ and then increase sharply from 187 to 350 kg N $ha^{-1}$ based on a meta-analysis of 19 agricultural systems with highly variable properties (Van Groenigen et al., 2010). More field control experiments are needed to test whether or not the yield-scaled $N_2O$ emission indicator is suitable to optimize fertilizer managements. We tested the responses of crop yield and yield-scaled $N_2O$ emissions to long-term N fertilization rates in a typical winter wheat–summer corn double-cropping system in the North China Plain.

The results showed that the crop yields increased with increasing N fertilizer rates, but significant difference occurred only between control and fertilized treatments (Qin et al., 2012). The responses of soil $N_2O$ emissions to N fertilizer rates were dependent on cropping seasons. Generally, the yield-scaled $N_2O$ emissions of the summer-corn season were not significantly different among N fertilizer rates, whereas yield-scaled $N_2O$ emission of the winter-wheat season exponentially increased with increasing of N fertilizer rates (Fig. 5).

With the increase of N fertilizer rates, the yield-scaled $N_2O$ emissions of wheat grain N and wheat aboveground N were shown to be cubic functions of N fertilizer rates (Fig. 5). In contrast, the yield-scaled $N_2O$ emissions of corn grain N and corn aboveground N did not show specific relationships with N fertilizer rates (Fig. 5). For the 2007 corn season, when the N fertilizer rates were 0, 200, 400, and 600 kg N $ha^{-1}$ $yr^{-1}$, the grain N yield-scaled $N_2O$ emissions were 17.51, 9.72, 21.87, and 21.14 g $N_2O$-N $kg^{-1}$ N, respectively; the aboveground N yield-scaled $N_2O$ emissions were 12.47, 6.56, 14.61, and 15.46 g $N_2O$-N $kg^{-1}$ N, respectively. For the 2008 corn season, when the N fertilizer rates were 0, 200, 400, and 600 kg N $ha^{-1}$ $yr^{-1}$, the grain N yield-scaled $N_2O$ emissions were 7.15, 6.66, 26.03, and 26.76 g $N_2O$-N $kg^{-1}$ N, respectively; the aboveground N yield-scaled $N_2O$ emissions were 3.59, 3.37, 12.96, and 14.42 g $N_2O$-N $kg^{-1}$ N, respectively (Fig. 5). The optimized N

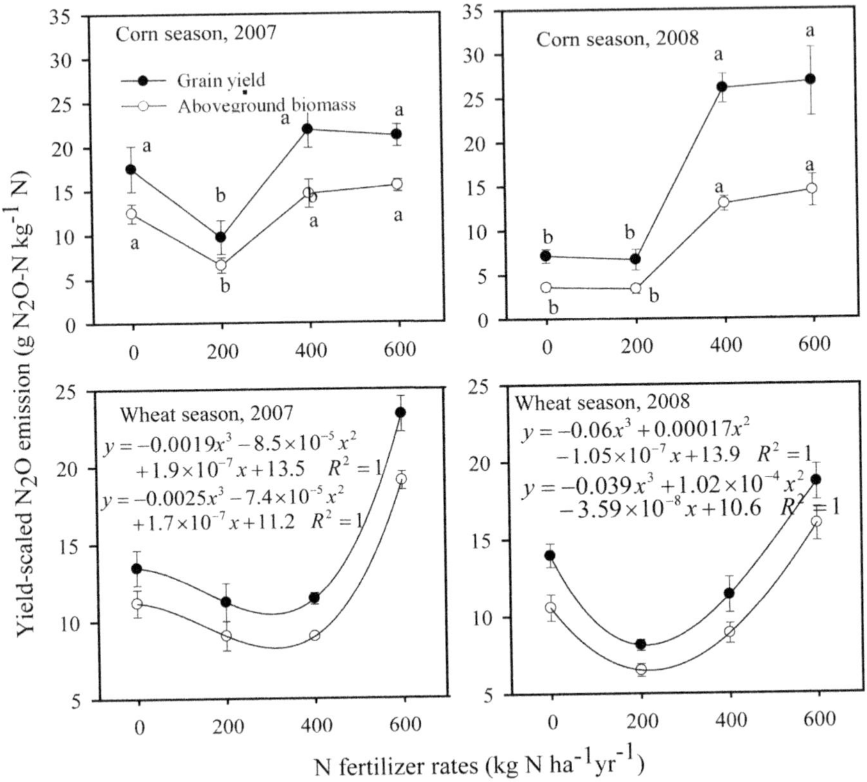


**Fig. 5. Yield-scaled $N_2O$ emissions as functions of N fertilizer rates in a typical area of North China. Different letters (a, b) near the error bars indicate significant difference at $p < 0.05$ between different N fertilizer rates.**

fertilizer rates suggested by the minimal yield-scaled $N_2O$ emissions of wheat grain N and aboveground N were from 211 to 312 kg N $ha^{-1}$ $yr^{-1}$.

Nitrogen is still one of the limiting factors for world grain production (Vitousek and Howarth 1991; Hodge and Fitter 2010). The responses of crop yields to N fertilizer indicate that when the soil N is limited, N fertilizer will lead to dramatic increase in crop yield, while when the N in soil is superabundant, increasing N fertilizer has limited effects on crop yield increase (Ma et al., 2007; Essah and Delgado 2009), as was shown in this study (Fig. 5). Based on the results of linear-plateau regression, the minimum levels of N fertilizer needed to maximize grain yield differed significantly between the year of 2007 and 2008 (Fig. 5), probably due to the fact that, apart from N fertilizer, other meteorological elements (precipitation and temperature) also influenced crop yields. As far as the data extends, 310 kg N $ha^{-1}$ $yr^{-1}$ was enough to reach maximum crop yield in the study area (Fig. 5). Contrary to crop yield, the annual $N_2O$ emissions were generally linear functions of N fertilizer rate, as is shown in Fig. 3. These results indicate that N fertilizer rate >310 kg N $ha^{-1}$ $yr^{-1}$ probably resulted in $N_2O$ emission coupled with little yield benefit.

In consideration of the huge food demand of China, relatively high yield is the precondition for the Chinese Government to extend an environmentally friendly agricultural strategy. Therefore, the yield-scaled $N_2O$ emissions are superior for evaluating agricultural management effects on the environment in this region. Supposing the food demand of corn and wheat in China is stable and the cultivated area of corn and wheat is adjustable, minimal yield-scaled $N_2O$ emissions will achieve minimal total $N_2O$ emissions from the wheat–corn double-cropping systems. Our results indicated that when the N fertilizer rates were 305 and 211 kg N $ha^{-1}$ $yr^{-1}$, the wheat grain N yield-scaled $N_2O$ emissions of 2007 and 2008 achieved minimums of 10.4 and 8.1 g $N_2O$-N $kg^{-1}$ N, respectively. To be on the safe side, N fertilizer rate of 310 kg N $ha^{-1}$ $yr^{-1}$ seems to be suitable for both crop production and mitigation of $N_2O$ emission under wheat–corn double-cropping rotation in the North China Plain. This N rate was decreased as much as 25% of the farmers' practice. Further study is needed to consider the yield-scaled nitrate leaching, ammonia volatilization, and $CO_2$ emission to evaluate the multiple impacts of this cropping system on environment.

## Nitrous Oxide Emission Factor as a Function of Nitrogen Fertilization Rate, Precipitation, and Temperature

The $N_2O$ emission factor, defined as the ratio of the $N_2O$ emission divided by the amount of N fertilizer applied, is a simple way to estimate $N_2O$ emission from agricultural soil (Toma et al., 2007). The IPCC has proposed a default $N_2O$ emission factor value of 1.0%. However, $N_2O$ emission factors have been reported to show large variation among different field managements (Dobbie and Smith 2003). For the summer corn–winter wheat double-cropping system in northern China, the $N_2O$ emission factors range from 0.2 to 2.6% (average 0.90 ± 0.56%; Fig. 6). For the cotton and soybean farmlands, the $N_2O$ emission factors range from 0.75 to 4.80% (average 1.92 ± 1.38%) and from 1.2 to 3.2% (average 1.72 ± 1.02%), respectively (Fig. 6). The $N_2O$ emission factor has a decreasing trend with the increasing of N fertilization rate (Fig. 6). Soil water-filled pore space was reported to be one of the most important parameters stimulating $N_2O$ emission rate (Dobbie and Smith, 2003). Soil water-filled pore space is significantly higher at N fertilization rate of 0 kg N $ha^{-1}$ $yr^{-1}$ than at 200, 400, and 600 kg N $ha^{-1}$ $yr^{-1}$ in the summer corn–winter wheat double-cropping system in northern China (Wang et al., 2013) as a result of

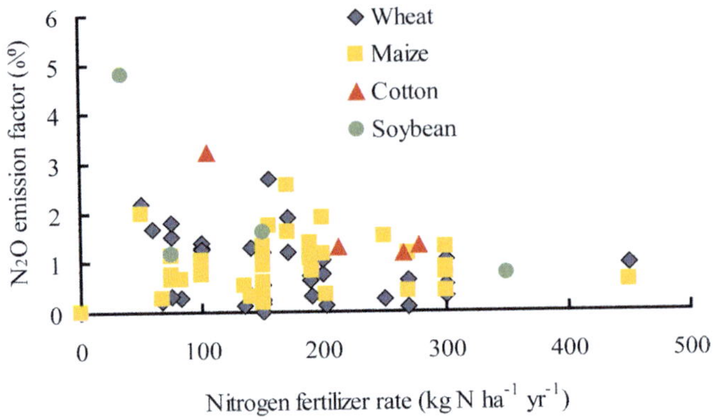


**Fig. 6. Nitrous oxide emission factor as function of N fertilization rate in North China. The flux data was collected from the literatures (Guohong et al., 1995; Huang et al., 1995; Yu et al., 1995; Xing, 1998; Jin et al., 1999; Xu et al., 2000; Ding et al., 2001; Shen et al., 2002; Xiong and Xing, 2002; Zhou et al., 2002; Zhang et al., 2003; Chen et al., 2005; Lu et al., 2005; Ye et al., 2005; Hu et al., 2006; Wang et al., 2006a,b, 2008; Ding et al., 2007; Gale and Huang, 2007; Li et al., 2007; Toma et al., 2007; Sun et al., 2008; Dong et al., 2009; Wan et al., 2009; Cui et al., 2010; Liu et al., 2010; Qin et al., 2010; Ding et al., 2011; Hoben et al. 2011, Ju et al., 2011; Zhang et al., 2011; Cui et al., 2012; Fu and Li, 2012; Meng et al., 2012; Pei et al., 2012; Qin et al., 2012; Tian et al., 2012; Yan, 2012; Zhang et al., 2012; Cui et al., 2013; Gao et al., 2013; Hu et al., 2013; Shi et al., 2013; Wang et al., 2013; Yuan and Shen, 2013).**

low transpiration rate under fertilization rate of 0 kg N $ha^{-1}$ $yr^{-1}$. This fact accounts for the relatively higher $N_2O$ emission factor at low N fertilization rate compared with moderate and high N fertilization rate (Fig. 6). However, annual precipitation has no significant effect on $N_2O$ emission factor (Fig. 7). On the contrary, the $N_2O$ emission factor first decreases and then increases with increasing annual mean temperature (Fig. 7). These results have limited indication for other places, since the ranges of annual precipitation and mean temperature are narrow.

Modeling approaches are helpful for $N_2O$ emission budgets at various temporal and spatial scales. The emission factor approach proposed by the IPCC is widely used to estimate $N_2O$ emissions associated with agricultural practices and land-use change at large temporal and spatial scales (Del Grosso et al., 2008). However, the emission factor approach has been criticized for its inaccuracy at small spatial and temporal scales and for not taking account of agricultural management practices aimed to reduce $N_2O$ emissions from agriculture, such as controlled-release fertilizers (Butterbach-Bahl et al., 2013). Additionally, our results indicate that the $N_2O$ emission factor is affected by the N fertilizer rates, probably as a result of the fact that mineralization and N deposition contribute a lot to $N_2O$ emission under low fertilization rate.

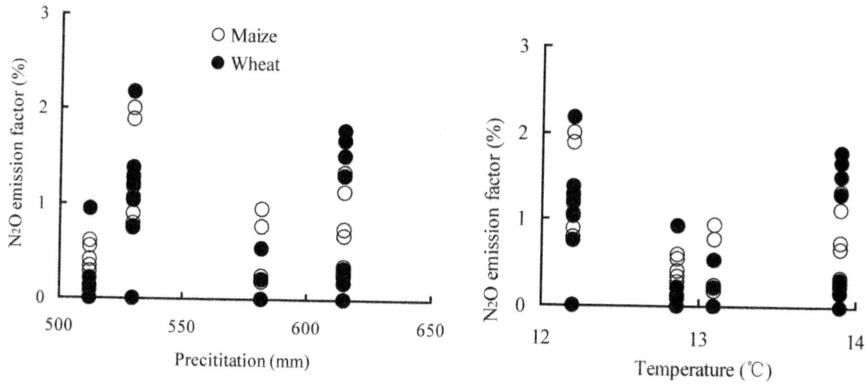


**Fig. 7. Nitrous oxide emission factor as functions of annual precipitation and annual mean temperature under summer maize–winter wheat double-cropping system in North China. The flux data was collected from the literatures (Guohong et al., 1995; Huang et al., 1995; Yu et al., 1995; Xing 1998; Jin et al., 1999; Xu et al., 2000; Ding et al., 2001; Shen et al., 2002; Xiong and Xing, 2002; Zhou et al. 2002; Zhang et al., 2003; Chen et al., 2005; Lu et al., 2005; Ye et al., 2005; Hu et al., 2006; Wang et al., 2006a,b; Ding et al., 2007; Gale and Huang, 2007; Li et al., 2007; Toma et al., 2007; Sun et al., 2008; Wang et al., 2008; Dong et al., 2009; Wan et al., 2009; Cui et al., 2010; Liu et al., 2010; Qin et al., 2010; Ding et al., 2011; Hoben et al., 2011; Ju et al., 2011; Zhang et al., 2011; Cui et al., 2012; Fu and Li, 2012; Meng et al., 2012; Pei et al., 2012; Qin et al., 2012; Tian et al., 2012; Yan, 2012; Zhang et al., 2012; Cui et al., 2013; Gao et al., 2013; Hu et al., 2013; Shi et al., 2013; Wang et al., 2013; Yuan and Shen, 2013).**

## A Conceptual First-Order Approach to Interpret Seasonal Nitrous Oxide Emissions in Future Studies, Enhance Emission Factors, and Incorporate Management Effects

A first-order reaction approach is commonly used in detailed process-based modeling of many soil chemical reactions, including greenhouse gas emissions. Such a first-order approach for $N_2O$ emissions has to be based on the chemical mechanism for this process. As described above, the $N_2O$ emissions occur in two ways. Under aerobic conditions, $N_2O$ is emitted mainly during the microbial process of nitrification of ammonium ions to nitrate. The source of ammonium ions is the organic matter decomposition by microbes or the applied ammonium containing fertilizers. Under anaerobic conditions, it is released during the denitrification process. However, even under aerobic conditions, there are commonly pockets of high soil moisture in the soil where some denitrification and, hence, the $N_2O$ release can take place. For simplicity, we will combine both the modes of production into one first-order approach. With this background, we formulate the following first-order approach for simulating seasonal emissions of $N_2O$:

$$N_2O \text{ emissions} = f_1[NH_4\text{-N (urea-N) applied} + \text{N mineralized during the period}] \times f_w \times f_{temp} \times f_{till} \times f_{fs} \times f_{upt}$$

where $f_1$ is the first-order rate constant, and $f_w$, $f_{temp}$, $f_{till}$, $f_{fs}$, and $f_{upt}$ are the factors to account for the effects of seasonal average soil water content, soil temperature, tillage, fertilizer source, and plant N uptake, respectively. The factors $f_w$, $f_{temp}$, $f_{till}$, and $f_{fs}$ will be ratios of actual average seasonal field conditions to optimal reference values. For example, the $f_w$ will be a ratio of seasonal average soil moisture to optimum average soil moisture content for maximum emission for moisture contents below the optimum. The $f_{temp}$ will be a ratio of seasonal average temperature to optimum temperature for temperatures below the optimum and the reverse of this ratio for temperatures above the optimum. The $f_{till}$ will be a ratio of emission from a conventionally tilled soil to no-till soil, and $f_{fs}$ a ratio of emission from a certain fertilizer source to emission from reference fertilizer source, such as urea. The $f_{upt}$ will represent the degree (0–1) to which plant uptake of N influences the source available for $N_2O$ emission. The plant uptake will not be expected to affect emission from applied fertilizer, since the ammonium source in the fertilizer is generally nitrified very quickly before any significant uptake will take place. The uptake of ammonium ions may influence the $N_2O$ emission from organic matter decomposition, and the uptake of nitrate may affect the release from denitrification in wet soil pockets.

To illustrate the use of the above approach, we present here the results of Mosier et al. (2006, Fig. 9) for $N_2O$ emissions in our Fig. 8. The authors used urea ammonium nitrate (UAN) as fertilizer injected 5 cm below the soil surface in bands along the crop rows. The experiment contained three cropping systems (conventionally tilled continuous corn [CT-CC], no-till continuous corn [NT-CC], and no-till corn–soybean rotation [NT-CB]) with three levels of fertilizer in 2002 (0, 134, and 202 kg N ha$^{-1}$) and four levels in 2003 and 2004 (0, 67, 134, and 234 kg ha$^{-1}$). The crops were irrigated fully to nearly optimum level, so the soil moisture factor can be assumed equal to 1.0. The soil temperature in all 3 yr was pretty close to the same, so for comparison among the years the soil temperature can also be assumed to be 1.0. The fertilizer source factor was also set equal to 1.0, since there was only one fertilizer, the UAN.

Figure 8 shows the seasonal average flux rate for $N_2O$ as a function of N application rates for each cropping system for each year separately as an average rate of 3 yr and as all data combined for all years in one graph. There are very strong linear relationships fitted in the first three plots and fairly strong in the fourth. The slope of the fitted functions represents the value of first-order constant $f_1 \times f_{upt}$ for no-till systems (where $f_{till}$ was 1.0) and $f_1 \times f_{till} \times f_{upt}$ for the tilled system. The intercepts of the linear fits represent the values of [N mineralized during the season] $\times f_1 \times f_{upt}$ for no-till systems (where $f_{till}$ was 1.0) and [N mineralized during the season] $\times f_1 \times f_{till} \times f_{upt}$ for the tilled system.

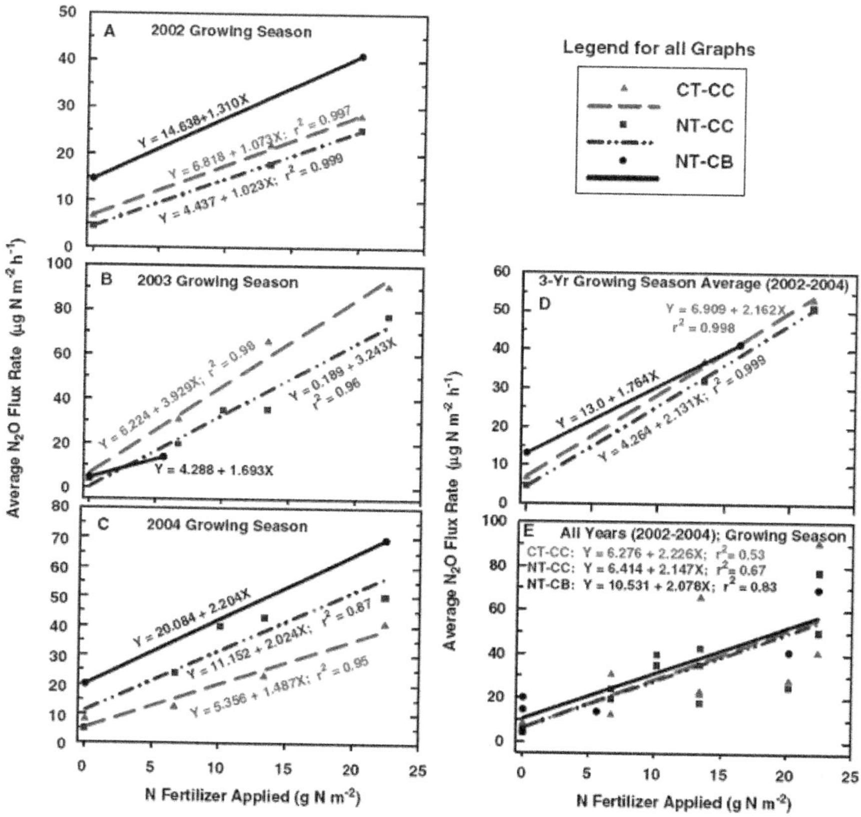


**Fig. 8. Effect of N fertilizer rate on growing season $N_2O$ emissions in (A) 2002, (B) 2003, (C) 2004, (D) 3-yr average data, and (E) all 3 yr of data. Note the different scales. CT-CC, conventionally tilled continuous corn; NT-CC, no-till continuous corn; NT-CB, no-till corn–soybean rotation. (Reproduced from Mosier et al., 2006).**

The results vary greatly among individual years. This is most likely a result of the very high spatial and temporal variability of data commonly encountered in such measurements but may also be caused by variations in timing of rainfall and other weather factors. The relationships in the 3-yr average data also show a very high $R^2$. The conventional till had higher slope and higher intercept than the no-till. The tillage factor in the above first-order equation can be obtained from these relations. The NT-CB (corn–soybean) system has higher slope and intercept than both the continuous corn (CC) systems. We think this is caused by the fact that soybeans added an organic N source to the soil, which resulted in higher mineralization of N and, hence, higher release of $N_2O$ during the corn season.

This example illustrates that the above first-order approach enhances the simple emission factor approach by including the effects of tillage and crop rotations and has the potential to include the effects of the fertilizer source and plant-uptake effects, if any. Halvorson and DelGrosso (2012) provide the data on emissions from several fertilizer sources to calculate relative fertilizer source factors ($f_{fs}$). The approach can also explain how, in some cases, the functional relationship between the emission and level of fertilizer applied shows faster than a linear increase at higher levels of fertilizers so as to be fit better by an exponential function, rather than a linear function. At very high rates of N application, beyond the plant-uptake needs, more of the N is available for denitrification in wetter soil pockets, resulting in more $N_2O$ emission.

## Conclusions

For uplands with high pH, low groundwater table, and low dissolved organic C content, such as northern China, nitrification may play an important role in soil $N_2O$ emission. The microbial mechanisms and controls seem to be different in soils with high groundwater table, for example, in northern Europe. Although N fertilization inevitably increases $N_2O$ emission, the good news here is that the lowest yield-scaled $N_2O$ emissions were obtained at modest N application rates (and not at zero application). Modeling soil $N_2O$ flux is still difficult because of its large temporal and spatial variation. More attention should be paid on the process-based models for accurate estimation of soil $N_2O$ emission.

## References

Abalos, D., A. Sanz-Cobena, L. Garcia-Torres, J.W. van Groenigen, and A. Vallejo. 2013. Role of maize stover incorporation on nitrogen oxide emissions in a non-irrigated Mediterranean barley field. Plant Soil 364:357–371. doi:10.1007/s11104-012-1367-4

Bateman, E., and E. Baggs. 2005. Contributions of nitrification and denitrification to $N_2O$ emissions from soils at different water-filled pore space. Biol. Fertil. Soils 41:379–388. doi:10.1007/s00374-005-0858-3

Braker, G. and R. Conrad. 2011. Diversity, structure, and size of $N_2O$-producing microbial communities in soils—What matters for their functioning? In: S.S. Allen, I. Laskin, and M.G. Geoffrey, editors, Advances in applied microbiology. Academic Press, Orlando, FL. p. 33–70.

Butterbach-Bahl, K., E.M. Baggs, M. Dannenmann, R. Kiese, and S. Zechmeister-Boltenstern. 2013. Nitrous oxide emissions from soils: How well do we understand the processes and their controls? Philos. Trans. R. Soc. Lond. B Biol. Sci. 368:20130122. doi:10.1098/rstb.2013.0122

Butterbach-Bahl, K., and M. Dannenmann. 2011. Denitrification and associated soil $N_2O$ emissions due to agricultural activities in a changing climate. Current Opinion Environ. Sustainability 3:389–395. doi:10.1016/j.cosust.2011.08.004

Chen, S., Y. Huang, X. Zheng, and Y. Chen. 2005. Nitrous oxide emission from cropland and its driving factors under different crop rotations. Scientia Agricultura Sinica 38:2053–2060.

Cui, F., G. Yan, Z. Zhou, X. Zheng, and J. Deng. 2012. Annual emissions of nitrous oxide and nitric oxide from a wheat–maize cropping system on a silt loam calcareous soil in the North China Plain. Soil Biol. Biochem. 48:10–19. doi:10.1016/j.soilbio.2012.01.007

Cui, Z., X. Chen, and F. Zhang. 2010. Current nitrogen management status and measures to improve the intensive wheat–maize system in China. Ambio 39:376–384. doi:10.1007/s13280-010-0076-6

Cui, Z., S. Yue, G. Wang, F. Zhang, and X. Chen. 2013. In-season root-zone N management for mitigating greenhouse gas emission and reactive N losses in intensive wheat production. Environ. Sci. Technol. 47:6015–6022. doi:10.1021/es4003026

Davidson, E.A. 2009. The contribution of manure and fertilizer nitrogen to atmospheric nitrous oxide since 1860. Nat. Geosci. 2:659–662. doi:10.1038/ngeo608

Davidson, E. A., J. Rogers, and W. Whitman. 1991. Fluxes of nitrous oxide and nitric oxide from terrestrial ecosystems. Microbial production and consumption of greenhouse gases: Methane, nitrogen oxides, and halomethanes.:219-235.

Davidson, E.A., and L.V. Verchot. 2000. Testing the Hole-in-the-Pipe Model of nitric and nitrous oxide emissions from soils using the TRAGNET Database. Global Biogeochem. Cycles 14:1035–1043. doi:10.1029/1999GB001223

Del Grosso, S.J., T. Wirth, S.M. Ogle, and W.J. Parton. 2008. Estimating agricultural nitrous oxide emissions. Eos Trans. AGU 89:529. doi:10.1029/2008EO510001

Ding, H., G. Cai, Y. Wang, and D. Chen. 2001. $N_2O$ emission from different crop-fluvo-aquic soil systems in the North China Plain. Agroenviron. Prot 20:7–9.

Ding, W., Y. Cai, Z. Cai, K. Yagi, and X. Zheng. 2007. Nitrous oxide emissions from an intensively cultivated maize–wheat rotation soil in the North China Plain. Sci. Total Environ. 373:501–511. doi:10.1016/j.scitotenv.2006.12.026

Ding, W., H. Yu, and Z. Cai. 2011. Impact of urease and nitrification inhibitors on nitrous oxide emissions from fluvo-aquic soil in the North China Plain. Biol. Fertil. Soils 47:91–99. doi:10.1007/s00374-010-0504-6

Dobbie, K.E., and K.A. Smith. 2003. Nitrous oxide emission factors for agricultural soils in Great Britain: The impact of soil water-filled pore space and other controlling variables. Glob. Change Biol. 9:204–218. doi:10.1046/j.1365-2486.2003.00563.x

Dong, W., C. Hu, S. Chen, and Y. Zhang. 2009. Tillage and residue management effects on soil carbon and $CO_2$ emission in a wheat–corn double-cropping system. Nutr. Cycl. Agroecosyst. 83:27–37. doi:10.1007/s10705-008-9195-x

Essah, S.Y.C. and J.A. Delgado. 2009. Nitrogen management for maximum potato yield, tuber quality, and environmental conservation. In: E. Yanful, editor, Appropriate technologies for environmental protection in the developing world. Springer, Netherlands. p. 307–315.

Firestone, M., and E. Davidson. 1989. Microbiological basis of NO and $N_2O$ production and consumption in soil. In: M.O. Andreae, D.S. Schimel, and G. P. Robertson, editors, Exchange of trace gases between terrestrial ecosystems and the atmosphere. Wiley, New York. p. 7–21.

Foley, J.A., N. Ramankutty, K.A. Brauman, E.S. Cassidy, J.S. Gerber, M. Johnston, N.D. Mueller, C. O'Connell, D.K. Ray, and P.C. West. 2011. Solutions for a cultivated planet. Nature 478:337–342. doi:10.1038/nature10452

Fu, X.Q. and Y. Li. 2012. Spatiotemporal variability of soil nitrous oxide emission and its study methods: A review. (In Chinese with English abstract.) Chinese J. Ecol. 31:724–730.

Gale, H. and K. Huang. 2007. Demand for food quantity and quality in China. Economic Research Report No. 32. USDA-ERS. USDA, Washington DC.

Gao, X., M. Tenuta, A. Nelson, B. Sparling, D. Tomasiewicz, R.M. Mohr, and B. Bizimungu. 2013. Effect of nitrogen fertilizer rate on nitrous oxide emission from irrigated potato on a clay loam soil in Manitoba, Canada. Can. J. Soil Sci. 93:1–11. doi:10.4141/cjss2012-057

Giltrap, D.L., A.G.E. Ausseil, K.P. Thakur, and M.A. Sutherland. 2013. Investigating a method for estimating direct nitrous oxide emissions from grazed pasture soils in New Zealand using NZ-DNDC. Sci. Total Environ. 465:7–16. doi:10.1016/j.scitotenv.2013.03.053

Granli, T., and O.C. Boeckman. 1994. Nitrous oxide from agriculture. Agric. Univ. of Norway, Advisory Service, Ås, Akershus, Norway.

Guohong, H., C. Guanxiong, W. Jie, H. Bin, and Y. Kewei. 1995. $N_2O$ and $CH_4$ fluxes from typical upland fields in northeast China. Chin. J. Appl. Ecol. 6:383–386.

Halvorson, A.D., and S.J. DelGrosso. 2012. Nitrogen source and placement effects on soil nitrous oxide emissions from no-till corn. J. Environ. Qual. 41:1349–1360. doi:10.2134/jeq2012.0129

Hoben, J., R. Gehl, N. Millar, P. Grace, and G. Robertson. 2011. Nonlinear nitrous oxide ($N_2O$) response to nitrogen fertilizer in on-farm corn crops of the US Midwest. Glob. Change Biol. 17:1140–1152. doi:10.1111/j.1365-2486.2010.02349.x

Hodge, A., and A.H. Fitter. 2010. Substantial nitrogen acquisition by arbuscular mycorrhizal fungi from organic material has implications for N cycling. Proc. Natl. Acad. Sci. USA 107:13754–13759. doi:10.1073/pnas.1005874107

Hu, C., S.A. Saseendran, T.R. Green, L. Ma, X. Li, and L.R. Ahuja. 2006. Evaluating nitrogen and water management in a double-cropping system using RZWQM. Vadose Zone J. 5:493–505. doi:10.2136/vzj2005.0004

Hu, X.K., F. Su, X.T. Ju, B. Gao, O. Oenema, P. Christie, B.X. Huang, R.F. Jiang, and F.S. Zhang. 2013. Greenhouse gas emissions from a wheat–maize double cropping system with different nitrogen fertilization regimes. Environ. Pollut. 176:198–207. doi:10.1016/j.envpol.2013.01.040

Huang, G., G. Chen, and J. Wu. 1995. $N_2O$ and $CH_4$ fluxes from typical upland fields in northeast China. (In Chinese with English abstract.) Chinese J. Applied Ecology 6:383–386.

Intergovernmental Panel on Climate Change (IPCC). 2006. Chapter 10. Emissions from livestock and manure management. guidelines for national greenhouse inventories. Vol. 4. Agriculture, forestry and other land use. IPCC, Geneva, Switzerland. p. 10.1–10.87.

Jin, M., R. Zhang, L. Sun, and Y. Gao. 1999. Temporal and spatial soil water management: A case study in the Heilonggang region, PR China. Agric. Water Manage. 42:173–187. doi:10.1016/S0378-3774(99)00039-6

Ju, X., X. Lu, Z. Gao, X. Chen, F. Su, M. Kogge, V. Römheld, P. Christie, and F. Zhang. 2011. Processes and factors controlling $N_2O$ production in an intensively managed low carbon calcareous soil under sub-humid monsoon conditions. Environ. Pollut. 159:1007–1016. doi:10.1016/j.envpol.2010.10.040

Keeney, D., I. Fillery, and G. Marx. 1979. Effect of temperature on the gaseous nitrogen products of denitrification in a silt loam soil. Soil Sci. Soc. Am. J. 43:1124–1128. doi:10.2136/sssaj1979.03615995004300060012x

Li, X., C. Hu, J.A. Delgado, Y. Zhang, and Z. Ouyang. 2007. Increased nitrogen use efficiencies as a key mitigation alternative to reduce nitrate leaching in north china plain. Agric. Water Manage. 89:137–147. doi:10.1016/j.agwat.2006.12.012

Liu, C., X. Zheng, Z. Zhou, S. Han, Y. Wang, K. Wang, W. Liang, M. Li, D. Chen, and Z. Yang. 2010. Nitrous oxide and nitric oxide emissions from an irrigated cotton field in Northern China. Plant Soil 332:123–134. doi:10.1007/s11104-009-0278-5

Lu, Y., Y. Huang, and X. Zheng. 2005. $N_2O$ emission factor for agricultural soils. (In Chinese with English abstract.) Chinese J. Applied Ecology 16:1299-1302.

Ma, L., R.W. Malone, P. Heilman, D.L. Karlen, R.S. Kanwar, C.A. Cambardella, S.A. Saseendran and L.R. Ahuja. 2007. RZWQM simulation of long-term crop production, water and nitrogen balances in Northeast Iowa. Geoderma 140:247–259.

Meng, Q., Q. Sun, X. Chen, Z. Cui, S. Yue, F. Zhang, and V. Römheld. 2012. Alternative cropping systems for sustainable water and nitrogen use in the North China Plain. Agric. Ecosyst. Environ. 146:93–102. doi:10.1016/j.agee.2011.10.015

Mosier, A.R., A.D. Halvorson, C.A. Reule, and X.J. Liu. 2006. Net global warming potential and greenhouse gas intensity in irrigated cropping systems in Northeastern Colorado. J. Environ. Qual. 35:1584–1598. doi:10.2134/jeq2005.0232

Mosier, A., C. Kroeze, C. Nevison, O. Oenema, S. Seitzinger, and O. Van Cleemput. 1998. Closing the global $N_2O$ budget: Nitrous oxide emissions through the agricultural nitrogen cycle. Nutr. Cycl. Agroecosyst. 52:225–248. doi:10.1023/A:1009740530221

Pei, S., Y. Zhang, J. Liu, X. Lun, and Y. Mu. 2012. $N_2O$ exchange fluxes from wheat-maize crop rotation system in the North China Plain. (In Chinese with English abstract.) Chinese J. Environ. Sci. 33:3641–3646.

Qin, S., C. Hu, and W. Dong. 2010. Nitrification results in underestimation of soil urease activity as determined by ammonium production rate. Pedobiologia 53:401–404. doi:10.1016/j.pedobi.2010.06.001

Qin, S., Y. Wang, C. Hu, O. Oenema, X. Li, Y. Zhang, and W. Dong. 2012. Yield-scaled $N_2O$ emissions in a winter wheat–summer corn double-cropping system. Atmos. Environ. 55:240–244. doi:10.1016/j.atmosenv.2012.02.077

Reay, D.S., E.A. Davidson, K.A. Smith, P. Smith, J.M. Melillo, F. Dentener, and P.J. Crutzen. 2012. Global agriculture and nitrous oxide emissions. Nat. Clim. Change 2:410–416. doi:10.1038/nclimate1458

Rees, R., J. Augustin, G. Alberti, B. Ball, P. Boeckx, A. Cantarel, S. Castaldi, N. Chirinda, B. Chojnicki, and M. Giebels. 2013. Nitrous oxide emissions from European agriculture–an analysis of variability and drivers of emissions from field experiments. Biogeosciences 10:2671–2682. doi:10.5194/bg-10-2671-2013

Ryden, J., and L. Lund. 1980. Nitrous oxide evolution from irrigated land. J. Environ. Qual. 9:387–393. doi:10.2134/jeq1980.00472425000900030012x

Sanz-Cobena, A., L. Sánchez-Martín, L. García-Torres, and A. Vallejo. 2012. Gaseous emissions of $N_2O$ and NO and $NO_3$ leaching from urea applied with urease and nitrification inhibitors to a maize (*Zea mays*) crop. Agric. Ecosyst. Environ. 149:64–73. doi:10.1016/j.agee.2011.12.016

Schellenberg, D.L., M.M. Alsina, S. Muhammad, C.M. Stockert, M.W. Wolff, B.L. Sanden, P.H. Brown, and D.R. Smart. 2012. Yield-scaled global warming potential from $N_2O$ emissions and $CH_4$ oxidation for almond (*Prunus dulcis*) irrigated with nitrogen fertilizers on arid land. Agric. Ecosyst. Environ. 155:7–15. doi:10.1016/j.agee.2012.03.008

Shen, Y., A. Kondoh, C. Tang, Y. Zhang, J. Chen, W. Li, Y. Sakura, C. Liu, T. Tanaka, and J. Shimada. 2002. Measurement and analysis of evapotranspiration and surface conductance of a wheat canopy. Hydrol. Processes 16:2173–2187. doi:10.1002/hyp.1149

Shi, Y., W. Wu, F. Meng, Z. Zhang, L. Zheng, and D. Wang. 2013. Integrated management practices significantly affect $N_2O$ emissions and wheat–maize production at field scale in the North China Plain. Nutr. Cycl. Agroecosyst. 95:203–218. doi:10.1007/s10705-013-9558-9

Smith, K.A., I.P. McTaggart, K.E. Dobbie, and F. Conen. 1998. Emissions of $N_2O$ from Scottish agricultural soils, as a function of fertilizer N. Nutr. Cycl. Agroecosyst. 52:123–130. doi:10.1023/A:1009781518738

Sun, Y., P. Lu, J. Li, Q. Yu, S. Sun, J. Wang, and Z. Ouyang. 2008. Characteristics of soil $N_2O$ flux in a winter wheat–summer maize rotation system in North China Plain and analysis of influencing factors. (In Chinese with English abstract.) Chinese J. Agrometeorol. 29:1–5.

Syakila, A., and C. Kroeze. 2011. The global nitrous oxide budget revisited. Greenhouse Gas Meas. Manage. 1:17–26. doi:10.3763/ghgmm.2010.0007

Tian, S., T. Ning, H. Zhao, B. Wang, N. Li, H. Han, Z. Li, and S. Chi. 2012. Response of $CH_4$ and $N_2O$ emissions and wheat yields to tillage method changes in the North China Plain. PLoS ONE 7:e51206. doi:10.1371/journal.pone.0051206

Tilman, D., K.G. Cassman, P.A. Matson, R. Naylor, and S. Polasky. 2002. Agricultural sustainability and intensive production practices. Nature 418:671–677. doi:10.1038/nature01014

Toma, Y., S.D. Kimura, Y. Hirose, K. Kusa, and R. Hatano. 2007. Variation in the emission factor of $N_2O$ derived from chemical nitrogen fertilizer and organic matter: A case study of onion fields in Mikasa, Hokkaido, Japan. Soil Sci. Plant Nutr. 53:692–703. doi:10.1111/j.1747-0765.2007.00184.x

Van Groenigen, J., G. Velthof, O. Oenema, K. Van Groenigen, and C. Van Kessel. 2010. Towards an agronomic assessment of $N_2O$ emissions: A case study for arable crops. Eur. J. Soil Sci. 61:903–913. doi:10.1111/j.1365-2389.2009.01217.x

Venterea, R.T., M. Bijesh, and M.S. Dolan. 2011. Fertilizer source and tillage effects on yield-scaled nitrous oxide emissions in a corn cropping system. J. Environ. Qual. 40:1521–1531. doi:10.2134/jeq2011.0039

Vitousek, P. and R. Howarth. 1991. Nitrogen limitation on land and in the sea: How can it occur? Biogeochemistry 13:87–115.

Wan, Y., X. Ju, J. Ingwersen, U. Schwarz, C.F. Stange, F. Zhang, and T. Streck. 2009. Gross nitrogen transformations and related nitrous oxide emissions in an intensively used calcareous soil. Soil Sci. Soc. Am. J. 73:102–112. doi:10.2136/sssaj2007.0419

Wang, C.Y., J. Zheng, J.X. Gu, Y. Shi, and X. Chen. 2006a. $CO_2$ and $N_2O$ emission fluxes and impact factors of soybean field in lower reaches of Liaohe Plain. Soils 38:708–711.

Wang, C., J. Zheng, J. Gu, S. Wang, Y. Shi, and X. Chen. 2006b. $N_2O$ emission flux from several dry crops and impact factor s in the lower reaches of Liaohe plain. J. Agro-Environ. Sci. 25:657–663.

Wang, L., H. Li, and J. Qiu. 2008. Characterization of emissions of nitrous oxide from soils of typical crop fields in Huang-Huai-Hai Plain. Scientia Agricultura Sinica 41:1248–1254.

Wang, Y.Y., C.S. Hu, H. Ming, Y.M. Zhang, X.X. Li, W.X. Dong, and O. Oenema. 2013. Concentration profiles of $CH_4$, $CO_2$ and $N_2O$ in soils of a wheat–maize rotation ecosystem in North China Plain, measured weekly over a whole year. Agric. Ecosyst. Environ. 164:260–272. doi:10.1016/j.agee.2012.10.004

Xing, G. 1998. $N_2O$ emission from cropland in China. Nutr. Cycl. Agroecosyst. 52:249–254. doi:10.1023/A:1009776008840

Xiong, Z., and G. Xing. 2002. The effects of summer legume crop cultivation on nitrous oxide emissions from upland farmland. Scientia Agricultura Sinica 35:1104–1108.

Xu, H., G. Xing, Z. Cai, and H. Tsuruta. 2000. Effect of soil texture on $N_2O$ emissions from winter wheat and cotton fields. Agro-Environ. Protect. 19:1–3.

Yan, G. 2012. The $N_2O$ and NO emissions of three types of agricultural system in North China Plain. In: Abstracts EGU General Assembly 2012, Vienna, Austria, 22–27 Apr. 2012.

Ye, X., J. Li, Y.H. Wang, E.M. Liu, R.X. Li, Q. Yu, and B.X. Chen. 2005. Characterization of emissions of nitrous oxide from soils of typical crop fields in North China Plain. J. Agro-Environ. Sci. 24:1186–1191.

Yu, K., G. Chen, S. Yang, J. Wu, B. Huang, G. Huang, and H. Xu. 1995. Role of several upland crops in $N_2O$ emission from farmland and its response to environmental factors. (In Chinese with English abstract.) Chinese J. Appl. Ecol. 6:387–391.

Yuan, Z., and Y. Shen. 2013. Estimation of agricultural water consumption from meteorological and yield data: A case study of Hebei, North China. PLoS ONE 8:e58685. doi:10.1371/journal.pone.0058685

Zhang, X.Y., D. Pei, and C.S. Hu. 2003. Conserving groundwater for irrigation in the North China Plain. Irrig. Sci. 21:159–166.

Zhang, Y., J. Liu, Y. Mu, Z. Xu, S. Pei, X. Lun, and Y. Zhang. 2012. Nitrous oxide emissions from a maize field during two consecutive growing seasons in the North China Plain. Chin. J. Environ. Sci. 24:160–168. doi:10.1016/S1001-0742(10)60594-3

Zhang, Y., Y. Shen, H. Sun, and J.B. Gates. 2011. Evapotranspiration and its partitioning in an irrigated winter wheat field: A combined isotopic and micrometeorologic approach. J. Hydrol. 408:203–211. doi:10.1016/j.jhydrol.2011.07.036

Zhou, S.L., F.S. Zhang, and X.R. Wang. 2002. The spatio-temporal variations of soil $NO_3$–N and apparent budget of soil nitrogen II. Summer maize. Acta Ecol. Sin. 1:48–53.

# Comparison of DayCent and DNDC Models: Case Studies Using Data from Long-Term Experiments on the Canadian Prairies

**Brian B. Grant,* Ward N. Smith, Con A. Campbell, Raymond L. Desjardins, Reynald L. Lemke, Roland Kröbel, Brian G. McConkey, Elwin G. Smith, and Guy P. Lafond**


## Abstract

Two prominent process-based models, DNDC (i.e., Denitrification–Decomposition) and DayCent, were investigated for their ability to capture the inter-annual variability in spring wheat grain yields, soil carbon change, and trace gas emissions for three long-term experiments conducted on the Canadian prairies. The cropping systems included continuous spring wheat (*Triticum aestivum* L.) receiving N and P fertilizer [Cont-W(N+P)], continuous wheat receiving P [Cont-W(+P)], and fallow–wheat receiving N and P [F-W(N+P)]. These studies were conducted on three distinct soils: a Brown Chernozen (silt loam), a Dark Brown Chernozem (sandy clay loam) and a thin Black Chernozem (heavy clay). Both models were effective in estimating variations in grain yield for the Cont-W(N+P) rotation, particularly for drier to normal periods, with coefficients of determination ($R^2$) ranging from 0.56 to 0.73 and 0.51to 0.70 for DayCent and DNDC, respectively. When crop N requirements were not met fully by fertilizer inputs [i.e., rotations Cont-W(+P) and F-W(N+P)], the models demonstrated decreased accuracy in predicting yields, which suggests deficiencies in the estimated N balance. Crop water use was generally well estimated for both models (AREs < 6% and < 15% for DayCent and DNDC) considering the marked differences in each models conceptualization of the soil profile and water movement. Nitrous oxide estimates by both models compared well with measurements and captured the relative differences between fertilized and unfertilized N systems. The utilization of this long-term dataset proved to be highly informative in assessing each model's characterization of C and N dynamics while highlighting further development needs.


Abbreviations: ARE, average relative error; DNDC, Denitrification–Decomposition model; HI, harvest index; SOC, soil organic carbon; WFPS, water-filled pore space.

Brian B. Grant, Ward N. Smith (ward.smith@agr.gc.ca), Con A. Campbell (con.campbell@agr.gc.ca), and Raymond L. Desjardins (ray.desjardins@agr.gc.ca), Eastern Cereal and Oilseed Research Centre, Agriculture and Agri-Food Canada, 960 Carling Ave., Ottawa, ON K1A 0C6, Canada. *Corresponding author (brian.grant@agr.gc.ca). Reynald L. Lemke, Agriculture and Agri-Food Canada, 107 Science Place, Saskatoon, SK S7N 0X2, Canada. (reynald.lemke@agr.gc.ca). Roland Kröbel (Roland.Kroebel@agr.gc.ca) and Elwin G. Smith (Elwin.Smith@agr.gc.ca), Lethbridge Research Centre, Agriculture and Agri-Food Canada, 5403 1st Ave. South, Lethbridge, AB T1J 4B1, Canada. Brian G. McConkey, Semiarid Prairie Agricultural Research Centre, Agriculture and Agri-Food Canada, 1 Airport Road, P.O. Box 1030, Swift Current, SK S9H 3X2, Canada. (Brian.McConkey@agr.gc.ca). Guy P. Lafond, deceased, formerly Indian Head Research Farm, Agriculture and Agri-Food Canada, PO Box 760, Indian Head, SK S0G 2K0, Canada.

doi:10.2134/advagricsystmodel6.2013.0035


Published by ASA, CSSA, SSSA, 5585 Guilford Rd., Madison, WI 53711-5801, USA.
Synthesis and Modeling of Greenhouse Gas Emissions and Carbon Storage in Agricultural and Forest Systems to Guide Mitigation and Adaptation
S. Del Grosso, L. Ahuja, and W. Parton, Editors
Advances in Agricultural Systems Modeling, Volume 6. Lajpat R. Ahuja, Series Editor

One of the long-term experimental plots in Swift Current, Saskatchewan selected for this study.

With global populations continuing to increase, and with new competition for energy from biomass, the demand on agriculture to increase productivity per unit area will become even more critical. The agricultural sector has successfully implemented management strategies to increase production. This trend should continue provided that climate change does not adversely affect agriculture. Soil organic carbon (SOC) plays an important role in limiting global atmospheric $CO_2$ levels while providing significant agronomic benefits at the localized scale (Lal, 2004). Greater productivity, reduced erosion, and improvements in nutrient balance and water retention are some of the many benefits that increased SOC can provide (Carter et al., 1997). Conversely, the decomposition of SOC negatively affects the environment by releasing $CO_2$ to the atmosphere, which in turn may enhance climate change and increase temperatures (Abdalla et al., 2011). As nearly 60% of $N_2O$ attributed to anthropogenic activities originates from agriculture (Smith et al., 2007), with an associated global warming potential of nearly 310 times that of $CO_2$, there is an obvious imperative to conduct assessments with a wider scope than just $CO_2$.

Agricultural farming systems are complex, managed environments where nutrient and energy flows are tightly coupled. For this reason, any change to the farming system, such as the introduction of a new management practice, or a changing climate, has the potential to alter these nutrient and energy flows and impact the net greenhouse gas emissions that result from these changes. Biogeochemical models are formulated to consider the dynamics and mass balance of these flows and thus are useful in the quantification of the net radiative forcing required to assess a wide range of ecosystem services. Additionally, they are used to evaluate the impact of changes to climate on crop productivity and changes to soil quality factors while considering an array of agricultural management practices. As opposed to field measurements that are restricted in their assessment value because of their cost/time investment, these models can be employed over a multitude of temporal and spatial scales. Their conceptual framework facilitates synthesizing the current state of knowledge; however, rigorous validation of their outputs is required to ensure confidence in their predictions. Validating carbon change can be particularly challenging as changes in soil carbon induced by management and crop residue carbon inputs are often within measurement variability because of their slow rate of change, the small size of change relative to the total SOC pool, the variability of weather, the heterogeneity of the soil, and possible imprecisions in measurement methodologies. Datasets for experiments conducted over long durations are invaluable for model testing as it usually takes many decades of measurements to reveal significant trends for soil carbon change. Campbell et al. (2005) state that long-term studies provide a greater probability of experiencing an array of diverse weather conditions, thereby allowing us to better define the outcomes one might expect under a wide spectrum of growing conditions. Models are often employed to investigate carbon change (Smith et al., 1997, 2012b; Campbell et al., 2007; Lemke et al., 2012), crop yields (Touré et al., 1995; Roloff et al., 1998a; Kersebaum et al., 2008; Kröbel et al., 2011), or trace gas emissions (Del Grosso et al., 2008; Smith et al., 2008), but few studies have been conducted that aggregate all of these together (Zhang et al., 2002; Del Grosso et al., 2005). By comparing two prominent trace gas models, DNDC and DayCent, we hope to identify model mechanisms or concepts that could be improved as well as investigating and modeling the impact of climate variability on multiple agronomic outcomes.

The objectives of this study were to use the DNDC and DayCent models together with results from three long-term crop rotation studies on the Canadian prairies to (i) estimate grain production of spring wheat in response to inter-annual changes in climate-induced stresses, (ii) compare against measured data and systematically test the ability of the models to estimate crop water use, crop yields, carbon inputs, soil carbon change, and nitrous oxide emissions, and (iii) compare the performance of the models in relation to the mechanisms they embody and suggest possible model improvements.

## Materials and Methods

### Site Description

Three long-term crop rotation experimental sites were selected for this study. They characterize some of the diversity in climate and soil conditions that exist across the Canadian prairies. Commonalities in experimental cropping systems among the sites provided an opportunity to isolate the impact of climate and soil characteristics, independent of crop type, on inter-annual crop yields, SOC, and nitrous oxide emission. The sites ranged from the dry semiarid soils at Swift Current, SK (Brown Chernozem) and Lethbridge, AB (Dark Brown Chernozem) to a sub-humid soil at Indian Head, SK (thin Black Chernozem) (Table 1). The rotations investigated consisted of three spring wheat (*Triticum aestivum* L.) based systems: continuous wheat receiving N and P fertilizer [Cont-W(N+P)], continuous wheat receiving P only [Cont-W(+P)], and fallow–wheat receiving N and P fertilizer [F-W(N+P)]. Annual yields were available for a minimum of 33 yr along

**Table 1. Selected experimental sites, climate, and crop rotation characteristics.**

| Location and duration of experiment | Soil group (Canadian Classification) | Soil texture | Soil C† (15-cm depth) | Mean growing season‡ temp. | Total growing season‡ precip. | Mean growing season‡ wind speed | Mean growing season‡ PET§ |
|---|---|---|---|---|---|---|---|
| | | | Mg C $ha^{-1}$ | °C | mm | km $h^{-1}$ | mm |
| Swift Current, SK (1967–2005) | Brown Chernozem | silt loam | 30.5 | 15.5 | 212 | 18.1 | 459 |
| Lethbridge, AB (1972–2004) | Dark Brown Chernozem | sandy clay loam | 35.0 | 15.6 | 209 | 16.7 | 430 |
| Indian Head, SK (1957–1996) | thin Black Chernozem | heavy clay | 30.5 | 15.9 | 251 | 13.7 | 367 |

† Initial soil C at the start of each experimental period.

‡ Growing Season length is from 1 May to 31 August.

§ PET, potential evapotranspiration as estimated by Penman–Montieth (FAO approach).

with soil carbon measurements taken at each of the study sites. Detailed descriptions of each of the study sites are as follows.

The Swift Current long-term site was established in 1966 at the Agriculture Canada Research station in southwestern Saskatchewan (50°17′ N, 107°48′ W). Before the experiment, the fields had been cropped as F-W with minimal fertilizer since 1922. The experimental plots are located on an Orthic Brown Chernozem (Aridic Haploboroll) soil and are characterized as a Swinton loam (Ayres et al., 1985). For the rotations containing spring wheat, the timing of seeding typically occurred in May with harvest operations generally performed in late August. Fertilizer N was applied as $NH_4NO_3$ with amounts based on soil N tests conducted each fall, and P was applied as ammonium phosphate at seeding at an annual rate of 10 kg P $ha^{-1}$. For the period of 1967 to 1989, N rates were based on the recommendations of the soil-testing laboratory of the University of Saskatchewan (65 kg $ha^{-1}$ for cereals); after 1990 the recommended total N rate changed to 90 kg $ha^{-1}$ for cereals on fallow and 73 kg $ha^{-1}$ for cereals grown on stubble (Campbell et al., 2005). The increase in N rates coincided with improved growing season weather conditions for the period 1989 to 2002. Tillage operations consisted of a single pass using a cultivator once in the spring and with multiple passes during fallow years. Swift Current had the most extensive measurement array of any of the three selected sites with periodic measurements of soil C, grain/straw yields, grain/straw N, soil moisture, and nitrous oxide emissions (Campbell and Zentner 1993; Campbell et al., 2005, 2007). Growing season precipitation (1 May to 31 August) averaged 212 mm coupled with relatively high estimates of potential evapotranspiration (PET) resulting from windy conditions.

Nearly 30% of the annual precipitation originates as snowfall. The study experienced a number of severe droughts during the years 1968, 1980, 1984, 1985, and 1988 that were reflected in the observed wheat yields. A more detailed description of the design and management of this experiment can be found in previous manuscripts (Campbell et al., 1983, 1992, 2007; Zentner and Campbell, 1988; Campbell and Zentner, 1993).

Long-term plot studies at Agriculture Canada Research Station in Lethbridge (49°42′ N, 112°50′ W) are located in the southwest portion of Alberta on an Orthic Dark Brown Chernozem (Typic Haploboroll) with a sandy clay loam textured soil. The plots were first established in 1910 when the native prairies (short and mid grass species) were tilled to investigate various wheat cropping systems for their feasibility for possible broadscale application across the region. Initially none of the established rotations received fertilizer until 1967 when they were divided into one set of sub-plots receiving N fertilizer and another set without N fertilizer. Fertilizer N was broadcast at a constant rate of 45 kg N $ha^{-1}$ (ammonium nitrate). In 1972, the plots were further split to include P fertilizer, applied with the seed, at a fixed rate of 20 kg P $ha^{-1}$. Tillage operations were the same as in Swift Current, but in the late 1990s no-till was adopted for all plots. Planting times for spring wheat were normally in May with harvest operations conducted in late August. Growing season precipitation for the site is approximately 209 mm with annual precipitation amounts of about 380 mm. Precipitation received as snowfall accounts for nearly 30% of the annual precipitation received. Droughts were experienced periodically (mid 1930s, early 1960s) and more recently in 1984, 1985, and 2000. A more detailed description of the design and management of the study has been reported (Janzen, 1987, 1995; Monreal and Janzen, 1993; Janzen et al., 1997, 1998; Smith et al., 2012a).

The third long-term rotation experiment was initiated in 1957 at the Agriculture Canada Experimental Farm in Indian Head, SK (50°32′ N, 103°31′ W). The fields had previously been managed as either fallow–wheat or fallow–wheat–wheat rotations that received minimum fertilizer. The soil is classified as heavy clay textured on a thin Black Chernozem (Udic Boroll) that is moderately well drained. Fertilizer N and P was initially applied at recommend rates for spring wheat in the region and then modified to be based on soil test levels of N and P in 1978. Average annual N rates for Cont-W(N+P) increased from 24 kg N $ha^{-1}$ for the period 1958 to 1977 to 76 kg N $ha^{-1}$ for the years 1978 to 1991, while wheat following fallow increased from 6 to 10.5 kg N $ha^{-1}$ for the same time periods. Seedbed preparation was conducted with a single operation of a heavy-duty cultivator with an attached harrow. Stubble plots received an additional harrow treatment before planting using a double disc press drill. Weed control on summer fallow was managed using five operations of heavy cultivator tillage. Conventional tillage was employed until 1990 and then changed to no-till practices. Growing season precipitation (1 May to 31 August) is 250 mm with an annual precipitation of around 430 mm. Annual PET (670 mm) was lower at Indian Head than the other two locations because of lower wind speeds (Campbell et al., 1990). A more detailed description discussing the establishment of the plots and their management history has previously been reported (Campbell et al., 2001; Lafond et al., 2009, 2012; Zentner et al., 1987; Lemke et al., 2012).

## Model Description—DNDC

The DNDC model by Li et al. (1992, 1994, 2000) is a biogeochemical model designed to estimate carbon and N dynamics and trace gas emissions ($NO_x$, $NH_3$, and $CH_4$) for a diverse set of agricultural management cropping systems. It is conceptualized into four main sub-models: soil/climate, crop vegetation, decomposition, and denitrification. The soil/climate sub-model is built as a layered water balance to allow prediction of soil moisture, soil water flow, and crop water uptake on an hourly basis. The soil profile is conceptualized to a 50-cm depth together with a single pool for water storage beyond 50 cm. By default, the size of this latter pool is determined by soil properties defined for the top 50 cm as DNDC. This study employed a slight modification so that soil properties for the 50- to 100-cm depth could be utilized to better represent the water holding capacity of this segment of the soil. Various crops can be simulated with the empirically defined crop growth sub-model to predict biomass C and N content of grain, stalk, and roots. Crop growth is limited by a combination of heat unit, nitrogen, and water availability. The decomposition sub-model is composed of four main carbon pools (litter, labile humus, passive humus, and microbial biomass). Each of these has their own fixed C to N ratio and base decomposition rate. These rates are influenced by a combination of soil texture, soil moisture, soil temperature, and nitrogen. The DNDC model's framework for N gas emissions adopts a dynamic anaerobic balloon concept to regulate the rate of the nitrification and denitrification reactions. This balloon is controlled using soil redox potential (Eh), which is estimated using the Nernst equation (Li et al., 2000). The DNDC model is not conceptualized to explicitly track populations of different types of bacteria and assumes that the initial enzyme activity (i.e., ammonia monooxygenase) is proportional with the initial microbial biomass but not any specific bacterium species. The scientific basis for calculating nitrification (or other microbial reactions) on the basis of total microbial biomass is the hypothesis of "parallel gene transfer." To predict growth and death rates of microorganisms, Li et al. (2000) chose to adopt equations from Blagodatsky and Richter (1998) that regulate bacterial growth/death through soil dissolved organic carbon concentration, temperature, and moisture. It is these equations that are used in DNDC for calculating nitrifier activity. The nitrification reaction itself is a function of $NH_4^+$ concentration and nitrifier biomass and is limited by temperature and moisture stress (Li et al., 2012). The formation of N gas, including $N_2O$, is a function of water-filled pore space (WFPS) and the quantity of N nitrified. Denitrification is handled by estimating the quantity of denitrifier bacteria using a multi-nutrient dependent (Michaelis–Menten) growth function that is regulated by soil temperature, soil water, Eh, and pH (Li et al., 2000). Denitrification is then calculated as a stepwise ($NO_x$, $N_2O$, $N_2$) transformation process as a function of microbial growth and pH. All $N_2O$ produced in the soil through either mechanism is further regulated by a diffusion algorithm that considers the soil properties of the layer before the gas is emitted to the atmosphere. The denitrification sub-model is utilized when soil moisture increases and/or soil oxygen decreases. Further details regarding the mechanisms in DNDC in comparison to DayCent are described in Table 2.

## Model Description—DayCent

The DayCent model (Kelly et al., 2000; Parton et al., 1998; Del Grosso et al., 2001) represents the iterative development of a daily version of the CENTURY model (Parton et al., 1994). Its development was initiated to allow for the improved quantification of trace gas fluxes ($CO_2$, $N_2O$, $NO_x$, and $CH_4$) across varying agroecosystems. It is a one-dimensional model designed to simulate crop growth, carbon and nutrient dynamics (N, P, and K), and soil water transport. It differs from CENTURY, however, in that it characterizes many processes at a daily time step rather than monthly and has a more spatially explicit soil profile. The shift to a daily time step was necessary as the estimation of trace gas fluxes requires a finer temporal resolution as these fluxes are typically associated with short-term precipitation or irrigation events (Frolking et al., 1998). Pools of C and N are conceptualized across a number of pools (surface active, soil active, slow, and passive) and are regulated by their size, C to N ratio, lignin content, soil moisture, and soil temperature. The microbial processes of nitrification and denitrificiation are characterized by a concept known as the "hole-in the pipe" (Firestone and Davidson, 1989), which gives rise to N gas emissions. Nitrification in DayCent is a function of $NH_4^+$ concentration, water content, temperature, pH, and soil texture with the $N_2O$ produced being a constant fraction of the total N nitrified. On the denitrification side, the maximum rate is estimated as being a function of labile C (acts as the $e^-$ donor) and $NO_3^-$ concentration ($e^-$ acceptor) with $O_2$ availability ($e^-$ acceptor) also playing a role. Simulated heterotrophic respiration acts to regulate labile C availability and $O_2$ availability. This rate is further reduced by a

**Table 2. Overview of processes used to simulate soil hydrology, carbon and nitrogen dynamics, biomass production, and trace gas emissions in the DayCent and DNDC models†**

| | DNDC | DayCent |
|---|---|---|
| | Desired outcomes and major model inputs | |
| Major model outputs | N gas emissions ($N_2O$, NO, $NH_3$, $CH_4$, $CO_2$), soil organic matter change, crop production, N leaching, and runoff | N gas emissions ($N_2O$, NO, $NH_3$, $CH_4$, $CO_2$), soil organic matter change, crop production, nutrient leaching, and runoff (N, P, S) |
| Agricultural management and events simulated | Plant type (annual or perennial), planting and harvest date, cultivation, organic matter or fertilizer additions, N inhibitor, slow release N, grazing, intercropping, rice paddies, and irrigation | Plant type (annual or perennial), planting and harvest date, cultivation, organic matter or fertilizer additions, N inhibitor, grazing, irrigation, fire scheduling |
| Model inputs: soil properties | Soil properties of surface layer (bulk density, texture, field capacity, wilting point, hydraulic conductivity), can modify SOM C to N ratios, organic matter partitioning is determined by model as a function of SOM concentration and soil properties | Soil properties of multiple layers (bulk density, texture, field capacity, wilting point, hydraulic conductivity, residual water content, plant rooting depth), organic matter partitioning based on equilibrium simulation. |
| Model inputs: environmental drivers | Max and min air temperature, precipitation, solar radiation, wind speed, humidity | Max and min air temperature, precipitation, solar radiation, wind speed, humidity |
| | Soil–water–plant dynamics | |
| Soil profile | Homogenous profile to 50-cm depth, underlying deepwater pool to below 50-cm depth | Heterogeneous profile to any depth, user-defined layers |
| Soil water transport | Cascade flow where water drains to the next layer if above field capacity, layered water balance, empirical capillary rise | Saturated water flow occurs when there is sufficient water inputs (precipitation, irrigation, snow melt) for the soil profile to go above field capacity. Unsaturated flow occurs on days that the profile does not reach saturation and is dependent on matric and gravitational potential (can flow up or down the profile). A layered water balance is calculated. |
| Potential evapotranspiration | Penman–Montieth FAO approach | Penman–Montieth |
| Transpiration | Function of crop biomass | Function of leaf biomass |
| Evaporation | Function of PET, temperature, available soil moisture | Function of PET, temperature, available soil moisture |
| Infiltration | Water loss occurs first as canopy interception followed by infiltration limited by surface saturated hydraulic conductivity | Water loss occurs first as canopy interception followed by infiltration limited by surface saturated hydraulic conductivity. Excess water is assumed to be runoff. |

Table continued.

**Table 2. Continued.**

| | DNDC | DayCent |
|---|---|---|
| Erosion | Estimated by the Soil Conservation Service Curve Number (SCS-CN) methodology and the Modified Universal Soil Loss Equation (MUSLE) | Erosion is not simulated, but wind and water erosion can be estimated externally and included as input |
| Soil thermal properties and freeze/thaw | Changes in soil temperature are estimated as the rate of change in heat flux from layer to layer divided by the soil volumetric heat capacity. Heat flux from layer to layer are estimated as a function of thermal conductivity and change in temperature between layers. Heat release due to water freezing is estimated as a function of water volume (Li et al., 2006). | Soil heat flux is based on the Fourier heat flow equation and is approximated using an explicit finite difference scheme. Daily maximum and minimum temperature and soil thermal diffusivity (method of De Vries) are estimated for each soil layer. To incorporate the discontinuous freeze–thaw process, a tipping-bucket routine using a calculated fusion energy pool was employed (Eitzinger et al., 2000). |
| Snow accumulation/insulation | Estimates daily snow accumulation, snow melt, snow thermal insulation, and sublimation | Estimates daily snow accumulation, snow melt, snow melt infiltration |
| Nutrient leaching | A small portion of $NO_3^-$ is leached to groundwater as preferential flow and the remainder is limited in mobility through the profile by the adsorptive or buffering capacity of the soil. $NH_4^+$ is simulated through the complete profile with the adsorption and desorption of ammonium ions is determined by the Langmuir equation (Li et al., 2006). Nutrients that leach below 50 cm are assumed to be leached into groundwater or the sub-soil. | A fraction of $NO_3^-$ is leached to the next soil layer during saturated events. Ammonium is simulated in the top 15 cm and is assumed to be immobile. Nutrients that leave the deepest soil layer are assumed to be leached into groundwater or the sub-soil. |
| | Biomass production | |
| Biomass production (NPP) | Function of N availability, soil water, soil temperature (additional stress during anthesis reducing harvest index), vegetation type, shading, and plant phenology (empirical updated for growth rates in Canada) | Function of nutrient availability (N, P, K), soil water (additional stress at time of grain filling reducing harvest index), soil temperature, shading, vegetation type, and plant phenology (empirical or growing degree day sub-models) |
| Biomass partitioning | Root to shoot ratio is a function of plant type, phenology soil water, N availability | Root to shoot ratio is a function of plant type, phenology, soil water, and nutrient availability |
| | Yield formation is a function of harvest index, total aboveground biomass affected by N, and temperature stress | Yield formation is a function of harvest index with grain filling/dynamic carbon allocation affected by N and water stress |

| | DNDC | DayCent |
|---|---|---|
| Plant phenology | Empirical Canada specific growth curves affected by growing degree day, photoperiod, vernalization, and senescence | Function of growing degree day, photoperiod, vernalization, senescence, translocation of nutrients. Germination and harvest dates can be defined by user or made a function of soil temperature, or accumulated growing degree days, respectively. |
| Effect of $CO_2$ fertilization on biomass | Assimilation in C3 crops, water use efficiency, N use efficiency | Assimilation in C3 crops, water use efficiency, N use efficiency |
| Nitrogen uptake | Based on plant C to N ratio from biomass production | Upper and lower boundaries for adequate and deficient C:N |
| | Soil organic matter and N dynamics | |
| Live carbon | Leaf + stem, grain, root | Leaf + stem, grain, juvenile and mature roots |
| Litter | Standing dead, very labile, labile, and resistant litter | Standing dead, metabolic, and structural surface littler partitioned by lignin to N ratio |
| SOM pools | Compartmental by layer. Main SOM pools: microbial biomass (labile and resistant) and humus (labile, resistant, and passive) | Compartmental by layer. Main SOM pools: surface active, soil active, slow, and passive |
| Decomposition of SOM | First order kinetics. Influenced by soil texture, soil moisture, temperature, clay content, type and quantity of residue inputs, tillage, depth in the soil profile, and nitrogen limitations (Li et al., 1994). Pools have different C to N ratios and decomposition rates. | First order kinetics. Influenced by soil texture, soil moisture, temperature, clay content, quality and quantity of litter inputs, tillage, and nutrient limitations. Pools have different C to N ratios and decomposition rates. |
| Nitrification rate | Nitrifier-bacteria growth and death rate are functions of DOC and a temperature factor. Nitrification rate is a function of $NH_4^+$ concentration, nitrifier biomass, a temperature reduction factor, and a moisture reduction factor (Li et al., 2012). | Nitrification rate is a function of $NH_4^+$, water content, temperature, pH, and texture |
| | N gas emissions | |
| General N gas formulation | Anaerobic balloon concept. Eh estimated using Nernst equation. Anaerobic balloon increases or shrinks depending on Eh regulating size of aerobic (nitrification) and anaerobic (denitrification fractions). | Uses a "hole-in-the-pipe" formulation and sums emissions from nitrification and denitrification. |
| N gas from nitrification | Function of water-filled pore space and quantity of N nitrified | Fraction of N nitrified |

Table continued.

**Table 2. Continued.**

| | DNDC | DayCent |
|---|---|---|
| N gas from denitrification | Quantity of denitrifier bacteria is estimated using a multi-nutrient dependent (Michaelis–Menten) growth function dependent on temperature, soil water, Eh, pH (Li et al., 2000). Denitrification calculated as stepwise transformation process as a function of microbial growth and pH. | DayCent estimates a maximum denitrification rate as a function of the electron donor (labile C) or acceptor ($NO_3^-$) that is most limiting. Heterotrophic respiration acts as a proxy for labile C availability and $O_2$ demand. The denitrification rate is further reduced by a soil moisture factor that is an arctangent function of water-filled pore space, respiration, and gas diffusivity. A gas loss ratio determines the ratio of $N_2$ to $N_2O$ (Parton et al., 2001; Chianese et al., 2009). |
| $NH_3$ volatilization | The model calculates the hydrolysis rate as a first order function of urease enzyme activity and urea concentration. The urease enzyme activity is estimated as a linear function of DOC, moisture, and temperature (Li et al., 2012). Hydrolysis of urea consumes $H^+$ and shifts the $NH_4^+/NH_3$ equilibrium increasing $NH_3$ volatilization. $NH_4^+$ produced reaches equilibrium with dissolved $NH_3$ and $NH_3$ concentration in soil gas phase is estimated. The DNDC model estimates $NH_3$ volatilization as a function of $NH_3$ concentration in soil gas phase and soil air-filled porosity. | A portion of N excreted from animals is volatilized based on soil texture (more emissions for coarser texture). A portion of harvested or senesced biomass N is volatilized. |
| Methane oxidation/hydrolysis | Methane oxidation rate is a function of $CH_4$ concentration and redox potential.<br>Methane production from manure is a function of redox potential, DOC, $CO_2$, microbial activity, manure temperature, and manure pH. | $CH_4$ oxidation rate is calculated as the product of a site specific maximum rate (varies with soil gas diffusivity as a function of soil bulk density and field capacity), a normalized beta function for soil water effects, a temperature multiplier, and an agricultural management adjustment factor. In the beta function for water effects, gas diffusivity limits $CH_4$ emissions when soil water is high and $CH_4$ is limited biologically when soil water is low (Del Grosso et al., 2000). |
| Anaerobic $CH_4$ production (methanogenisis) | Methanogen activity drives $CH_4$ production under anaerobic conditions in DNDC and is estimated based several factors including temperature, redox potential, pH, DOC, a microbial activity index, and $CO_2$. | A version of DayCent with $CH_4$ production in flooded systems is currently being tested. |

† DNDC, Denitrification–Decomposition model; DOC, dissolved organic carbon; Eh, soil redox potential; SOM, soil organic matter.

soil moisture factor, which is a function of WFPS, respiration, and gas diffusivity (Del Grosso et al., 2000). In DayCent, the $N_2$ to $N_2O$ ratio is a function of the ratio of soil $NO_3$ concentration to heterotrophic $CO_2$ emission and an index of soil gas diffusivity (Del Grosso et al., 2000). DayCent assumes that all gases produced at any depth are emitted into the atmosphere as the diffusivity of soil is only used to infer the $O_2$ status for a given depth and does not act to limit the amount of N gas emitted to the atmosphere from soil. Crop growth can be estimated using either an empirical growth curve model or a growing degree day (phenologically based) sub-model and is regulated by a combination of genetic potential (energy to biomass conversion factor), phenology, and solar radiation and is reduced by any nutrient, water, and temperature stresses. Additional source code modifications were not required as the default model structure allows for a robust parameterization of temperature and water stress via model input files.

## Model Simulation Methods

Simulations for both models were performed in a consistent manner ensuring, when possible, that parameterization between the models remained uniform. The small amount of regionalized development work performed with DNDC version 9.5 to improve simulation of the deepwater pool was necessary as it lacked a detailed characterization of the soil profile already included in DayCent that was considered to be important for this type of analysis, particularly in relation to simulating crop water stress. This study was coordinated with another study investigating the impact of climate change on production by Smith et al. (2013) and as such included additional modifications employed through collaboration with the model developer (i.e., temperature stress at anthesis, influence of $CO_2$ on carbon assimilation and water and nitrogen use efficiency, crop growth rate as a function of cardinal temperature). We decided that the Swift Current dataset would be used as the baseline site to calibrate crop water requirements for both models because the water balance was best characterized at this location (Campbell et al., 1997a; de Jong et al., 2011). Previously published literature (Campbell et. al., 1997a) indicated that water limitation was a dominant factor in determining spring wheat yields at this site. The C to N ratios for biomass partitions were calibrated using measured N in grain and straw (data not shown) at Swift Current (Campbell et al., 2004) and detailed in Kröbel et al. (2011). In DNDC these parameterizations pertained to the crop inputs defined as maximum biomass production (kg C $ha^{-1}$), biomass fraction, and biomass C to N ratio for the grain, leaf + stem, and root fractions as well as the crop water demand (g water/kg dry matter). For DayCent the modified crop parameters included those that represent the upper and lower boundaries for biomass C to N ratios during growth (PRAMN,

PRAMX), the coefficient for potential aboveground monthly production as a function of solar radiation (PRDX), the fraction of aboveground N removed with grain (EFRGRN), the effect of water stress on harvest index (HIWSF), and the relative water content required for 50% of maximum production (WSCOEFF). As DayCent defines the biomass C to N ratios as a dynamic minimum and maximum boundary during growth, parameterization was formulated to ensure the measured values for grain/straw and root C:N were described by the minimum N requirement boundary. The crop parameterizations (grain/straw/root fractions, C to N ratios of biomass fractions, water requirements) formulated from the Swift Current data were extended to the other two locations. Daily climate data, including maximum and minimum temperature, daily precipitation, and daily wind speed (when available), were obtained for each of the study sites using the CANSIS weather database (Agriculture and Agri-Food Canada, 2006). Soil information for each site including bulk density, initial organic carbon content, and soil texture was obtained from published literature cited in the site description. Field capacities (i.e., moisture held at −0.3 MPa) and wilting points (moisture held at −1.5 MPa) for each of the sites were determined for the Swift Current site (Roloff et al., 1998b) or estimated by using the pedotransfer functions described by Saxton and Rawls (2006) when measurements were not available. The DNDC and DayCent simulations were both formulated to include spin-up periods so as to stabilize the partitioning of the C and N pools. For DNDC the spin-up period typically is 10 yr to stabilize the N pools, while in DayCent a 1000-yr spin-up simulation is normally performed (Del Grosso et al., 2011). In both models the partitioning of the carbon pools is important as it regulates the resistance of these pools to decomposition. Each pool is defined as having different base decomposition rates, and these are then modified by soil moisture/temperature and management.

Comparisons against measured yields for each study site and for all rotations were calculated using the following statistical evaluators: average relative error (ARE, %) (Gold, 1977) (also known as "mean percentage error"), RMSE (Loague and Green, 1991), and the coefficient of determination ($R^2$). Regression slopes were tested using 95% confidence intervals to see whether they differed statistically from the 1:1 line. When plot replicate data were available, we calculated ARE and RMSE while considering measurement variability using the probable error range method (Harmel and Smith 2007; Harmel et al., 2010). This method intends to remove the measurement variability component from statistical measures, such as ARE and RMSE, to provide a better indicator of the modeling error component. The probable error range was employed using the 95% confidence intervals for the upper and lower bounds by adding or subtracting 1.96 times the standard error to the mean observation. These values were then substituted into the observed mean values to calculate revised RMSE and ARE statistics. To determine if model performance was biased toward specific weather conditions, the models were further evaluated under "HIGH," "NORMAL," and "LOW" periods of PET for the Cont-W(N+P) cropping system. Each of these periods was calculated by evaluating which climate years were less than the 25th percentile (HIGH) and more than the 75th percentile (LOW) for PET as calculated using Penman–Montieth during the measured periods.

Crop water use was calculated from the models as the total sum of evaporation, transpiration, runoff, leaching, and interception by the crop canopy and compared against measured spring soil water minus harvest soil water. Soil organic carbon comparisons were made for the 0- to 15-cm depth, the depth to which measurements were taken, with weighted layer conversions for DNDC and DayCent to shift from a 20-cm output to 15 cm. For the Lethbridge comparisons of SOC, measurements were only available for N unfertilized systems Cont-W(+P) and F-W(+P) for the period 1911 to 1990, and as such an additional simulation for the F-W(+P) system was performed. Crop residue carbon inputs were calculated using measured harvest indices (HIs) and measured grain yields applying the Bolinder et al. (2007) methodology. Nitrous oxide emissions were compared between modeled and measured values only for days where data were available with no gap filling being employed at Swift Current for the years 2001 to 2003.

## Results and Discussion

### Estimating Inter-annual Spring Wheat Yields

Understanding soil carbon dynamics is important if we are to assess the impact that agricultural management and climate have on the terrestrial carbon balance and the global carbon cycle. As soil carbon change is to a large part determined by the magnitude of crop residue carbon inputs, it is prudent to first assess the ability of the models to simulate crop production before evaluating soil carbon estimates. Both models were employed to estimate spring wheat yields on the three investigated rotations—Cont-W(N+P), Cont-W(+P), and F-W(N+P)—at the following three sites: Swift Current, Lethbridge, and Indian Head (Fig. 1, Fig. 2, and Fig. 3). Overall, both models performed well for simulating yields in the Cont-W(N+P) system with AREs of 3.8 and 6.3% for DayCent and DNDC, respectively. As neither model accounts for the impact of insects nor disease, and assuming all other stresses are simulated correctly, a slight overestimation of yields might be expected. DayCent captured the overall inter-annual variation for Cont-W(N+P) marginally better than DNDC, with lower RMSEs (460–744 vs. 559–778 kg $ha^{-1}$) and similar coefficients of determination (0.56–0.73 vs. 0.51–0.70) (Table 3). Neither model showed a particular bias toward any of the three locations.

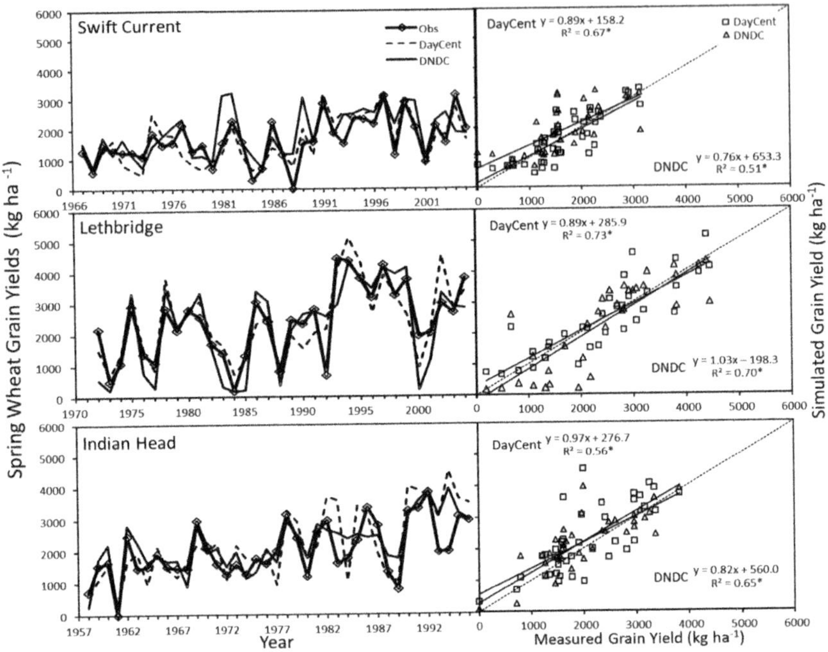


**Fig. 1. Observed and estimated inter-annual grain yields for Cont-W(N+P) at three locations using DNDC and DayCent. * LSD at *P* < 0.05.**

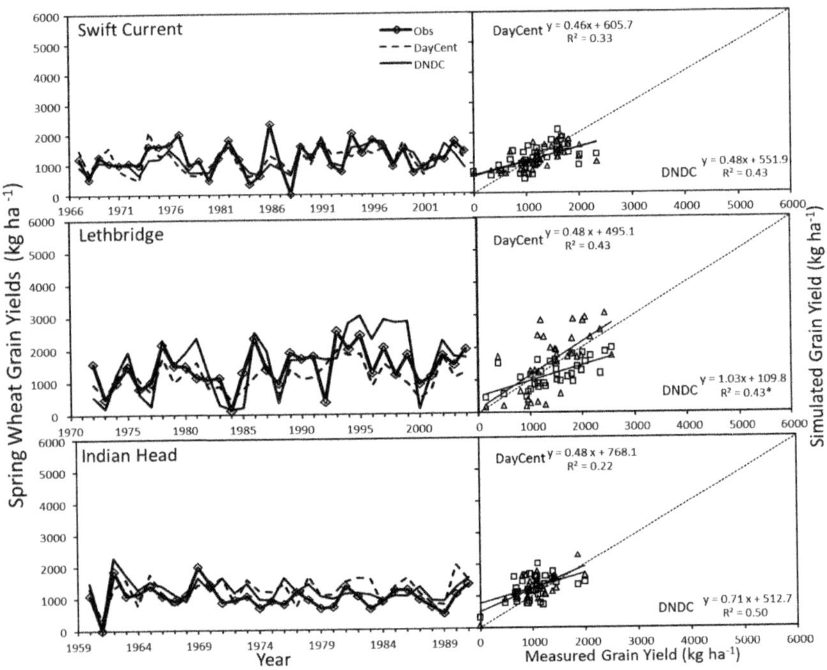


**Fig. 2. Observed and estimated inter-annual grain yields for Cont-W(+P) at three locations using DNDC and DayCent. * LSD at *P* < 0.05.**

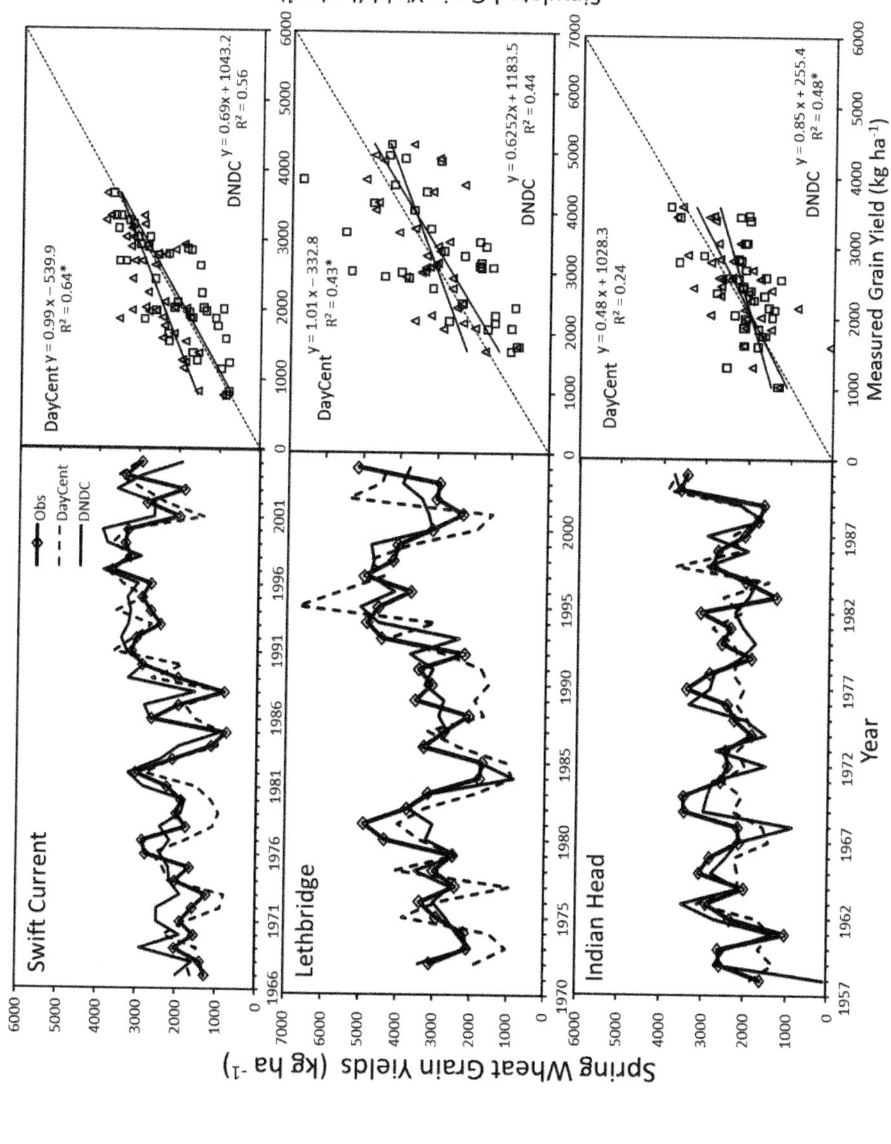


**Fig. 3. Observed and estimated inter-annual grain yields for F-W(N+P) at three locations using DNDC and DayCent. * LSD at *P* < 0.05.**

**Table 3. Evaluation of inter-annual spring wheat grain yield using DNDC and DayCent for three cropping systems at three locations.†**

| | Cont-W(N+P) | | | Cont-W(+P) | | | F-W(N+P) | | |
|---|---|---|---|---|---|---|---|---|---|
| | Swift Current | Lethbridge | Indian Head | Swift Current | Lethbridge | Indian Head | Swift Current | Lethbridge | Indian Head |
| | Observed yield‡, kg ha$^{-1}$yr$^{-1}$ | | | | | | | | |
| | 1637 | 2403 | 2020 | 1211 | 1411 | 1029 | 2264 | 3289 | 2477 |
| | ARE, % | | | | | | | | |
| DNDC | 15.8 | −4.9 | 9.8 | −6.0 | 10.6 | 20.8 | 15.4 | −1.5 | −4.7 |
| DayCent | −0.9 | 1.3 | 10.2 | −4.5 | −17.8 | 22.2 | −3.3 | −9.5 | −10.1 |
| | RMSE, kg ha$^{-1}$ | | | | | | | | |
| DNDC | 624 | 778 | 559 | 379 | 698 | 356 | 617 | 770 | 573 |
| DayCent | 460 | 631 | 744 | 416 | 503 | 452 | 559 | 1162 | 676 |
| | $R^2$ | | | | | | | | |
| DNDC | 0.51 | 0.70 | 0.65 | 0.43 | 0.43 | 0.50 | 0.56 | 0.44 | 0.48 |
| DayCent | 0.67 | 0.73 | 0.56 | 0.33 | 0.43 | 0.22 | 0.64 | 0.43 | 0.24 |

† ARE, average relative error; DNDC, Denitrification–Decomposition model.

‡ Mean observed yields over the experiment duration.

To determine if the models had any tendency to simulate specific moisture stress periods more accurately, comparisons between measured and simulated yields were made within defined soil moisture regimes using the Cont-W(N+P) system (Table 4). The periods were defined as years when the PET was "HIGH," "LOW," and "NORMAL." In general, both DayCent and DNDC simulated yield variation under HIGH and NORMAL conditions better than years under LOW conditions. RMSE for DayCent ranged from 359 to 637 kg ha$^{-1}$ for HIGH and NORMAL years, respectively, and 363 to 958 kg ha$^{-1}$ for LOW years. The RMSE for DNDC ranged from 419 to 766 kg ha$^{-1}$ for HIGH and NORMAL years and from 705 to 853 kg ha$^{-1}$ for LOW years. The observation that HIGH and NORMAL years were better simulated by the models was further supported by the $t$ test at $p < 0.05$, which indicated significant trends for HIGH and NORMAL years compared with the LOW years for both models. Campbell et al. (1997a) determined that water use in semiarid Canadian prairies accounts for nearly 65% of the variability in wheat yields, N (soil test N and fertilizer) was responsible for 13% of the yield variability, and degree days was responsible for only 3%. With water playing such an important role in determining the inter-annual yield variability, it is logical that yields in the driest years would be most influenced by water stress when N is not limiting. Henry et al. (1986) has shown that yield responses to N availability are usually greater as water availability increases, suggesting that under more humid conditions spring wheat yields are influenced more by a combination of nitrogen and water availability than just water alone. We concluded that both

**Table 4. Impact of climate induced stress on estimated grain yield for Cont-W(N+P) at three locations using DayCent and DNDC. †**

| | | DayCent | | | DNDC | | |
|---|---|---|---|---|---|---|---|
| | Statistical evaluator | HIGH‡ | NORMAL‡ | LOW‡ | HIGH‡ | NORMAL‡ | LOW‡ |
| Swift Current | ARE, % | −8.9 | 5.6 | −5.5 | 25.9 | 12.2 | 0.0 |
| Lethbridge | ARE, % | −6.6 | −5.7 | 12.6 | −24.1 | −4.9 | 2.2 |
| Indian Head | ARE, % | −18.2 | 4.1 | 24.4 | 31.4 | 7.6 | 9.8 |
| Swift Current | RMSE, kg $ha^{-1}$ | 564 | 359 | 363 | 722 | 448 | 822 |
| Lethbridge | RMSE, kg $ha^{-1}$ | 515 | 526 | 839 | 766 | 701 | 853 |
| Indian Head | RMSE, kg $ha^{-1}$ | 421 | 637 | 958 | 419 | 477 | 705 |
| Swift Current | $R^2$ | 0.33 | 0.82* | 0.83 | 0.42* | 0.74* | 0.00 |
| Lethbridge | $R^2$ | 0.77* | 0.80* | 0.55* | 0.73* | 0.85* | 0.30 |
| Indian Head | $R^2$ | 0.89** | 0.54* | 0.49* | 0.98* | 0.68* | 0.50* |

*Trend is significant at $p < 0.05$.

** Trend is significant at $p < 0.10$.

† ARE, average relative error; DNDC, Denitrification–Decomposition model.

‡ Growing season potential evapotranspiration values that were below the 25% percentile or above the 75% were characterized as HIGH and LOW, respectively, for the measured period.

models performed best when water represented the dominant stress; they had greater difficulty in estimating yields when this stress was alleviated or was no longer the dominant factor influencing yields. Results of the rotations Cont-W(+P) and F-W(N+P) further supported this conclusion as the model's ability to simulate yields declined when the influence of nitrogen on yield determination became more pronounced (Fig. 2 and Fig. 3).

The Cont-W(+P) treatment, contrasted with the Cont-W(N+P) system, represented a cropping system in which N stress was the dominant factor influencing crop growth. In the models, P was not limiting. While both models performed adequately in simulating the overall magnitude of the yields for Cont-W(+P) (ARE ranging from −6.0 to 22.2%) (Table 3), they both had difficulties simulating the annual yield variations with an $R^2$ of 0.22 to 0.43 for DayCent compared with an $R^2$ of 0.43 to 0.50 for DNDC (Table 3). As N inputs were either zero or very low in Cont-W(+P), with only a nominal amount (5 kg $ha^{-1}$) applied as part of the P source for Swift Current, crop demands could only be met by N supplied primarily from mineralization. DayCent and DNDC underestimated mean spring $NO_3$–N by 11.6 and 20.9 kg $ha^{-1}$ $yr^{-1}$ in Cont-W(+P) and by 24.9 and 43.4 kg $ha^{-1}$ $yr^{-1}$ for F-W(N+P) systems, respectively, in the 0- to 50-cm depth, while the measured $NO_3$–N in Cont-W(+P) appeared to suggest that some unknown external N input to the system was occurring or that mineralization might be higher than was simulated.

Observed yields suggest that N availability was important in determining spring wheat yield, as indicated by yields of Cont-W(+P) that were 51 to 74% of the Cont-W(N+P) yields for the three locations. In a lysimeter study conducted on a Swinton Loam at the Swift Current Research Station, Campbell et al. (1977) demonstrated the importance of the positive interaction of soil moisture and N fertilizer in determining spring wheat yield. Assuming fertilizer N rates similar to the average N rate of this Cont-W(N+P) study (41 kg N $ha^{-1}$), nitrogen was calculated as being responsible for approximately 39% of the yield variability with the remaining 61% being related to water. Both models have previously been reported to underestimate N supplied through mineralization (Smith et al., 2008). Kröbel et al. (2011) found that DNDC 9.3, and a version of DNDC that was developed for Canadian spring wheat, underestimated available N content in a semiarid soil in western Canada. When simulating F-W in Oregon with CENTURY, Parton and Rasmussen (1994) added extra N to compensate for underestimated deep soil N mineralization. For this study, it may not be prudent to just increase the model's rate of N mineralization through the adjustment of the base decomposition rates since this would result in greater rates of SOC decomposition and too much loss of SOC at the studied sites, though there could be some feedback to the system through increased C inputs. In the F-W(N+P) system where moisture and N or P fertility would be less limiting, DayCent simulated yields well for the semiarid sites (Table 3) but not at Indian Head. The sub-humid climate at Indian Head does not require fallow, and the resulting high soil moisture in the F-W(N+P) can result in $NO_3^-$ leaching (Campbell et al., 1994). In contrast to DayCent, DNDC performed less well at simulating yield in the semiarid sites, but better than DayCent in the sub-humid site (Fig. 3). DayCent simulated the overall magnitude of yields well with ARE ranging from –3.3 to –10.1%, similar to DNDC's –1.5 to 15.4%, but both models had difficulty capturing the variability at Indian Head as evident by the $R^2$ of 0.24 for DayCent and 0.48 for DNDC (Table 3). Inclusion of fallow in the rotation is a well-known practice in semiarid agricultural regions aimed at improving water availability for subsequent cropping years (de Jong et al., 2011), but in the Black Chernozem fallow is used only rarely to control disease or weeds. Crops grown on fallow will have greater yield than their stubble cropped counterparts as was indicated by both the observations and the DayCent and DNDC models where measured yields were 22 to 38% higher for F-W(N+P) than for Cont-W(N+P). We suspect that similar to the Cont-W(+P) system, an underestimation of available N by the models influenced their ability to capture the proper inter-annual variations.

In some studies in western Canada where mass balance of N has been assessed, it was found that there was higher soil N than expected. It has been suggested there is a need for more research to determine the source of this additional N (Campbell et al., 2004; Janzen et al., 1997; Biederbeck et al., 1984). In some cases large amounts of atmospheric N deposition may have occurred (Bremer et al., 1995). Other studies have shown that large quantities of N can accumulate over time at deep soil depths (Thorup-Kristensen, 2006) and may become available when deep rooted crops are included in a rotation, or through capillary rise. There is also research indicating that asymbiotic N fixation may occur and thereby provide an N source, particularly in systems with low anthropogenic N inputs (Groffman, 2008). Orr et al. (2011) measured the activity of free-living N-fixing bacteria and found that increases in N uptake by barley (*Hordeum vulgare* L.) created conditions that were more favorable for abundant and diverse bacterial community than a bean crop. Studies have indicated that there may be a very large range in asymbiotic N fixation for various soils and climatic and management conditions with fixation ranging from 0 to 90 kg N $ha^{-1}$ (Raju et al., 1972; Burgmann et al., 2004; Kahindi et al., 1997; Dobereiner et al., 1972). This problem of possible missing N sources is not isolated to western Canadian soils. Findings by Jenkinson (1991) for the Rothamsted long-term experimental plots suggest that nearly 30 kg N $ha^{-1}$ $yr^{-1}$ was supplied to the unfertilized wheat plots from somewhere else, most likely the atmosphere. Since the activity of free-living bacteria across various soils is so diverse, it is difficult to formulate a generalized mechanism in biophysical models to characterize their growth. Our results indicate that both the DayCent and DNDC models are unable to quantify the sources of N inputs in all three cropping systems, but particularly so for the N limited Cont-W(+P) system and the F-W(N+P) system. To further improve the performance of these models, a better understanding of N dynamics needs to be achieved through targeted field research.

Another factor that may have contributed to differences between measured and modeled yields might be related to changes in wheat cultivar over the course of the studies (Campbell et al., 2005). Each of the selected long-term studies had multiple changes in cultivar since their inception (Campbell et al., 1997b). The idea was to reflect cultivar improvements over the course of the study to better represent current farming practices. Unfortunately, there was not sufficient information available to differentiate between cultivars in terms of crop parameterization; therefore, a single generalized cultivar was employed in the model. While we do not believe this would have contributed to a gross overestimation/underestimation of crop yields by the model, some influence was possible.

In general, the DayCent and DNDC models performed as good or better at estimating wheat production than other modeling studies conducted on the semiarid prairies. For example, Kersebaum et al. (2008) used the HERMES model to simulate grain yields at Swift Current for Cont-W and F-W rotations between

1967 and 1991 and found it performed reasonably well with a $R^2$ of 0.57. HERMES performed even better when simulating aboveground plant biomass ($R^2$ of 0.7). The EPIC model, using two PET methods, was also employed in simulating wheat production in the Swift Current rotation study (Roloff et al., 1998a). This resulted in an $R^2$ of 0.45 to 0.49, 0.60 to 0.66, and 0.18 to 0.56 for Cont-W(+P), Cont-W(N+P), and F-W(N+P), respectively. Touré et al. (1995) evaluated five models (EPIC, CENTURY, CERES, Stewart, and Sinclair) for their ability to simulate spring wheat yields in the Lethbridge ABC long-term crop rotation study. Similar to our findings, the models generally performed better on the Cont-W(N+P) and F-W(N+P) systems than for less well N-fertilized systems, with $R^2$ ranging from 0.07 to 0.62 for Cont-W(N+P) compared with the Cont-W(+P) system ($R^2$ of 0.08–0.42). They concluded that none of the models was able to capture the full range of yield variability and suggested that model selection should be based on other constraints (i.e., data input intensity, and type of system being modeled).

## Simulating Crop Water Use

On the Canadian prairies, yields are primarily a function of water use and available N. Crop water use was estimated using measurements of soil water over a 35-yr period at Swift Current (de Jong et al., 2008). DayCent and DNDC effectively simulated the appropriate inter-annual fluctuations in water use (Fig. 4). DayCent simulated AREs of −1.7, 1.7, and 5.6% and $R^2$'s of 0.76, 0.75, and 0.78 for Cont-W(N+P), Cont-W(+P), and F-W(N+P), while DNDC simulations resulted in AREs of −2.0, −5.5, and 14.8% and $R^2$'s of 0.75, 0.75, and 0.84 for the same rotations. We conclude that the models' ability to simulate water availability was not a major factor limiting accurate estimation of crop production although both models showed a slight tendency to overestimate water use on the F-W(N+P) rotation. Predicting an accurate water balance in the models is also instrumental for determining rates of SOC decomposition, mineralization, nitrification, and denitrification.

Both DayCent and DNDC employ a cascade layered water balance approach to estimate unidirectional downward flow of water under saturated conditions. DayCent extends its water flow routine to include unsaturated flow via Darcy's law using modeled hydraulic conductivity and hydraulic potential (Del Grosso et al., 2001). Note that the DNDC model does not include unsaturated flow but does include a simple empirical routine to simulate capillary rise. Transpiration in DayCent is a function of leaf biomass (regulated by soil water potential and root biomass), whereas in DNDC transpiration is a function of crop biomass (regulated by defined crop water requirement). The simulation results were encouraging for DNDC considering that the model does not simulate unsaturated flow and that it has a simplified conceptualization of the soil profile beyond the charac-

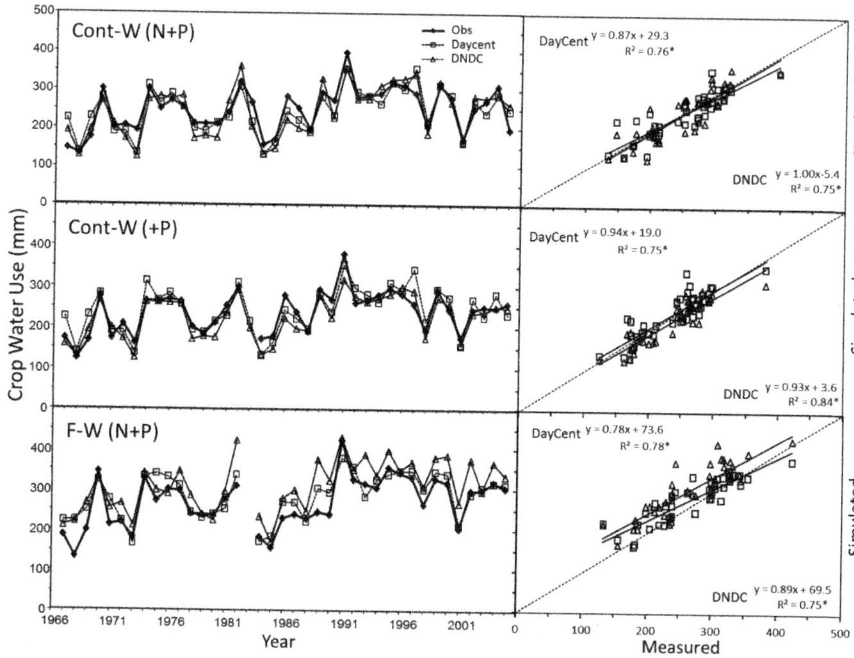


**Fig. 4. Observed and estimated crop water use for three cropping systems at Swift Current, SK using DNDC and DayCent. * denotes LSD at $P < 0.05$.**

terized 50-cm depth. The model assumes that all water that moves below this depth is captured in a flexible water pool container with its size determined by soil parameters. In our analysis we modified the model to read in soil parameters for the underlying 50- to 100-cm depth rather than use the default function of surface parameters. This improved results at the Swift Current site as the original implementation of the water pool did not induce the proper crop water stress, because the soil water holding capacity below 50 cm was low because of the high bulk density (1.78 g $cm^{-3}$) in the deep profile. In DNDC, the crop is allowed to tap into this reservoir once the rooting depth reaches beyond 50 cm to supplement its transpiration demands.

Precipitation received outside of the growing season is significant in terms of water availability for crops grown on fallow, and especially for stubble crops in semiarid regions. The models require the capacity to hold this water for later use by the crop. An analysis of spring soil water at the Swift Current site indicated that both models predicted more water availability after fallow than after stubble (as expected) with average differences between these systems being generally similar to observations (Fig. 5). Using 18 yr of measured data at Swift Current, de Jong et al. (2011) determined that stubble cropping systems have, on average, 40

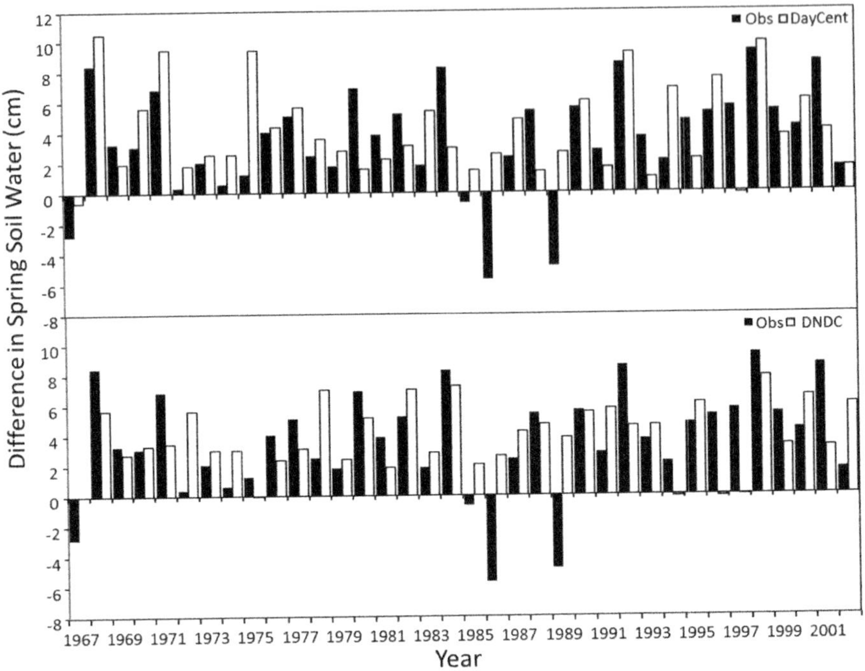


**Fig. 5. Difference in spring soil water (0–120 cm) between fallow and stubble cropped years at Swift Current, SK (1967–2002)—measured and modeled.**

mm less soil water reserves than after fallow. The DNDC and DayCent models estimated this difference to be on average 38 mm and 41 mm, respectively, over the 39-yr study. There were a number of discrepancies between measured and simulated results for some specific years, particularly in years where measurements indicated that less water was stored under fallow than in stubble. These measured anomalies coincided with years where the growing season precipitation was very low (e.g., drought years 1985 and 1988) and resulted in severely depressed yields in the stubble cropped systems. Both models indicated that, regardless of the climatic conditions, soils under fallow had nearly the same, or significantly more, stored soil water than continuously cropped soils. It is important to note that the DayCent model performed very well using its default water balance framework and was parameterized to define the soil depth up to 210 cm. No recommendations for additional modifications to the layered water balance in DayCent seem necessary within the scope of our analysis.

## Estimating Soil Carbon Change and Crop Residue Carbon Inputs

Changes in soil C will depend primarily on the balance between changes in C inputs from crop or animal residues (positive effect) and rate of decomposition of organic matter (negative effect). Both of these processes depend on factors such as moisture, temperature (thus weather), soil texture, and nutrients, especially in the case of C inputs (Campbell et al., 2007). In cases of prolonged periods of above average moisture (e.g., 1989–2007 at Swift Current), crop production and thus C inputs will increase, especially if fertility increases as was the case in the Swift Current study for the treatments receiving N. An increase in available water will also result in an increase in the decomposition of soil C and thus tend to result in lost C. It is the balance between these processes that will dictate the final observed SOC level at any time. There is another factor that must be considered in assessing changes in SOC over time: changes in SOC are normally slow and small relative to the total SOC pool along with spatial variability usually being large, which makes significant differences difficult to detect in systems that are usually very heterogeneous in C content (Campbell et al., 2007). Consequently, reliable measured changes in SOC require that many measurements are taken regularly over decades (Conant et al., 2001). Alternatively, it may be possible to use models to make such determinations.

The DayCent and DNDC models are considered to be compartmental in nature, whereby SOC is divided into conceptual C pools of similar rates of turnover (age) and degradability (Table 2). The age of a particular carbon molecule in each of the conceptual pools is not tracked (Manzoni and Porporato, 2009), and SOC decomposition is instead controlled by a specific rate constant defined for each pool, which represents the average decomposition rate of that C pool. This decomposition rate is a function of water and temperature factors, nitrogen limitations, and soil texture, whereby finer textured soils act to physically and chemically reduce the rate of decomposition (Del Grosso et al., 2011).

At the Swift Current site, the DNDC model accurately predicted carbon inputs for all three cropping systems (Fig. 6A). Both the carbon input estimated (Bolinder et al., 2007) and results from the DNDC model showed that there was an increase in crop yields and C inputs in the 1990s due to improved weather and higher rates of fertilizer N for Cont-W(N+P) and F-W(N+P) (Campbell et al., 2007). Soil organic carbon was underestimated by the DNDC model and thus did not reflect the increase in C inputs or the measured gain in SOC (Fig. 6B). In DNDC the decomposition rate of the active SOC pools is faster than in DayCent, resulting in a larger portion of the initial C inputs being decomposed and lost as $CO_2$. This explains the lower SOC in the two Cont-W systems for DNDC than for DayCent (Fig. 6B), relative to estimated C inputs. The fraction of active SOC stabilizes in DNDC at about 1% in comparison to the 3 to 4% in DayCent. This is compensated in DNDC with a higher fraction of passive SOC that decays slowly. In a future study, it would be beneficial to better characterize the size and rate of decomposition of the active and passive

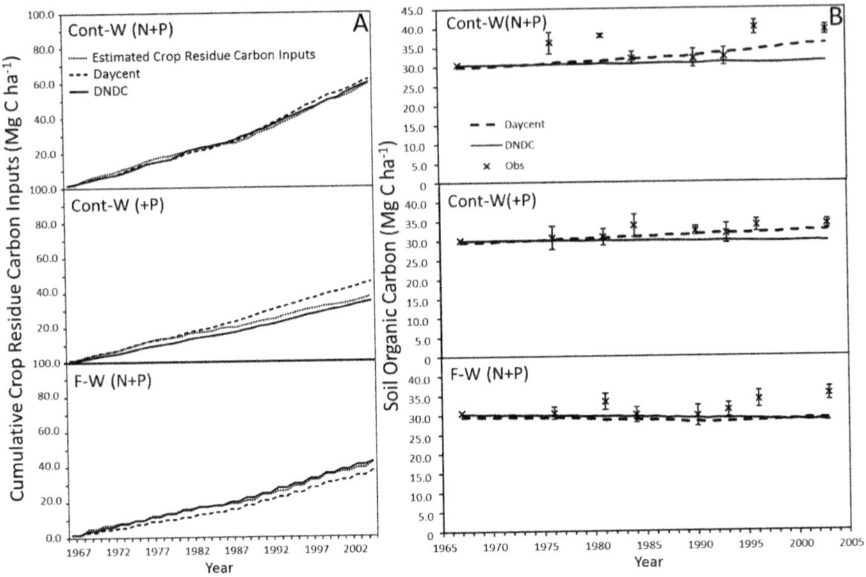


**Fig. 6. Estimated soil carbon change and crop residue carbon inputs for three cropping systems at Swift Current, SK (1967–2003).**

SOC pools using experimental studies done in Canada. For our study, we chose to employ default rate constants for estimating the SOC.

DayCent estimated C input well only for Cont-W(N+P), overestimated it for Cont-W(+P), and slightly underestimated it for F-W(N+P) (Fig. 6A). In further output analyses, it was determined that the DayCent model overestimated C input for Cont-W(+P) because it predicted a lower HI (0.36) than measured (0.42) and also a higher fraction of roots (0.23) than was employed in the Bolinder et al. (2007) method (0.21) used to estimate roots from measured aboveground biomass. DayCent adjusts biomass fractions dynamically depending on water and nitrogen stress, indicating that the low N availability was responsible for increasing root mass and thus C inputs. Conversely, estimated HIs compared well, within 0.01 for both models, to measured HIs for the Cont-W(N+P) and F-W(N+P) systems of 0.37 and 0.38, respectively. DayCent came close to estimating SOC in both Cont-W(N+P) and Cont-W(+P), but like DNDC it underestimated SOC in F-W(N+P). Soil organic carbon was underpredicted because of an underestimation of crop production with an ARE of −3.3% (Table 3), and thus C inputs. Interestingly, measured SOC increased even in the F-W(N+P) system as measured crop yields increased primarily because of improvements in weather in the 1990s and a marked increase in fertilizer N rates after 1990 (8 kg N $ha^{-1}$ $yr^{-1}$ before 1991 and 41 kg N $ha^{-1}$ $yr^{-1}$; after Campbell et al., 2007).

At Lethbridge and Indian Head, both models generally performed well in estimating changes in SOC although DNDC performed slightly better than DayCent (Fig. 7). There were limited SOC data available at Indian Head, and Lethbridge only had carbon data available for P treatments (Monreal and Janzen, 1993). At the Lethbridge site the DNDC model estimated a slower decline in SOC and response to management than DayCent, and fit the measured data better. Estimated changes in SOC by DNDC generally occurred more slowly than in DayCent, primarily because the size of the passive pool in DNDC was much larger. The fraction of passive SOM (humus) in the DNDC model stabilizes at a higher value than in DayCent, resulting in slower responses to changes in C inputs. In a modeling study addressing the effects of residue removal on SOC, Smith et al. (2012b) reported that DNDC produced less SOC change than DayCent, which was credited largely to the higher humus fraction assumed in the DNDC model.

## Estimating Nitrous Oxide Emissions

The evolution of nitrous oxide fluxes from agricultural soils represents a multipart interaction of biophysical and physical processes along with the compounding impact of climate and agricultural management on these processes. While the fundamental processes (nitrification, denitrification, and chemodenitrification) that are responsible for N gas formation are well understood, it is the complex interactions of the regulators of these processes that make the quantification across temporal and spatial scales extremely challenging (Robertson and Tiedje, 1987; Davidson et al., 1991). Both the DayCent and DNDC models are designed to

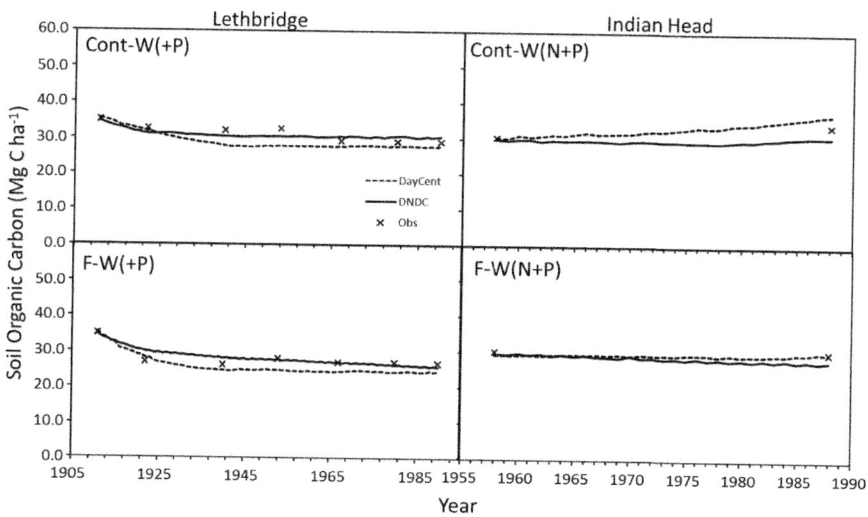


**Fig. 7. Observed and estimated soil carbon change in the 0- to 15-cm depth at Lethbridge, AB and Indian Head, SK using DNDC and DayCent.**

encapsulate the interaction of these elements with varying complexity and are able to estimate $N_2O$ emissions over a multitude of scales.

Measurements of nitrous oxide emissions were available for 3 yr of the crop rotation study at the Swift Current location for two of the rotations [Cont-W(N+P) and Cont-W(+P)]. Observations were taken using a chamber technique described by Lemke et al. (2002) for both the spring burst and frost-free time period during the years 2001 to 2003. Emission intensity was low overall and differences between the two rotations (fertilized vs. unfertilized) suggest that available nitrogen was partially responsible for limiting $N_2O$ emissions from nitrification and denitrification with the lack of available water being the primary regulator. Over the course of the 3 yr, the fertilized system Cont-W(N+P) had total non-continuous measured emissions of 217 g $N_2O$–N $ha^{-1}$, which was more than double the 99 g $N_2O$–N $ha^{-1}$ measured in the N unfertilized system of Cont-W(+P). These emissions corresponded with the 3-yr mean measured soil $NO_3$–N levels of 45 and 18 g $ha^{-1}$ in the 0- to 120-cm depth for Cont-W(N+P) and Cont-W(+P). Both models provided reasonable estimates of the total emission period, with values of 278 and 107 g $N_2$–N $ha^{-1}$ for DayCent and 275 and 56 g $N_2O$–N $ha^{-1}$ for DNDC over the Cont-W(N+P) and Cont-W(+P) rotations, respectively. Lemke et al. (2005) suggested that for drier regions in Saskatchewan a mean annual $N_2O$–N emission rate of less than 500 g $ha^{-1}$ $yr^{-1}$ was representative of the region. The DayCent model estimated emissions events that corresponded well with the timing of measurements for both rotations (Fig. 8). Calculated RMSE values when considering measurement variability ranged from 0.0 to 5.5 g $N_2O$–N $ha^{-1}$ $d^{-1}$ for the frost-free and spring burst periods. The DNDC model also compared well to total measured emissions during the periods, but differed from DayCent results in that it predicted more sizeable emission events in non-measured periods during 2001 and 2002 and no emissions for the 2003 spring burst period. Calculated RMSE when considering measurement variability was similar to DayCent, ranging from 1.1 to 5.0 g $N_2O$–N $ha^{-1}$ $d^{-1}$ for the frost-free and spring burst periods. The large emission peaks in DNDC during the non-measured period corresponded with a single large precipitation event (~19 mm) on Day 196 in 2001 and the melting of snowpack on Day 123 of 2002. Even though soil water contents were relatively low (less than 0.36 WFPS in the top 30 cm), the denitrification routine acted to produce significant $N_2O$ emission events in the DNDC model. This proved to be an interesting difference as DayCent estimated that all emission events were related to the nitrification process except for the spring 2003 (Fig. 9). Only when soil water contents were high in the spring months of 2003 (above 0.55 WFPS) did the soil climate move to a more anaerobic state to drive the denitri-

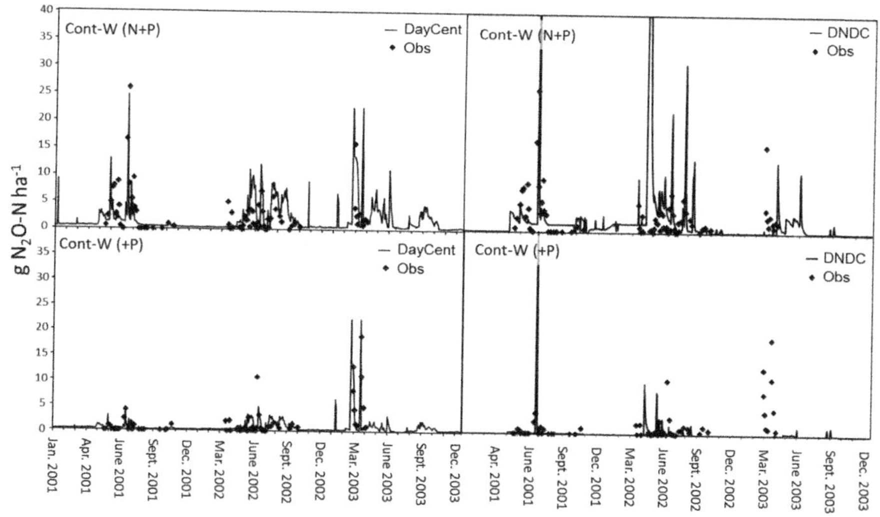


**Fig. 8. Observed and estimated $N_2O$ emissions using DayCent and DNDC for Cont-W(N+P) and Cont-W(+P) at Swift Current (2001–2003).**

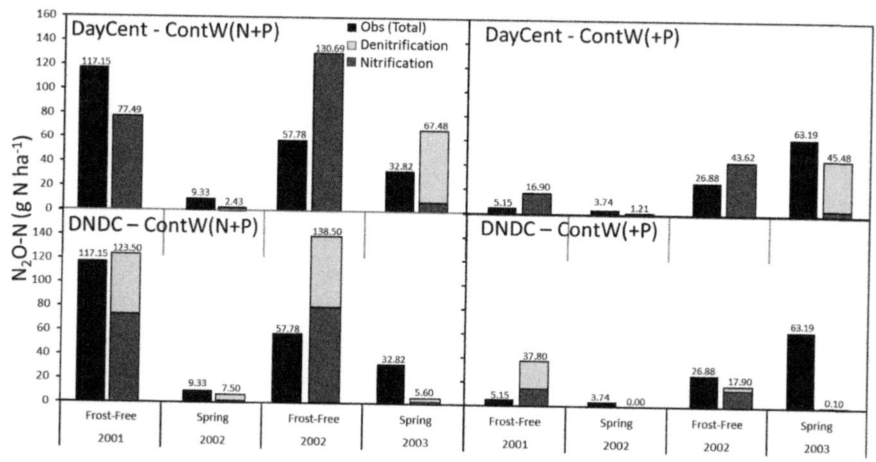


**Fig. 9. Estimated contribution of denitrification and nitrification processes to $N_2O$–N emissions for Cont-W(N+P) at Swift Current, SK (2001–2003).**

fication process for DayCent. Estimated water contents for the 7.5- and 15-cm depths on measured days averaged less than 0.30 WFPS for the years 2001 to 2002 and 0.67 and 0.47 WFPS for the 2003 spring period in Cont-W(N+P). In DayCent, denitrification is formulated to only occur at WFPS above 0.55 with peak nitrification rates occurring near 0.50 WFPS (Parton et al., 2001). Results from DNDC indicated a more balanced contribution of the two mechanisms (nitrification, denitrification) toward $N_2O$ gas formation (Fig. 9). We attributed this result partly to the linear approximation of redox potential (Eh) between wilting point and field capacity. In DNDC the redox potential acts to regulate the magnitude of the denitrification side of the anaerobic balloon. Future modification of this mechanism to a nonlinear approach would be prudent so as to put more emphasis on the nitrification routine for semiarid conditions and would reduce the frequency of large denitrification events that are not as common for this predominantly dry region. Bateman and Baggs (2005) concluded from analysis with a combination of stable isotope and acetylene techniques, on a silt loam soil, that denitrification was the process responsible for $N_2O$ production when soil moisture was at 70% WFPS and nitrification accounted for nearly 81% of all $N_2O$–N emissions between 35 and 60% WFPS over a 24-d period. Similar relationships between WFPS and relative N gas emissions from nitrification and denitrification have been well documented by others (Linn and Doran, 1984; Bouwman, 1998; Dalal et al., 2003). Measured soil water contents ($n = 108$) for Cont-W(N+P) between 1967 and 2002 indicated that soil moisture only once went above 0.7 WFPS and only in the 90- to 120-cm depth. More than 85% of the measurements at Swift Current were in the range of 0.2 to 0.5 WFPS, suggesting that nitrification would be the primary process involved in N gas loss. Low estimated spring $N_2O$ emissions by DNDC in the year 2003 were another notable difference from DayCent results and measurements. Water contents were estimated to be higher in 2003 (~0.45 WFPS) than in 2002 (~0.23 WFPS) in the 0- to 30-cm depth by DNDC, which suggests this was not the main cause of the low predicted emissions. Estimated soil $NO_3^-$ and $NH_4^+$ concentrations in the spring 2003 measured period did, however, provide insight as to why N gas formation was minimal in DNDC. Total soil N (ammonium and nitrate) for the top 50 cm was estimated to be only 6.9 kg ha$^{-1}$ for Cont-W(N+P) at the beginning of the spring melt and only 0.2 kg ha$^{-1}$ for Cont-W(+P). These low soil $NO_3$–N concentrations acted to limit the microbial processes that result in $N_2O$ emissions and are a product of the previously mentioned issues regarding missing N inputs in the models. Estimated annual average $N_2O$ emissions indicated that higher soil water availability resulted in greater $N_2O$ emissions (Table 5). Both DNDC and DayCent estimated the lowest annual emissions for the semiarid Brown Chernozem and the highest for the sub-humid thin Black Chernozem. Fertilizer N inputs were also positively correlated with $N_2O$ emissions for both models with the Cont-W(N+P) treatment always emitting more $N_2O$ than ContW(+P). Overall, annual emissions were higher for DayCent than for DNDC, which was also the case for the 3-yr measured comparison at Swift Current. Without additional measurements, it is difficult to ascertain which model provides a better estimate although both models appeared to agree on the general magnitude of emissions.

**Table 5. Estimated mean annual $N_2O$ emissions for three cropping systems using DNDC and DayCent at three locations.†**

| | Brown Chernozem (Swift Current) | | | Dark Brown Chernozem (Lethbridge) | | | Thin Black Chernozem (Indian Head) | | |
|---|---|---|---|---|---|---|---|---|---|
| | Cont-W(N+P) | Cont-W(+P) | F-W(N+P) | Cont-W(N+P) | Cont-W(+P) | F-W(N+P) | Cont-W(N+P) | Cont-W(+P) | F-W(N+P) |
| | Average $N_2O$–N kg $ha^{-1}$ | | | | | | | | |
| DNDC | 0.47 | 0.28 | 0.36 | 0.61 | 0.37 | 0.89 | 0.56 | 0.23 | 1.16 |
| DayCent | 0.72 | 0.35 | 0.58 | 0.93 | 0.48 | 0.76 | 1.60 | 1.03 | 1.36 |
| Avg. Fert-N, kg N $ha^{-1}$ | 37.7 | 4.8 | 18.7 | 45.0 | 0.0 | 45.0 | 49.4 | 4.9 | 18.4 |

† DNDC, Denitrification–Decomposition model.

## Conclusions

Biogeochemical models have proven useful in helping us determine the impacts of agricultural management and soil and climate regimes on soil carbon change and trace gas fluxes while accounting for a full mass balance (C, N, and water). The strength of such models lies in their ability to facilitate assessments at different spatial and temporal scales and in representing the current state of knowledge as an integrated framework. By utilizing a full-system accounting approach, one limits the need for further upscaling/downscaling of results, which will help to reduce uncertainty in assessments. Further model improvement is dependent on the continued investigation of processes and their interactions with one another that influence carbon change and trace gas fluxes in both short-term and long-term experimental studies.

For this study, two prominent North American biogeochemical models (DNDC and DayCent) were investigated for their ability to estimate crop production, carbon change, and nitrous oxide fluxes using data from three long-term experiments in three locations on the Canadian prairies. Both models simulated average grain yield well for Cont-W(N+P) across three distinct soils types: a Brown Chernozem (silt loam) at Swift Current, a Dark Brown Chernozem (sandy clay loam) at Lethbridge, and a thin Black Chernozem (heavy clay) at Indian Head. They also simulated year-to-year variations well, particularly for normal to dry periods where water stress was the primary influence on yield. The models had increased difficulty simulating both Cont-W(+P) and F-W(N+P) systems, which was attributed primarily to their underestimation of available N. Further research is warranted to discern whether experiments in the region can determine a full mass balance of nitrogen, which would help identify the deficiencies in the estimated mass balance. Crop water use was well estimated by both DayCent and DNDC, even though their conceptualization of the soil profile and water movement had some striking differences (50-cm profile depth in DNDC and unsaturated flow in DayCent). Soil carbon change and crop residue carbon inputs generally agreed with measurements from the three studies; however, DNDC may require some adjustments to the base decomposition rates of its pools along with adjustments to the size of the pool fractions to allow for a better soil C to crop residue carbon inputs fit at Swift Current. Presently, the weighted distribution of carbon toward the slow humus pool is too high in DNDC and restricts its response to changes in crop residue input carbon. Nitrous oxide measurements compared well with estimates by both models and captured the differences between N fertilized and unfertilized continuous wheat. The distribution between which microbial processes (nitrification or dentrification) were responsible for the majority of the $N_2O$ emissions differed substantially between the models. While the majority of the emissions were derived from nitrification in both models, DNDC weighted the distribution more evenly between nitrification and denitrification, while DayCent heavily favored the nitrification process except for a single spring burst period. Modification of the linear approximation of the redox potential for low WFPS needs to be considered in DNDC to improve this allocation since nitrification is likely the dominant term for these regions. Overall, this assessment indicated that both models provided good results when used to investigate inter-annual crop production, soil carbon change, and nitrous oxide emissions for N fertilized wheat systems in the semiarid and sub-humid regions of the Canadian prairies.

## References

Abdalla, M., S. Kumar, M. Jones, J. Burke, and M. Williams. 2011. Testing DNDC model for simulating soil respiration and assessing the effects of climate change on the $CO_2$ gas flux from Irish agriculture. Global Planet. Change 78(3):106–115. doi:10.1016/j.gloplacha.2011.05.011

Agriculture and Agri-Food Canada. 2006. Cansis weather database. Agriculture and Agri-Food Canada. http://10.117.206.21/wthrarch/index.html (accessed 8 Aug. 2006).

Ayres, K.W., D.F. Acton, and J.G. Ellis. 1985. The soils of the Swift Current map area 72J Saskatchewan. Publ. S6. Saskatchewan Inst. of Pedology, Univ. of Saskatchewan, Saskatoon, SK, Canada.

Bateman, E.J., and E.M. Baggs. 2005. Contributions of nitrification and denitrification to $N_2O$ emissions from soils at different water-filled pore space. Biol. Fertil. Soils 41(6):379–388. doi:10.1007/s00374-005-0858-3

Biederbeck, V.O., C.A. Campbell, and R.P. Zentner. 1984. Effect of crop rotation and fertilization on some biological properties of a loam in southwestern Saskatchewan. Can. J. Soil Sci. 64:355–367. doi:10.4141/cjss84-037

Blagodatsky, S.A., and O. Richter. 1998. Microbial growth in soil and nitrogen turnover: A theoretical model considering the activity state of microorganisms. Soil Biol. Biochem. 30(13): 1743–1755.

Bolinder, M.A., H.H. Janzen, E.G. Gregorich, D.A. Angers, and A.J. VandenBygaart. 2007. An approach for estimating net primary productivity and annual carbon inputs to soil for common agricultural crops in Canada. Agric. Ecosyst. Environ. 118(1–4):29–42. doi:10.1016/j.agee.2006.05.013

Bouwman, A.F. 1998. Nitrogen oxides and tropical agriculture. Nature 392:866–867. doi:10.1038/31809

Bremer, E., H.H. Janzen, and C. Gilbertson. 1995. Evidence against associative $N_2$ fixation as a significant N source in long-term wheat plots. Plant Soil 175:13–19. doi:10.1007/BF02413006

Burgmann, H., F. Widmer, W. Von Sigler, and J. Zeyer. 2004. New molecular screening tools for analysis of free-living diazotrophs in soil. Appl. Environ. Microbiol. 70:240–247. doi:10.1128/AEM.70.1.240-247.2004

Campbell, C.A., H.R. Davidson, and F.G. Warder. 1977. Effects of fertilizer N and soil moisture on yield, yield components, protein content and N accumulation in the aboveground parts of spring wheat. Can. J. Soil Sci. 57(3):311–327. doi:10.4141/cjss77-036

Campbell, C.A., H.H. Janzen, and N.G. Juma. 1997b. Case studies of soil quality in the Canadian Prairies: Long-term field experiments. Dev. Soil Sci. 25:351–397. doi:10.1016/S0166-2481(97)80044-X

Campbell, C.A., G.P. Lafond, R.P. Zentner, and Y.W. Jame. 1994. Nitrate leaching in a Udic Haploboroll as influenced by fertilization and legumes. J. Environ. Qual. 23(1):195–201. doi:10.2134/jeq1994.00472425002300010030x

Campbell, C.A., D.W.L. Read, V.O. Biederbeck, and G.E. Winkleman. 1983. First 12 years of a long term crop rotation study in southwestern Saskatchewan nitrate N distribution in soil and N uptake by the plant. Can. J. Soil Sci. 63:563–578. doi:10.4141/cjss83-058

Campbell, C.A., F. Selles, G.P. Lafond, V.O. Biederbeck, and R.P. Zentner. 2001. Tillage fertilizer changes: Effect on some soil quality attributes under long term crop rotations in a thin Black Chernernozem. Can. J. Soil Sci. 81:157–165. doi:10.4141/S00-085

Campbell, C.A., F. Selles, R.P. Zentner, B.G. McConkey, S.A. Brandt, and R.C. McKenzie. 1997a. Regression model for predicting yield of hard red spring wheat grown on stubble in the semiarid prairie. Can. J. Plant Sci. 77:43–52. doi:10.4141/P96-043

Campbell, C.A., A.J. VandenBygaart, B. Grant, R.P. Zentner, B.G. McConkey, R. Lemke, E.G. Gregorich, and M.R. Fernandez. 2007. Quantifying carbon sequestration in a conventionally tilled crop rotation study in southwestern Saskatchewan. Can. J. Soil Sci. 87:23–38. doi:10.4141/S06-015

Campbell, C.A., and R.P. Zentner. 1993. Soil organic matter as influenced by crop rotations and fertilization. Soil Sci. Soc. Am. J. 57:1034–1040. doi:10.2136/sssaj1993.03615995005700040026x

Campbell, C.A., R.P. Zenter, H.H. Janzen, and K.E. Bowren. 1990. Crop rotation studies on the Canadian prairies. No. 1841/E. Research Branch, Agriculture Canada, Ottawa.

Campbell, C.A., R.P. Zentner, F. Selles, V.O. Biederbeck, and A.J. Leyshon. 1992. Comparative effects of grain lentil-wheat and monoculture wheat on crop production, N economy and N fertility in a Brown Chernozemic soil. Can. J. Plant Sci. 72:1091–1107. doi:10.4141/cjps92-135

Campbell, C.A., R.P. Zentner, F. Selles, V.O. Biederbeck, B.G. McConkey, R. Lemke, and Y.T. Gan. 2004. Cropping frequency effects on yields of grain, straw, plant N, N balance and annual production of spring wheat in the semiarid prairie. Can. J. Plant Sci. 84:487–501. doi:10.4141/P03-078

Campbell, C.A., R.P. Zentner, F. Selles, P.G. Jefferson, B.G. McConkey, R. Lemke, and B.J. Blomert. 2005. Long-term effect of cropping system and nitrogen and phosphorus fertilizer on production and nitrogen economy of grain crops in a Brown Chernozem. Can. J. Plant Sci. 85:81–93. doi:10.4141/P03-215

Carter, M.R., E.G. Gregorich, D.W. Anderson, J.W. Doran, H.H. Janzen, and F.J. Pierce. 1997. Concepts of soil quality and their significance. Dev. Soil Sci. 25:1–19. doi:10.1016/S0166-2481(97)80028-1

Chianese, D.S., C.A. Rotz, and T.L. Richard. 2009. Simulation of carbon dioxide emissions from dairy farms to assess greenhouse gas reduction strategies. Trans. ASABE 52(4):1301–1312. doi:10.13031/2013.27780

Conant, R.T., K. Paustian, and E.T. Elliott. 2001. Grassland management and conversion into grassland: Effects on soil carbon. Ecol. Appl. 11:343–355. doi:10.1890/1051-0761(2001)011[0343:GMACIG]2.0.CO;2

Dalal, R.C., W. Wang, G.P. Robertson, and W.J. Parton. 2003. Nitrous oxide emission from Australian agricultural lands and mitigation options: A review. Aust. J. Soil Res. 41:165–195. doi:10.1071/SR02064

Davidson, E.A., J.E. Rogers, and W.B. Whitman. 1991. Fluxes of nitrous oxide and nitric oxide from terrestrial ecosystems. In: J.E. Rogers and W.B. Whitman, editors, Microbial production and consumption of greenhouse gases: Methane, nitrogen oxides, and halomethanes. American Society for Microbiology, Washington, DC. p. 219–235.

de Jong, R., C.A. Campbell, R.P. Zentner, P. Basnyat, H. Cutforth, and R. Desjardins. 2008. Quantifying soil water conservation in the semiarid region of Saskatchewan, Canada: Effect of fallow frequency and N fertilizer. Can. J. Soil Sci. 88:(4): 461–475.

de Jong, R., C. Campbell, R. Zentner, P. Basnyat, B. Grant, and R. Desjardins. 2011. Cropping frequency and N fertilizer effects on soil water distribution from spring to fall in the semiarid Canadian prairies. Agric. Sci. 2:220–237.

Del Grosso, S.J., A.D. Halvorson, and W.J. Parton. 2008. Testing DayCent model simulations of corn yields and nitrous oxide emissions in irrigated tillage systems in Colorado. J. Environ. Qual. 37(4):1383–1389. doi:10.2134/jeq2007.0292

Del Grosso, S.J., A.R. Mosier, W.J. Parton, and D.S. Ojima. 2005. DAYCENT model analysis of past and contemporary soil $N_2O$ and net greenhouse gas flux for major crops in the USA. Soil Tillage Res. 83(1):9–24. doi:10.1016/j.still.2005.02.007

Del Grosso, S.J., W.J. Parton, C.A. Keough, and M. Reyes-Fox. 2011. Special features of the DayCent modeling package and additional procedures for parameterization, calibration, validation, and applications. In: L.R. Ahuja and L. Ma, editors, Methods of introducing system models into agricultural research. ASA, CSSA, SSSA, Madison, WI. p. 155–176

Del Grosso, S., W.J. Parton, A.R. Mosier, M.D. Hartman, J. Brenner, D.S. Ojima, and D.S. Schimel. 2001. Simulated interaction of carbon dynamics and nitrogen trace gas fluxes using the DayCent model. In: M.J. Schaffer, L. Ma, and S. Hansen, editors, Modeling carbon and nitrogen dynamics for soil management. CRC Press, Boca Raton, FL. p. 303–332.

Del Grosso, S.J., W.J. Parton, A.R. Mosier, D.S. Ojima, A.E. Kulmala, and S. Phongpan. 2000. General model for $N_2O$ and $N_2$ gas emissions from soils due to dentrification. Global Biogeochem. Cycles 14(4):1045–1060. doi:10.1029/1999GB001225

Dobereiner, J., J.M. Day, and P.J. Dart. 1972. Nitrogenase activity and oxygen sensitivity of the *Paspalum notatum Azotobacter paspali* association. J. Gen. Microbiol. 71:103–116. doi:10.1099/00221287-71-1-103

Eitzinger, J., W.J. Parton, and M. Hartman. 2000. Improvement and validation of a daily soil temperature submodel for freezing/thawing periods. Soil Sci. 165(7):525–534. doi:10.1097/00010694-200007000-00001

Firestone, M.K., and E.A. Davidson. 1989. Microbiological basis of NO and $N_2O$ production and consumption in soil. In: M.O. Andreae and D.S. Schimel, editors, Exchange of trace gases between terrestrial ecosystems and the atmosphere. John Wiley & Sons, Chichester, UK p. 7–21.

Frolking, S.E., A.R. Mosier, D.S. Ojima, C. Li, W.J. Parton, C.S. Potter, E. Priesack, R. Stenger, C. Haberbosch, P. Dörsch, H. Flessa, and K.A. Smith. 1998. Comparison of $N_2O$ emissions from soils at three temperate agricultural sites: Simulations of year-round measurements by four models. Nutr. Cycling Agroecosyst. 52:77–105. doi:10.1023/A:1009780109748

Gold, H.J. 1977. Mathematical modeling of biological systems. An introductory guidebook. John Wiley & Sons, New York.

Groffman, P.M. 2008. Nitrogen balances of ecosystem, landscape, regional and global scales In: J. Schepers and W. Raun, editors, Nitrogen in agricultural soils. SSSA, Madison. p. 731–758.

Harmel, R.D., and P.K. Smith. 2007. Consideration of measurement uncertainty in the evaluation of goodness-of-fit in hydrologic and water quality modeling. J. Hydrol. 337(3–4):326–336. doi:10.1016/j.jhydrol.2007.01.043

Harmel, R.D., P.K. Smith, and K.W. Migliaccio. 2010. Modifying goodness-of-fit indicators to incorporate both measurement and model uncertainty in model calibration and validation. Trans. ASABE 53(1):55–63. doi:10.13031/2013.29502

Henry, J.L., J.B. Bole, and R.C. McKenzie. 1986. Effect of nitrogen-water interactions on yield and quality of wheat in Western Canada. In: A.E. Slinkard and D.B. Fowler, editors, Wheat production in Canada: A review. Proceedings of the Canadian Wheat Production Symposium, Saskatoon, SK, Canada. 3–5 Mar. 1986. Univ. of Saskatchewan, Saskatoon, SK, Canada. p. 165–191.

Janzen, H.H. 1987. Effect of fertilizer on soil productivity in long-term spring wheat rotations. Can. J. Soil Sci. 67:165–174. doi:10.4141/cjss87-014

Janzen, H.H. 1995. The role of long-term sites in agroecological research: A case study. Can. J. Soil Sci. 75:123–133. doi:10.4141/cjss95-016

Janzen, H.H., C.A. Campbell, R.C. Izaurralde, B.H. Ellert, N. Juma, W.B. McGill, and R.P. Zentner. 1998. Management effects on soil C storage on the Canadian prairies. Soil Tillage Res. 47(3–4):181–195. doi:10.1016/S0167-1987(98)00105-6

Janzen, H.H., A.M. Johnston, J.M. Carefoot, and C.W. Lindwall. 1997. Soil organic matter dynamics in long-term experiments in Southern Alberta. Soil organic matter in temperate agroecosystems: Long-term experiments in North America. CRC Press, Boca Raton, FL. p. 283–296.

Jenkinson, D.S. 1991. The Rothamsted long-term experiments: Are they still of use? Agron. J. 83(1):2–10. doi:10.2134/agronj1991.00021962008300010008x

Kahindi, J.H.P., P. Woomer, T. George, F.M. de Souza Moreira, N.K. Karanja, and K.E. Giller. 1997. Agricultural intensification, soil biodiversity and ecosystem function in the tropics: The role of nitrogen-fixing bacteria. Appl. Soil Ecol. 6:55–76. doi:10.1016/S0929-1393(96)00151-5

Kelly, R.H., W.J. Parton, M.D. Hartman, L.K. Stretch, D.S. Ojima, and D.S. Schimel. 2000. Intra and interannual variability of ecosystem processes in shortgrass steppe. J. Geophys. Res. 105:20,093–20,100. doi:10.1029/2000JD900259

Kersebaum, K.C., A. Wurbs, R. de Jong, C.A. Campbell, J. Yang, and R.P. Zentner. 2008. Long-term simulation of soil–crop interactions in semiarid southwestern Saskatchewan, Canada. Eur. J. Agron. 29(1):1–12. doi:10.1016/j.eja.2008.01.011

Kröbel, R., W.N. Smith, B.B. Grant, R.L. Desjardins, C.A. Campbell, N. Tremblay, C.S. Li, R.P. Zentner, and B.G. McConkey. 2011. Development and evaluation of a new Canadian spring wheat submodel for DNDC. Can. J. Soil Sci. 91:503–520. doi:10.4141/cjss2010-059

Lafond, G.P., C.A. Campbell, R.L. Lemke, W.E. May, and C.B. Holzapfel. 2012. Indian Head Long Term Crop Rotations: Indian Head Saskatchewan. Prairie Soils Crops 5:42–50.

Lafond, G.P., M. Stumborg, W.E. May, C.B. Holzapfel, and C.A. Campbell. 2009. Quantifying straw removal through baling and measuring the long-term impact on soil quality and wheat production. Agron. J. 101:529–537. doi:10.2134/agronj2008.0118x

Lal, R. 2004. Soil carbon sequestration impacts on global climate change and food security. Science 304(5677):1623–1627. doi:10.1126/science.1097396

Lemke, R.L., T.G. Goddard, F. Selles, and R.P. Zentner. 2002. Nitrous oxide emissions from wheat-pulse rotations on the Canadian prairies. In: Proceedings of the 4th Annual Canadian Pulse Research Workshop, Edmonton, AB, Canada. 8–10 Dec. 2002. Organizing Committee Canadian Pulse Research Workshop, Edmonton, AB, Canada. p. 95–98.

Lemke, R., B. Helgason, R. Farrell, D. Pennock, and B. McConkey. 2005. Agriculture and greenhouse gases: What have we learned? In: Proceedings of the Soils Crops Workshop 2004 [CD-ROM], Saskatoon, SK, Canada. 19–20 Feb. 2005. Univ. of Saskatchewan, Saskatoon, SK, Canada.

Lemke, R.L., A.J. Vandenbygaart, C.A. Campbell, G.P. Lafond, B.G. McConkey, and B. Grant. 2012. Long-term effects of crop rotations and fertilization on soil C and N in a thin Black Chernozem in southeastern Saskatchewan. Can. J. Soil Sci. 92(3):449–461. doi:10.4141/cjss2010-047

Li, C., J. Aber, F. Stange, K. Butterbach-Bahl, and H. Papen. 2000. A process-oriented model of $N_2O$ and NO emissions from forest soils: 1. Model development. J. Geophys. Res. 105(D4):4369–4384. doi:10.1029/1999JD900949

Li, C., N. Farahbakhshazad, D.B. Jaynes, D.L. Dinnes, W. Salas, and D. McLaughlin. 2006. Modeling nitrate leaching with a biogeochemical model modified based on observations in a row-crop field in Iowa. Ecol. Modell. 196:116–130. doi:10.1016/j.ecolmodel.2006.02.007

Li, C., S. Frolking, and T.A. Frolking. 1992. A model of nitrous oxide evolution from soil derived by rainfall events: I. Model structure and sensitivity. J. Geophys. Res. 97:9759–9776. doi:10.1029/92JD00509

Li, C., S. Frolking, and R.C. Harriss. 1994. Modeling carbon biogeo-chemistry in agricultural soils. Global Biogeochem. Cycles 8:237–254. doi:10.1029/94GB00767

Li, C., W. Salas, R. Zhang, C. Krauter, A. Rotz, and F. Mitloehner. 2012. Manure-DNDC: A biogeochemical process model for quantifying greenhouse gas and ammonia emissions from livestock manure systems. Nutr. Cycling Agroecosyst. 93(2):163–200. doi:10.1007/s10705-012-9507-z

Linn, D.M., and J.W. Doran. 1984. Effect of water-filled pore space on carbon dioxide and nitrous oxide production in tilled and non-tilled soils. Soil Sci. Soc. Am. J. 48:1267–1272. doi:10.2136/sssaj1984.03615995004800060013x

Loague, K., and R.E. Green. 1991. Statistical and graphical methods for evaluating solute transport models: Overview and application. J. Contam. Hydrol. 7:51–73. doi:10.1016/0169-7722(91)90038-3

Manzoni, S., and A. Porporato. 2009. Soil carbon and nitrogen mineralization: Theory and models across scales. Soil Biol. Biochem. 41:1355–1379. doi:10.1016/j.soilbio.2009.02.031

Monreal, C.M., and H.H. Janzen. 1993. Soil organic-carbon dynamics after 80 years of cropping a dark brown Chernozem. Can. J. Soil Sci. 73:133–136. doi:10.4141/cjss93-014

Orr, C.H., A. James, C. Leifert, J.M. Cooper, and S.P. Cummings. 2011. Diversity and activity of free-living nitrogen-fixing bacteria and total bacteria in organic and conventionally managed soils. Appl. Environ. Microbiol. 77(3):911–919. doi:10.1128/AEM.01250-10

Parton, W., M. Hartman, D. Ojima, and D.S. Schimel. 1998. DayCent and its land surface submodel: Description and testing. Global Planet. Change 19:35–48. doi:10.1016/S0921-8181(98)00040-X

Parton, W.J., E.A. Holland, S.J. Del Grosso, M.D. Hartman, R.E. Martin, A.R. Mosier, D.S. Ojima, and D.S. Schimel. 2001. Generalized model for $NO_x$ and $N_2O$ emissions from soils. J. Geophys. Res. Atmos. 106(D15):17403–17419. doi:10.1029/2001JD900101

Parton, W.J., D.S. Ojima, C.V. Cole, and D.S. Schimel. 1994. A general model for soil organic matter dynamics: Sensitivity to litter chemistry, texture and management. In: R.B. Bryant and R.W. Arnold, editors, Quantitative modeling of soil forming processes. Spec. Publ. 39. SSSA, Madison, WI. p. 147–167.

Parton, W.J., and P.E. Rasmussen. 1994. Long-term effects of crop management in wheat-fallow: II. CENTURY model simulations. Soil Sci. Soc. Am. J. 58(2):530–536. doi:10.2136/sssaj1994.03615995005800020040x

Raju, P.N., H.J. Evans, and R.J. Seidler. 1972. An asymbiotic nitrogen-fixing bacterium from the root environment of corn. Proc. Natl. Acad. Sci. USA 69(11):3474–3478. doi:10.1073/pnas.69.11.3474

Robertson, G.P., and J.M. Tiedje. 1987. Nitrous oxide sources in aerobic soils: Nitrification, denitrification and other biological processes. Soil Biol. Biochem. 19(2):187–193. doi:10.1016/0038-0717(87)90080-0

Roloff, G., R. de Jong, C.A. Campbell, R.P. Zentner, and V.M. Benson. 1998a. Estimating spring wheat yield variability with EPIC. Can. J. Soil Sci. 78:541–549. doi:10.4141/S97-063

Roloff, G., R. de Jong, C.A. Campbell, R.P. Zentner, and V.M. Benson. 1998b. EPIC estimates of soil water, nitrogen and carbon under semiarid temperate conditions. Can. J. Soil Sci. 78(3):551–562. doi:10.4141/S97-064

Saxton, K.E., and W.J. Rawls. 2006. Soil water characteristic estimates by texture and organic matter for hydrologic solutions. Soil Sci. Soc. Am. J. 70(5):1569–1578. doi:10.2136/sssaj2005.0117

Smith, E.G., B.H. Ellert, H.H. Janzen, and D.J. Nakonechny. 2012a. Rotation ABC—Lethbridge, Alberta. Prairie Soils Crops 5:147–154.

Smith, P., D. Martino, Z. Cai, D. Gwary, H. Janzen, P. Kumar, B. McCarl, S. Ogle, F. O'Mara, C. Rice, B. Scholes, and O. Sirotenko. 2007. Agriculture. In: B. Metz, O.R. Davidson, P.R. Bosch, R. Dave, and L.A. Meyer, editors, Climate change 2007: Mitigation. Contribution of Working Group III to the Fourth Assessment Report of the Intergovernmental Panel on Climate Change. Cambridge Univ. Press, Cambridge, UK. p. 497–540.

Smith, P., J.U. Smith, D.S. Powlson, W.B. McGill, J.R.M. Arah, O.G. Chertov, K. Coleman, U. Franko, S. Frolking, D.S. Jenkinson, L.S. Jensen, R.H. Kells, H. Klein-Gunnewiek, A.S. Komarov, C. Li, J.A.E. Molina, T. Mueller, W.J. Patron, J.H.M. Thornley, and A.P. Whitemore. 1997. A comparison of the performance of nine soil organic matter models using datasets from seven long-term experiments. Geoderma 81:153–225. doi:10.1016/S0016-7061(97)00087-6

Smith, W.N., B.B. Grant, C.A. Campbell, B.G. McConkey, R.L. Desjardins, R. Kröbel, and S.S. Malhi. 2012b. Crop residue removal effects on soil carbon: Measured and inter-model comparisons. Agric. Ecosyst. Environ. 161:27–38. doi:10.1016/j.agee.2012.07.024

Smith, W.N., B.B. Grant, R.L. Desjardins, R. Kröbel, C. Li, B. Qian, D.E. Worth, B.G. McConkey, and C.F. Drury. 2013. Assessing the effects of climate change on crop production and GHG emissions in Canada. Agric. Ecosyst. Environ. 179:139–150. doi:10.1016/j.agee.2013.08.015

Smith, W.N., B.B. Grant, R.L. Desjardins, P. Rochette, C.F. Drury, and C. Li. 2008. Evaluation of two process-based models to estimate soil $N_2O$ emissions in Eastern Canada. Can. J. Soil Sci. 88:251–260. doi:10.4141/CJSS06030

Thorup-Kristensen, K. 2006. Effect of deep and shallow root systems on the dynamics of soil inorganic N during 3-year crop rotations. Plant Soil 288:233–248. doi:10.1007/s11104-006-9110-7

Touré, A., D.J. Major, and C.W. Lindwall. 1995. Comparison of five wheat simulation models in southern Alberta. Can. J. Plant Sci. 75:61–68. doi:10.4141/cjps95-010

Zentner, R.P., and C.A. Campbell. 1988. First 18 years of a long-term crop rotation study in southwestern Saskatchewan—Yields, grain protein, and economic performance. Can. J. Plant Sci. 68:1–21. doi:10.4141/cjps88-001

Zentner, R.P., E.D. Spratt, H. Reisdorf, and C.A. Campbell. 1987. Effect of crop rotation and N and P fertilizer on yields of spring wheat grown on a Black Chernozemic clay. Can. J. Plant Sci. 67:965–982. doi:10.4141/cjps87-134

Zhang, Y., C. Li, X. Zhou, and B. Moore, III. 2002. A simulation model linking crop growth and soil biogeochemistry for sustainable agriculture. Ecol. Modell. 151(1):75–108. doi:10.1016/S0304-3800(01)00527-0

# CQESTR Simulation of Dryland Agroecosystem Soil Organic Carbon Changes under Climate Change Scenarios

**Hero Gollany**

Abstract

The potential effects of global climate change (CC) on C cycling and soil organic carbon (SOC) storage and loss in agroecosystems can be assessed by process-based models such as CQESTR. This model was used to simulate the effect of tillage and N fertilization on soil C storage and loss in three long-term experiments, where a wheat (*Triticum aestivum* L.)–fallow rotation under dryland is the predominant practice. Using a 30-yr simulation approach, the effects of biomass increases and decreases, intensifying cropping, and two decades of potential CC scenarios on SOC stock under 0 and 135 kg N $ha^{-1}$ (0N and 135N) in conventional tillage (CT) and no-till (NT) were examined. These SOC stocks were also compared with those of a pasture (GP). With a 30% increase in biomass, predicted SOC stock in the topsoil of the GP increased by 22% under current climate compared to CC scenario, and the topsoil was more affected by CC than the subsoil. A loss of SOC was predicted at the 0- to 30- and 30- to 60-cm depths for the CT–0N in all scenarios. The same was true for the CT–135N and NT–0N, except SOC gains of 0.26 and 1.37 Mg $ha^{-1}$ were predicted, respectively, in continuous winter wheat under CC scenarios. The NT–135N was the only treatment with SOC gains ranging from 1.35 Mg $ha^{-1}$ in wheat–fallow system with 30% biomass reduction to 8.61 Mg $ha^{-1}$ in continuous wheat. The C sink capacities of the NT–135N, NT–0N, CT–135N, and CT–0N were 12.69, 14.54, 13.46, and 18.57 Mg $ha^{-1}$, respectively, relative to the GP. Crop intensification under NT is a viable management system that could sequester up to 1.05 Mg $CO_2$ $ha^{-1}$ $yr^{-1}$ and improve SOC stock while reducing $CO_2$ in the atmosphere.

Abbreviations: CC, climate change; CT, conventional moldboard plow tillage; CT–0N, conventional moldboard plow tillage without N fertilization; CT–135N, conventional moldboard plow tillage with 135 kg N $ha^{-1}$ fertilization rate; GP, grass pasture; MAP, mean annual precipitation; MAT, mean annual temperature; NT, no-till; NT–0N, no-till without N fertilization; NT–135N, no-till with 135 kg N $ha^{-1}$ fertilization rate; SOC, soil organic carbon; and SOM, soil organic matter.

Hero Gollany, USDA-ARS-CPCRC, 48037 Tubbs Ranch Rd., Adams, OR 97810 (hero.gollany@ars.usda.gov).

doi:10.2134/advagricsystmodel6.2013.0004


Synthesis and Modeling of Greenhouse Gas Emissions and Carbon Storage in Agricultural and Forest Systems to Guide Mitigation and Adaptation
S. Del Grosso, L. Ahuja, and W. Parton, Editors
Advances in Agricultural Systems Modeling, Volume 6. Lajpat R. Ahuja, Series Editor

Long-term no-till experiments, established in 1982, at the Columbia Plateau Conservation Research Center in the foothills of the Blue Mountains in Pendleton, Oregon.

The impacts of global climate change on agriculture production and agroecosystem functions are some of the most pressing issues facing agriculture. The $CO_2$ concentration in the atmosphere is predicted to reach 700 ppmv by the end of this century (IPCC, 2001, 2014). Ecological disturbances due to shifts in land use from native forests and grasslands to agriculture have contributed to release of $CO_2$ to the atmosphere from the antecedent C as soil organic matter (Liang et al., 2008). Lal et al. (1998) estimated the total C release by land use change at ~117 Pg ($10^{15}$ g). The loss of C from US soils following conversion from native prairie or forest to agricultural cropping is well documented (Follett et al., 1997; Huggins et al., 1998; Janzen et al., 1998; Lal et al., 1999; Paustian et al., 1998; Rasmussen and Albrecht, 1998; Rasmussen and Collins, 1991). A significant land use change in the United States occurred in the early 1800s. An estimated 6.25 Tg ($10^{12}$ g) C $yr^{-1}$ was emitted as a result of shifting cultivation (Lal et al., 1998). The C losses were largest during the first 50 yr of land conversion due to soil disturbance then slowed as the soil approached a steady state condition. Liebig et al. (2005) reported an average relative soil organic matter (SOM) loss of 34 ± 14% in the top 30-cm depth in the northwestern United States and western Canada. Interest in abatement of climate change has prompted research in agricultural management strategies to reduce the increase in $CO_2$ levels in the atmosphere through sequestering $CO_2$ in a mineral soil as SOC (Follett et al., 2005; Franzluebbers, 2005; Lal, 2004; Liebig et al., 2005).

Agronomic management can play a major role in C cycling since soil can act as a source or sink of C. The amount of SOC in an agroecosystem is controlled by the difference between inputs of organic matter and outputs through mineralization, loss and deposition by erosion, and translocation of dissolved organic C through soil (Campbell et al., 1996; Mertens et al., 2007). The C sink capacity of the world's degraded soils and agricultural land is estimated at 66% of the historic C loss, or ~78 Pg ($10^{15}$ g) (Follett et al., 2012). There is the potential to reverse the historical C loss with proper management of agricultural land and forest while reducing $CO_2$ levels in the atmosphere. About 15% of the total $CO_2$ emitted in the United States is offset by the amount of C sequestered by land use, land-use change, and forestry (USEPA, 2014). The net C sequestration from land use and land-use change increased by ~17.8%, a net sequestration of 979.3 Tg $CO_2$ equivalents from 1990 to 2012 (USEPA, 2014). This C sequestration in soils was attributed to an increase in forest C stocks, while net C accumulation in cropland slowed, and emissions from land converted to cropland decreased (USEPA, 2014). This is not surprising, since management changes such as crop residue inputs, soil nutrient levels, water balance, soil temperature, and soil disturbance affect SOM and SOC stocks. Soil disturbance because of tillage operation affects soil physical and biochemical properties and $CO_2$ emission as SOC decomposes (Oorts et al., 2007). The SOC stocks may be increased by improved fertilization, crop residue inputs, organic amendments (i.e., green or organic waste), reduction in soil tillage, and irrigation. In general, a good N fertility program in continuously cropped management systems increases SOC stocks. However, N fertilization had a variable effect on SOC storage in several studies (Gregorich and Janzen, 1996; Halvorson et al., 2002; Jantalia and Halvorson, 2011; Khan et al., 2007; Russell et al., 2005). After 40 to 50 yr of N fertilization that exceeded grain N removal, a net decline in soil C occurred, despite residue returned under conventional tillage (Khan et al., 2007). Jantalia and Halvorson (2011) and Halvorson and Jantalia (2011) reported that N fertilization of irrigated continuous maize (*Zea mays* L.) near Fort Collins, Colorado, resulted in greater SOC stocks within the top 30 cm with no-till (NT) than with conventional tillage (CT). Dryland continuous cropping under NT increased soil C storage at 0 to 15 cm by 0.233 Mg $ha^{-1}$ $yr^{-1}$ compared to a loss of 0.141 Mg $ha^{-1}$ $yr^{-1}$ in crop–fallow under CT in the northern Great Plains (Halvorson et al., 2002). Conversely, loss of SOC occurred even with a high N fertilization rate in a fallow system under CT. In eastern Oregon, under a semiarid environment, loss of SOC occurred in CT wheat–fallow rotation across a range of N rates (Rasmussen and Parton, 1994). After 44 yr of conservation tillage (sweep tillage) and a high N rate (180 kg N $ha^{-1}$) SOC stocks increased by only 3% above that for the 45 kg N $ha^{-1}$ rate (Gollany et al., 2005). Conventional tillage fallow systems under dryland conditions are not conducive to SOC accretion.

The potential effects of global climate change on C cycling and SOC storage or loss in dryland agroecosystems can be assessed by process-based simulation models. These models integrate many soil properties and biogeochemical processes that control SOC turnover and allow for climate feedbacks. Soil C in its stable form as SOM responds gradually to agricultural management changes and requires very long-term monitoring to detect any changes. Because of this slow process and the complex interactions between management factors, climate, and edaphic properties, simulation models (Del Grosso et al., 2011; Gollany et al., 2011; Izaurralde et al., 2006; Liang et al., 2009; Parton et al., 1988; Paustian et al., 1992; Rickman et al., 2002) have been used to predict the short- and long-term effects of management practices on SOC dynamics and turnover. The crop production potential of a soil is strongly related to its organic matter content (Gollany et al., 1992; Lal et al., 1998; Mann et al., 2002), which in part is controlled by C inputs such as crop residue or organic amendments. Understanding the long-term effects of N fertilization, tillage and crop residue management practices, and climate change on SOC dynamics and C stocks is fundamental to sustainable crop production under the dryland cropping system of the semiarid region in the Pacific Northwest.

This chapter will present a CQESTR model overview and simulation results of the effect of soil tillage management and N fertilization on soil C storage or loss in three long-term experiments at the Columbia Plateau Conservation Research Center in Pendleton, OR, where dryland wheat cropping is the predominant agriculture practice. Using the simulation approach, we will also explore the effects of potential climate change scenarios on SOC stocks in a wheat–fallow system under two fertilizer rates in CT and NT management systems and compare these SOC stocks with the SOC stocks of a long-term grass pasture to examine the potential for C sequestration in this semiarid region.

## Site and Experiment Description

The data reported here are from two long-term winter wheat experiments at the Columbia Plateau Conservation Research Center (45°72′ N, 118°62′ W, elevation 458 m) and a long-term grass pasture (45°72′ N, 118°60′ W, elevation 453 m) at the Columbia Basin Agriculture Research Center, which are located 15 km northeast of Pendleton, OR. The site is in the intermediate precipitation zone (300–450 mm) of the US Pacific Northwest (Gollany et al., 2013). The climate is semiarid, with a 30-yr (1982–2012) mean annual temperature (MAT) of 10.4°C and mean annual precipitation (MAP) of 437 mm, 90% of which occurs mainly as winter rain during the period of 1 October to 30 June. Average air temperature ranges from −0.6°C in January to 21.2°C in July (Rasmussen and Smiley, 1997). The soil is a Walla Walla silt loam (coarse-silty, mixed, superactive, mesic Typic Haploxeroll) developed in loess from Pleistocene alluvial deposits overlying basalt flows of Miocene. The soil is well drained and depth to water table is greater than 50 m. The field was put into agriculture production in the mid 1880s, and was previously under conventional plow tillage winter wheat–fallow rotation before initiation of this NT experiment in 1982 (Albrecht et al., 2005). Selected soil chemical and physical properties for the Ap horizon (0–20 cm) were determined in the fall of 2007. The Ap horizon contained 720 g $kg^{-1}$ silt, 190 g $kg^{-1}$ sand, and 90 g $kg^{-1}$ clay, had a cation exchange capacity of 12.75 $cmol_c$ $kg^{-1}$, soil organic matter of 19 g $kg^{-1}$, and a pH (1:2 soil to 0.01 M $CaCl_2$ solution ratio) of 6.85.

The three long-term experiments used in this simulation study are a long-term conservation tillage experiment ($NT_A$), conventional moldboard tillage (CT), and a grass pasture (GP). The details of the experimental design and management operations were previously published (Rasmussen and Albrecht, 1998). Briefly, the $NT_A$ experiment consisted of three phases. The conservation tillage (direct-seed, NT) experiment with 40 plots (2.6 by 30.5 m) of a winter wheat–spring wheat rotation was initiated in 1982 (Rasmussen and Smiley, 1994). This experiment was strictly no-till, without disturbance other than seeding and stubble flailing. Since the establishment of the NT plots some treatment changes (i.e., residue management, fertilizer application method and rate, seeding dates, and cropping system) were made, but NT has been in place since 1982. The experiment was modified in 1985, 1988 (Rasmussen and Smiley, 1994), and 1997. In 1982, all west plots received 81 kg N $ha^{-1}$, and consisted of three replicates of treatments testing different drills in the winter wheat–spring wheat rotation, while east plots consisted of three replicates of seven treatments testing variable fertilizer rates and application methods. In phase I (1985–1988), a factorial experiment of four replicates of five N fertilization treatments with two residue management levels was established (Rasmussen et al., 1997). The fertilizer application treatments were four N fertilization rates (0, 56, 112, and 169 kg N $ha^{-1}$) banded at 2.5 to 3 cm below seed depth and one fertilizer rate (112 kg N $ha^{-1}$) surface broadcast. Residue treatments were standing stubble vs. finely chopped in 1984. From 1985 through 1988, residue treatments were changed to standing stubble vs. burned (after tilling a 30-cm-wide border around each odd numbered plot). In phase II (1989–1993), the winter wheat–spring wheat crop rotation was replaced with winter wheat–fallow rotation in 1989, and winter wheat was grown on even years. Two dates of seeding (10 September and 10 October) replaced the residue management treatments, with odd numbered plots seeded on 10 September and even numbered plots seeded on 10 October (Rasmussen and Smiley, 1994). In phase III (1994–1997), a randomized complete block design experiment was established with four replications of five N fertilization rates (0, 45, 90, 135, and 180 kg N $ha^{-1}$), with a minor change in fertilizer rates from previous rates (0, 56, 112, 112, and 169 kg N $ha^{-1}$). In phase IV (1997–present), the cropping system was retained as winter wheat–fallow rotation, and the seeding date variable was replaced by year cropped to represent both phases of the crop rotation each year.

A set of conventional moldboard plow tillage (CT) plots consisting of four replications of two N treatments (0 and 135 kg N $ha^{-1}$) was added to an adjacent area east of the $NT_A$ in the southeastern corner of the Research Center in 1997. The grass pasture (GP) experiment (45.7 by 109.8 m), divided by a waterway, with soil depth of ~1.22 m and slopes of 0 to 3%, is located to the west of CT and $NT_A$ and has been maintained since 1931. It approximates near-virgin grassland and serves as a baseline for changes in other long-term experiments maintained by Oregon State University. The dominant grass species is tall fescue (*Festuca arundi-*

*nacea* Schreb.) followed by bulbous bluegrass (*Poa Bulbosa* L.), green foxtail [*Setina viridis* (L.) P. Beauv.], and yellow foxtail [*Setina pumila* (Poir.) Roem. & Schult.]. Vegetation has been clipped once or twice during summer, periodically reseeded, occasionally fertilized, and occasionally irrigated and grazed from 1931 to 1985 (Rasmussen and Smiley, 1997).

## Crop Management

The details of the management operations for this study have been previously published (Rasmussen and Smiley, 1997). Briefly, the traditional winter wheat conservation tillage (direct-seed) fallow rotation was seeded (~4-cm deep) in 25-cm rows to soft-white winter wheat in September or early October using a John Deere (HZ 616) deep furrow drill, and Noble no-till drill capable of placing a fertilizer (Uran 32, Polyphos and Thiosul) solution 5 cm below the seed to reduce nutrient immobilization by residue. Nitrogen fertilizer was applied as urea-ammonium nitrate, P as ammonium polyphosphate, and S as ammonium thiosulfate. Winter wheat was harvested the following July or early August. Spring wheat was planted in 1985, 1987, 1997 (in rotation with winter wheat), and 2012 (due to winter wheat crop failure) in late March and early April. Wheat seeding rates were adjusted to ~250 plants $m^{-2}$. In the traditional CT experiment the cropping system was alternating winter wheat and fallow (15 mo fallow). Stubble remained standing after harvest until primary tillage (moldboard plowed ~22 cm deep) in April, followed by secondary tillage (field cultivator ~13 cm deep, tine harrow ~13 cm deep, and three or four rod-weeder passes depending on weather, with a set of three 3-m rod-weeders with 1.9-cm square rods set to a depth of 5 cm to kill weeds and seal the dust-mulch), and fertilized at seeding in September the following year. Weeds were controlled chemically in the $NT_A$ with various herbicides to prevent weed resistance. Herbicides were used to control weeds in both fallow and crop.

## Soil and Plant Biomass Sampling

Grain yield was determined by harvesting 1.5 by 18.2 m from the interior of the plots with a self-propelled harvester in the 1980s and later with a Wintersteiger-Delta plot combine. Whole plant samples (1.0 by 1.2 $m^{-2}$, 1982–2010) were collected from the inner four rows of each plot before combine harvest to determine straw and grain yield components. Plants were cut along the soil surface and dried at 60°C for at least 48 h. Grain and straw yield were adjusted to 120 g $kg^{-1}$ moisture content. Straw yield was calculated from combine grain yield and bundle straw/grain ratio. Stubble remained standing after harvest until flailed before seeding.

Four soil cores were collected following the fall harvests in 2002 and 2009 from each tillage and N treatment using a Giddings hydraulic probe 3.2 cm in diameter and 1.8 m in length (Giddings Machine Co.). Soil cores were sectioned at 0 to 5, 5 to 10, 10 to 20, 20 to 30, 30 to 60, 60 to 100, and 100 cm to a depth of 150 cm unless a restricted layer was reached. Previous soil samples, 60 cm in length, were taken in 1995 and 1999 and sectioned at 0 to 30 and 30 to 60 cm. Two, four, four, and three soil cores, 60 cm in length, were collected from grass pasture in 1990, 1995, 2000, and 2009, respectively, and sectioned at 0 to 30 and 30 to 60 cm. Two cores from each plot were composited. Soil samples were dried at 40°C, ground with a rolling pin, passed through nested 1- and 2-mm sieves, with all visible organic matter that was not collected on the sieve picked out with tweezers, and roller-milled for 4 h. These samples were used to determine total soil C and N. Total C was analyzed using a combustion analyzer (Thermo Finnigan FlashEA 1112 Elemental Analyzer). Total inorganic C was analyzed using a Skalar Primacs TOC Analyzer (Skalar Corp). Total organic carbon was determined by subtracting inorganic C from total C for each sample. The third and fourth cores from each plot were dried at 105°C and weighed to determine soil bulk density. These samples were then used to determine soil pH after they were ground and passed through a 1-mm sieve. The air-dried mass was converted to oven-dry basis, and sample volume was used to calculate soil bulk density ($\rho_b$). Mass of SOC by individual depth increment was based on soil bulk density. Initial SOC values for GP samples were determined in 1990 at 0- to 30- and 30- to 60-cm soil depths; therefore, weighted averages for four depths were calculated and used for the $NT_A$ and CT simulations in this chapter, and SOC was reported for the 0- to 30- and 30- to 60-cm soil depths.

## The CQESTR Model Description

The CQESTR model is a field-scale process-based soil C balance model of intermediate complexity. It has been in continuous development by USDA-ARS scientists at the Columbia Plateau Conservation Research Center since 2000 (Liang et al., 2009; Rickman et al., 2001). A brief description of the CQESTR model is reported here, and a detailed description can be found in Liang et al. (2009). It computes the rate of biological decomposition of crop residue or organic amendments as they convert to SOM/SOC (Liang et al., 2009; Rickman et al., 2001). It simulates the effects of climate (Gollany et al., 2013), tillage management practices (Liang et al., 2009; Leite et al., 2009; Rickman et al., 2002; Wilhelm et al., 2010), crop rotations (Gollany et al., 2012, 2013; Liang et al., 2008), amendment additions (Gollany et al., 2012; Liang et al., 2009; Plaza et al., 2012; Rickman et al., 2001), crop residue burning (Gollany et al., 2012; Liang et al., 2009; Rickman et al., 2001), or crop residue removal (Gollany et al., 2010, 2011) on SOC stocks in a soil profile of up to five layers (horizons). It works on a daily time-step and can perform long-term (e.g., 100-yr) simulations. Soil organic matter change is computed by maintaining a soil C budget for C additions as a result of atmospheric $CO_2$ sequestration during photosynthesis or added amendments like manure or municipal waste, and organic C losses as $CO_2$ through microbial decomposition. The total SOC budget can be represented by Eq. [1], using units of dry weight per unit area within each soil layer,

$$SOC = (C_{SOM} - C_{DOM}) + \sum_{l=1}^{\mu}(C_{S,l} - C_{DS,l}) + \sum_{m=1}^{v}(C_{R,m} - C_{DR,m}) + \sum_{n=1}^{w}(C_{A,n} - C_{DA,n}) \qquad [1]$$

where SOC is total soil organic carbon, $C_{SOM}$ is carbon in the stable soil organic matter, $C_{DOM}$ is decomposed organic matter lost as $CO_2$, $C_{S,l}$ is carbon in shoot residue $l$, $C_{DS',l}$ is carbon lost as $CO_2$ from decomposed shoot residue $l$, $C_{R,m}$ is carbon in root residue $m$, $C_{DR,m}$ is carbon lost as $CO_2$ from decomposed root residue $m$, $C_{A,n}$ is carbon in organic amendment $n$, $C_{DA,n}$ is carbon lost as $CO_2$ from decomposed amendment $n$, and $u$, $v$, $w$ are all applications of organic materials from the initial time to the current day (Liang et al., 2009).

The organic material decomposition is a three-phase process. After each organic material placement in the soil, decomposition occurs in two phases. Phase I is a rapid phase covering the first 1000 cumulative degree-days (thermal time), approximating the oxidation of readily metabolized substrate. Phase II is a slow decomposition phase representing oxidization of more recalcitrant materials (e.g., lignin). Crop residues and organic amendments are categorized by their placement in the soil and their identities are maintained throughout the two phase decomposition. Each organic residue addition is tracked separately according to its placement within distinct soil horizons. When Phase II is complete, the composted residue is transferred to the recalcitrant SOM pool (Phase III).

The CQESTR model uses readily available data. The major required model input data includes crop rotation and annual yields (above- and below-ground crop yields), grain to residue ratio, root distribution characteristics of crops, root biomass of crops used, dates of all operations (e.g., tillage, seeding, harvest, biomass addition, biomass removal, etc.), N content of both residue and organic amendments (i.e., manure or biosolid), weather data (mean monthly air temperature and monthly total precipitation), the number and thickness (depth) of soil layers (horizon), and the initial SOM or SOC content and bulk density of each soil layer. Soil horizons and depths are user defined. Crop rotation, grain and residue inputs, residue removal and tillage information are required explicitly. These consist of crop grain yields, shoot-to-grain ratios, depth of tillage and the fraction of the soil surface covered, and effects of tillage on residue (e.g., fraction of pre-tillage residue weight remaining on the soil surface after each tillage operation).

## Climate Change Scenarios Inputs

Climate change scenarios and crop biomass increase or reduction scenarios for fertilized ($N = 135$ kg N ha$^{-1}$) and nonfertilized ($N_0$) treatments were simulated for 20 yr, representing two potential rates of crop residue biomass improvement (10 or 30%) and two rates of crop biomass reduction (10 or 30%) representing anticipated changes in crop residue biomass associated with forecasted increases or decreases in the potential precipitation predicted to occur in the dryland Pacific Northwest region as a result of climate change (Littell et al., 2009). The historic MAP and MAT values were used until 2010, while new values for 2011 to 2020 and 2021 to 2030 were used to address climate change as described below. Climatic records at the Research Center indicate an increase in MAT from 10.2 to 10.4°C and MAP from 400 to 437 mm since the mid 1990s. There are many uncertainties in predicting long-term effects of potential climate change. Therefore, anticipated impacts of climate change on SOC due to increased MAT (1.1°C during 2011–2020 and 1.4°C during 2021–2030) and MAP (33 mm during 2011–2020 and 46 mm during 2021–2030) were simulated for 20 yr. All other input variables were kept constant. Actual annual grain and residue yields between 1981 and 2012 were used in simulations, with the resulting 10 and 30% of wheat straw under both tillage treatments being equivalent to each biomass harvest rate plus or minus the proposed increase or decrease from average yield for 2000 to 2010. In the continuous winter wheat scenarios, wheat straw yield equivalent to 75 and 90% of the projected 30 and 10% yield increase, respectively, to reflect the actual field relationship between biomass harvested when continuous cropping system was used relative to biomass yield in a wheat–fallow cropping system. No other changes in straw yield were assumed during the 30 yr of the simulations. The current C sink capacity of the top 0- to 30-cm depth for each experiment was determined relative to the GP experiment by subtracting current measured SOC stocks at the 0- to 30-cm depth in 2009 for each experiment from the measured GP SOC stocks values at the 0- to 30-cm depth in 1990.

## Statistical Procedures

Linear regression (PROC REG) and Pearson correlation (PROC CORR) procedures were used to assess the relationship between simulated and observed values and statistical significance (SAS Institute, 2008). The ANOVA was used to compute a standard error of the mean. Mean square deviation statistics were also used to evaluate the predictive performance of the model against observed data. According to Gauch et al. (2003), the mean square deviation (MSD) is partitioned into three components: non-unity slope (NU), lack of correlation (LC) or scatter, and squared bias (SB). All three components relate to terms of the linear regression equation ($Y = a + bX$) and the regression coefficient ($r^2$).

## CQESTR Model Evaluation

The CQESTR model was validated with long-term data from 13 sites across North America having a range of SOC (4.2–33.6 g SOC $kg^{-1}$) under diverse agronomic practices, climate, soil texture, and drainage classes (Liang et al., 2009). Regression analysis of observed and simulated SOC values resulted in a linear relation with an $r^2$ of 0.95 ($P < 0.0001$) and a 95% confidence interval of 2.5 g SOC $kg^{-1}$. Furthermore, regression analysis of simulated and observed SOC values from the three long-term experiments used in this chapter were closely related ($r^2 = 0.98$, $P < 0.0001$) with slope not significantly different from 1 ($P < 0.0001$), demonstrating the capacity of CQESTR to predict long-term SOC dynamics for these experiments under a range of management practices and climate change scenarios (Fig. 1). A significant ($P < 0.0001$) Pearson correlation coefficient ($r = 0.99$) and small MSD (3.722) indicated good model performance (Gauch et al., 2003). The contrib-

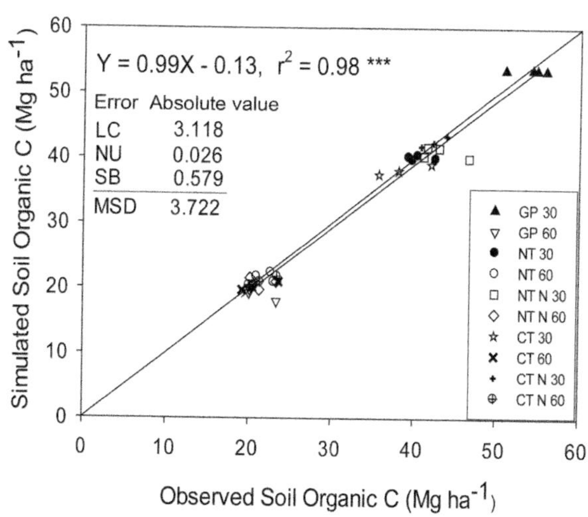


**Fig. 1. Scatter plots of measured total soil organic carbon versus CQESTR-predicted soil organic carbon in the 0- to 30-cm (30) and 30- to 60-cm (60) soil depths of the long-term conventional moldboard plow (CT) and no-till (NT) winter wheat–fallow cropping systems with nitrogen (N) and without fertilization and the grass pasture experiment. Superimposed is a linear regression line of all observed values vs. CQESTR-predicted soil organic carbon stocks for the experiments with slope and intercept; r, correlation coefficient; and ***, significant at the 0.0001 probability level.**

uting components of MSD were in the following order: scatter (LC = 3.118 or 83.8% of MSD) > translation (SB = 0. 579 or 15.6%) > rotation (NU = 0.026). The highest contributing component of MSD is the LC, which indicates that prediction errors are mainly associated with data scattering.

## Model Applications and Climate Change Scenarios

### Grass Pasture

Soil organic carbon dynamics for the long-term grass pasture were simulated for the 0- to 30-cm and 30- to 60-cm soil depths using the CQESTR model (Fig. 2). The model predicted SOC dynamics for the GP reasonably well at the 0- to 30-cm soil depth during 19 yr, although the CQESTR model predicted a slight SOC loss (0.16 Mg ha$^{-1}$) at a rate of 0.008 Mg ha$^{-1}$ yr$^{-1}$ while observed increased slightly (1.67 Mg ha$^{-1}$) at 0.086 Mg ha$^{-1}$ yr$^{-1}$. The simulation results as well as the observed data indicated SOC losses at the 30- to 60-cm depth, except in 2009 where the measured value indicated SOC gain. This discrepancy between predicted and measured SOC values could be due to sampling error or sample processing error, or contribution of dolomite that was not removed from the sample when calcium carbonate was removed while determining inorganic carbon in the soil samples, or underestimation of SOC by CQESTR due to the contribution of dissolved organic carbon to observed SOC stocks down in the soil profile, which CQESTR did not account for as a source of soil C sequestration in these arid soils. It seems most likely the discrepancy is from sample processing error since the previous three measured values indicated SOC losses at this depth. The rates of observed SOC stock gain and simulated SOC stock losses were 0.0139 and 0.131 Mg ha$^{-1}$ yr$^{-1}$, respectively, at

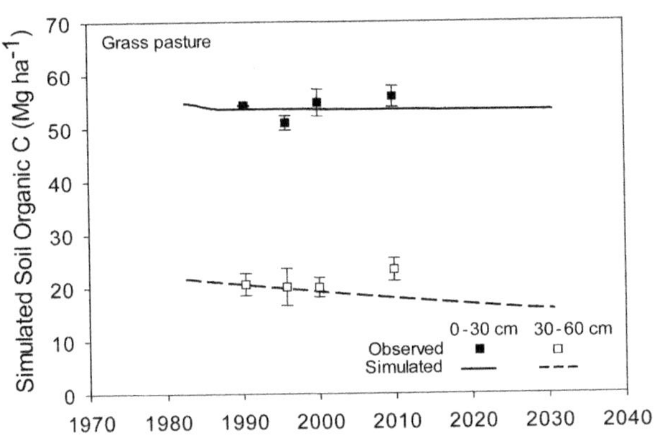


**Fig. 2. The CQESTR model simulated (lines) and observed (symbols) soil organic C stocks in the 0- to 30- and 30- to 60-cm soil depths of the long-term grass pasture experiment. Bars are standard errors of the mean (*n* = 2, 4, 4, and 3, in 1990, 1995, 2000, and 2009, respectively).**

the 30- to 60-cm depth. Greater predicted SOC loss at the 30- to 60-cm depth than at the 0- to 30-cm depth could be attributed either to limited root biomass growth at this depth or due to enhanced ecosystem respiration (Pendall et al., 2013) and decomposition of labile C and transport to lower depth.

## Long-Term No-till Wheat–Fallow Cropping System

Soil organic carbon dynamics for the long-term NT cropping system, with (NT–135N) and without fertilization (NT–0N), were simulated for the 0- to 30- and 30- to 60-cm soil depths using the CQESTR model (Fig. 3). The model predicted SOC dynamics for the NT–0N very well. Observed and simulated SOC values from 1995 to 2009 indicated SOC stock losses at both 0- to 30- and 30- to 60-cm soil depths in the unfertilized NT–0N plots (Fig. 3a). This is due to the low crop yield and amount of biomass produced and returned in the wheat–fallow system without N fertilization. Observed and simulated SOC stock losses were 0.62 and

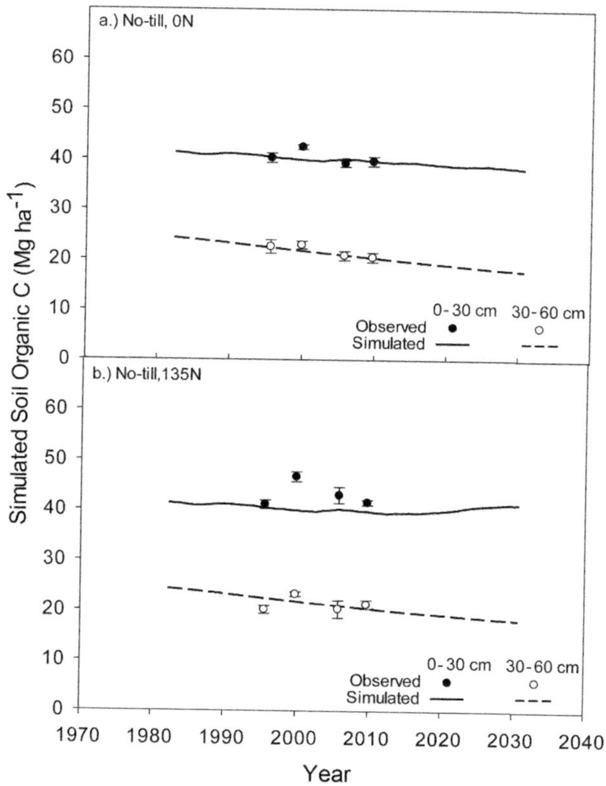


**Fig. 3. The CQESTR model simulated (lines) under current climate scenarios and observed (symbols) soil organic C stocks in the 0- to 30- and 30- to 60-cm soil depths of the long-term no-till winter wheat– fallow cropping systems under current climate (a) without nitrogen fertilization (0N) and (b) with 135 kg N ha$^{-1}$ nitrogen fertilization (135N). Bars are standard errors of the mean ($n$ = 4).**

0.68 Mg $ha^{-1}$ at rates of 0.04 and 0.05 Mg $ha^{-1}$ $yr^{-1}$ for the 0- to 30-cm soil depth, respectively. Observed and simulated SOC stock losses were 2.05 and 1.97 Mg $ha^{-1}$ at rates of 0.15 and 0.14 Mg $ha^{-1}$ $yr^{-1}$, respectively, for the 30- to 60-cm soil depth. The predicted rates of SOC losses at both depths were greater than in the GP. The wheat–fallow system is vulnerable to soil erosion, predominantly due to wind, and is not a biologically sustainable cropping system (Rasmussen and Parton, 1994) since loss of SOM and associated soil nutrients can reduce microbial populations and diversity with eventual crop failure due to soil nutrient depletion unless judicious supplementations of nutrients are made.

Observed SOC values in the fertilized NT–135N plots increased in the 0- to 30-cm depth in 1999 (Fig. 3b). An increase in measured SOC stocks for the 1999 sampling period most likely was due to undecomposed fine crop residue materials, light fraction, which passed through the 1-mm sieve during sample processing and was not removed with tweezers. The CQESTR model considers only C from decomposed residue to be incorporated into SOC stocks (Gollany et al., 2013). Crop residues are considered to be a transient C pool because light fraction organic material is not chemically or physically associated with soil minerals and is in a transitory pool between fresh residues and humified stable SOM (Gregorich et al., 1997; Gregorich and Beare, 2008). Observed and simulated SOC stocks in the fertilized NT–135N plots increased from 1995 to 2009 by 0.44 and 1.38 Mg $ha^{-1}$, respectively, at the 0- to 30-cm soil depth. This increase is probably due to increased wheat biomass as a result of N fertilization at the rate of 135 kg N $ha^{-1}$ and consequently increased the amount of wheat straw returned. Increased SOC stocks with increased N fertilization have been reported previously (Gregorich and Janzen, 1996; Halvorson et al., 2002; Halvorson and Jantalia, 2011; Omay et al., 1997; Rasmussen et al., 1980; Rasmussen and Rohde, 1988; Rasmussen and Smiley, 1997). A smaller observed SOC increase than predicted is most likely due to soil erosion which is not accounted for in the CQESTR model. At the 30- to 60-cm soil depth, observed SOC stocks increased at a rate of 0.08 Mg $ha^{-1}$ $yr^{-1}$ while CQESTR predicted SOC loss at a rate of 0.13 Mg $ha^{-1}$ $yr^{-1}$. This discrepancy between predicted and measured SOC values could be due either to the contribution of dissolved organic carbon to observed SOC stocks gained down in the soil profile, which CQESTR did not account for as a source of soil C sequestration in these arid soils, or could be from contribution of dolomite that was not removed from the sample when calcium carbonate was removed while determining inorganic carbon in the soil samples. Furthermore, there is a relatively poor understanding of the fate of C from root exudates. These mechanisms of soil C storage are poorly understood at this time; therefore, they are not accounted for in CQESTR.

## Conventional Wheat–Fallow System

Soil organic carbon dynamics for CT with N fertilization (CT–135N) and without (CT–0N) were simulated for the 0- to 30- and 30- to 60-cm soil depths using the CQESTR model (Fig. 4). The simulation results as well as the observed data indicated SOC loss at both soil depths for nonfertilized plots under the winter wheat–summer fallow cropping system (Fig. 4a). The simulated and observed SOC stock losses from 1999 to 2009 were 1.58 and 6.36 Mg ha$^{-1}$ for the 0- to 30-cm soil depth, respectively. Observed SOC decreased rapidly at a rate of 0.63 Mg ha$^{-1}$ yr$^{-1}$, while CQESTR predicted SOC stock loss at a rate of 0.16 Mg ha$^{-1}$ yr$^{-1}$ in the 0- to 30-cm soil depth. A similar pattern of SOC losses was observed and predicted by CQESTR but to a lesser degree for the 30- to 60-cm soil depth. Observed SOC decreased rapidly at a rate of 0.44 Mg ha$^{-1}$ yr$^{-1}$, while CQESTR predicted SOC

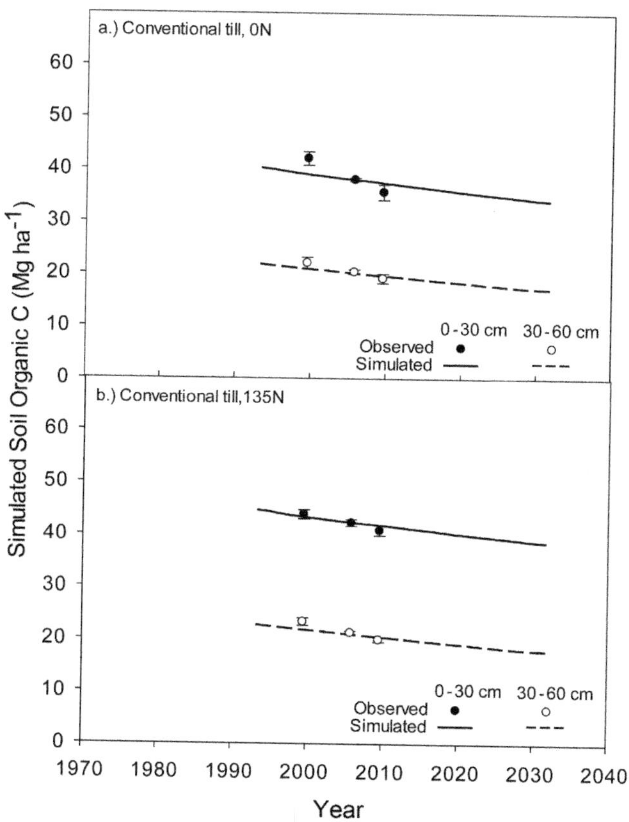


**Fig. 4. The CQESTR model simulated (lines) under current climate scenarios and observed (symbols) soil organic C stocks in the 0- to 30- and 30- to 60-cm soil depths of the long-term conventional tillage winter wheat–fallow cropping systems under current climate (a) without nitrogen fertilization (0N) and (b) with 135 kg N ha$^{-1}$ nitrogen fertilization (135N). Bars are standard errors of the mean (n = 4).**

stock loss at a rate of 0.13 Mg $ha^{-1}$ $yr^{-1}$ at the 30- to 60-cm soil depth. The model underestimated the rate of SOC stock losses by three and four times at the 30- to 60- and 0- to 30-cm soil depths, respectively. The reason for this discrepancy between observed and simulated values is unknown. A couple of limitations of CQESTR may have contributed to the simulation discrepancy. First, loss or deposition of C at the soil surface from wind and water erosion is not considered, nor is leaching of C through the soil profile in dissolved organic C form, or the physical transfer between soil layers or consumption of residue by worms, insects, small mammals, or macro fauna (Liang et al., 2009). A second limitation in CQESTR may be a missing parameter to account for the complex relationship between indigenous soil N level and plant N uptake.

The CQESTR model simulated SOC dynamics for the CT with N fertilization (CT–135N) very well for the 0- to 30- and 30- to 60-cm soil depths from 1999 to 2009 (Fig. 4b). The simulation results and observed SOC values indicated SOC losses at both soil depths, but the losses were less severe than those in the non-fertilized winter wheat–fallow cropping system. The SOC loss of 3.1 Mg $ha^{-1}$ in the CT–135N, from 1999 to 2009, was about half of that in the CT–0N at the 0- to 30-cm depth. This indicates that N fertilization either increased biomass and carbon input or decreased the decomposition of recalcitrant SOC stocks (Clapp et al., 2000; Varvel and Wilhelm, 2008). Observed SOC decreased rapidly at a rate of 0.31 Mg $ha^{-1}$ $yr^{-1}$ while CQESTR predicted SOC stock loss at a rate of 0.16 Mg $ha^{-1}$ $yr^{-1}$ at the 0- to 30-cm soil depth. The SOC losses at the 0- to 30-cm soil depth were 3.1 and 1.6 Mg $ha^{-1}$ for observed and predicted, respectively. The SOC losses in the dryland wheat–fallow cropping systems of the Pacific Northwest are attributed to low straw yield as a result of low precipitation, the biological oxidation associated with tillage, and erosion (Collins et al., 1992; Feng et al., 2011; Rasmussen and Parton, 1994; Thorne et al., 2003). The rates of SOC stocks lost at the 30- to 60-cm soil depth followed the same trend as in top soil and were 0.34 and 0.13 Mg $ha^{-1}$ $yr^{-1}$ for observed and predicted, respectively. The model underestimated SOC stock losses by two times for the topsoil, which was less than the 2.5 times for the 30- to 60-cm soil depth. The SOC losses we measured were greater than predicted and probably due to soil erosion and translocation that is not accounted for in CQESTR.

## Climate Change Scenarios and Wheat Yield Increase or Decrease

These scenarios were developed to represent anticipated changes in crop residue biomass associated with forecasted increases or decreases in the potential precipitation predicted to occur in the dryland Pacific Northwest region as a result of climate change.

## Grass Pasture under Current Climate with Biomass Increase or Decrease

The CQESTR model predicted SOC stock losses of 0.23 and 3.55 Mg ha$^{-1}$ at average annual rates of 0.008 and 0.118 Mg ha$^{-1}$ at the 0- to 30-cm and 30- to 60-cm soil depth, respectively, in the grass pasture (GP) long-term experiment, under current biomass yield and climatic conditions from 2000 to 2030 (Table 1). Increases in biomass yield of 10 and 30% (without CC) scenarios increased predicted SOC stocks during 30 yr of simulation by 0.35 and 1.44 Mg ha$^{-1}$, while reductions in biomass yield of 10 and 30% decreased SOC stocks by 0.74 and 1.84 Mg ha$^{-1}$, respectively, in the top 0- to 30-cm soil. This was expected since the rate of SOC stock change is highly correlated with the level of annual source C input into the system (Gollany et al., 2006; Rasmussen and Parton, 1994). Simulation results indicated that topsoil SOC stocks were more affected by biomass production than those in the subsoil. Soil organic C stock in the GP with climate change decreased from −0.23 to −0.64 Mg ha$^{-1}$ in the topsoil, while decreased from −3.55 to −3.81 Mg ha$^{-1}$ in the 30- to 60-cm depth and increased 1.21 Mg ha$^{-1}$ (from −0.23 to 1.44) in the topsoil vs. 0.04 Mg ha$^{-1}$ (from −3.55 to 3.51 Mg ha$^{-1}$) at the 30- to 60-cm depth, with 30% biomass yield increase (Table 1). Soil organic C stock losses were predicted for all biomass yield reductions or increases at the 30- to 60-cm soil depth. Soil organic C stock losses of 3.51 and 3.86 Mg ha$^{-1}$ were predicted for the 30- to 60-cm soil depth with 30 and 10% biomass yield reduction scenarios, respectively. These predicted SOC losses were close to the losses of 3.55 Mg C ha$^{-1}$ at the 30- to 60-cm depth under current climate conditions. This could be due to insufficient root biomass produced and C input below 30 cm because of prevalent dry summer and lack of precipitation from end of June until October.

## No-till Experiment under Current Climate with Biomass Increase or Decrease

Loss of SOC stock was predicted at both depths under the unfertilized wheat–fallow (NT–0N) cropping system for all scenarios including a 30% biomass increase (Table 1). A potential increase of biomass by 30% only decreased predicted SOC stock loss for the NT–0N treatments by 0.53 Mg ha$^{-1}$ (−0.82 vs. −1.35) at the 0- to 30-cm soil depth. An increase in SOC stocks was predicted for all NT–135N plots, including those with 30% biomass yield reduction. Soil organic C stock decreased by 0.66 Mg ha$^{-1}$ (from −1.35 to −2.01 Mg ha$^{-1}$) in the topsoil and decreased from by 0.32 Mg ha$^{-1}$ (from −3.94 to −4.26 Mg ha$^{-1}$) at the 30- to 60-cm depth, indicating that topsoil was more affected by CC than subsoil. The predicted SOC gain in the topsoil ranged from 1.93 Mg ha$^{-1}$ at an average annual rate of 0.064 Mg ha$^{-1}$ (0.234 Mg $CO_2$ yr$^{-1}$) with a potential 30% biomass yield reduction, to 4.62 Mg ha$^{-1}$ at a rate of 0.154 Mg ha$^{-1}$ (0.562 Mg $CO_2$ yr$^{-1}$) with a potential increase in biomass

**Table 1.** The CQESTR model simulated soil organic carbon changes (ΔSOC) from 2000 to 2030, in the 0- to 30- and 30- to 60-cm soil depths, for several anticipated changes in wheat biomass yields associated with forecasted increases or decreases in the potential precipitation predicted for the dryland Pacific Northwest region as a result of climate change, and climate change simulation scenarios for 20 yr under long-term grass pasture (GP), long-term no-till with 135 kg N ha-1 (NT–135N) and without fertilization (NT–0N), and conventional moldboard plow with (CT–135N) and without fertilization (CT–0N) in the wheat–fallow cropping systems at Pendleton, OR.

| | ΔSOC | | | | | | | | | |
|---|---|---|---|---|---|---|---|---|---|---|
| Soil depth: | 0–30 cm | | | | | 30–60 cm | | | | |
| Exp.: | GP | NT–0N | NT–135N | CT–0N | CT–135N | GP | NT–0N | NT–135N | CT–0N | CT–135N |
| Scenarios | | | | | | | | | | |
| | Mg $ha^{-1}$ | | | | | | | | | |
| Current yields and climate | −0.23 | −1.35 | 3.27 | −2.48 | −1.64 | −3.55 | −3.94 | −3.67 | −3.62 | −3.72 |
| Current climate, +10% yield increase | 0.35 | −1.18 | 3.72 | −2.31 | −1.39 | −3.54 | −3.94 | −3.61 | −3.61 | −3.71 |
| Current climate, −10% yield decrease | −0.74 | −1.61 | 2.82 | −2.66 | −1.89 | −3.56 | −3.95 | −3.68 | −3.63 | −3.73 |
| Current climate, +30% yield increase | 1.44 | −0.82 | 4.62 | −1.96 | −0.89 | −3.51 | −3.93 | −3.64 | −3.60 | −3.69 |
| Current climate, −30% yield decrease | −1.84 | −2.12 | 1.93 | −3.01 | −2.38 | −3.59 | −3.96 | −3.69 | −3.64 | −3.74 |
| Climate change and current yields | −0.64 | −2.01 | 2.74 | −3.14 | −2.43 | −3.81 | −4.26 | −3.97 | −3.90 | −4.01 |
| Climate change, +10% yield increase | −0.06 | −1.79 | 3.20 | −2.97 | −2.20 | −3.80 | −4.25 | −3.96 | −3.89 | −4.00 |
| Climate change, −10% yield decrease | −1.23 | −2.27 | 2.28 | −3.31 | −2.67 | −3.83 | −4.26 | −3.98 | −3.90 | −4.02 |
| Climate change, +30% yield increase | 1.12 | −1.34 | 4.12 | −2.64 | −1.70 | −3.77 | −4.24 | −3.95 | −3.88 | −3.99 |
| Climate change, −30% yield increase | −2.40 | −2.77 | 1.35 | −3.64 | −3.14 | −3.86 | −4.27 | −3.99 | −3.91 | −4.03 |
| Climate change and continuous wheat cropping | – | 1.37 | 8.61 | −1.09 | 0.26 | – | −3.49 | −3.12 | −3.49 | −3.60 |

yield of 30%. Many studies have shown that adoption of NT leads to an accretion of SOC at the soil surface (Angers and Eriksen-Hamel, 2008; Gollany et al., 2013; Liebig et al., 2005; West and Post, 2002). This was attributed to increased biomass or C source inputs, which promote SOC accretion (Follett et al., 2012; Gollany et al., 2006; 2013; Rasmussen et al., 1980). Gollany et al. (2006) reported that the soil C storage index doubled with winter wheat residue returned. Increases in biomass inputs to these dryland soils are expected to increase SOC stocks under NT and current climatic conditions.

## Conventional Tillage under Current Climate with Biomass Increase or Decrease

Soil organic C loss was predicted at both depths under unfertilized conventional wheat–fallow (CT–0N) cropping system for all scenarios including a 30% biomass increase (Table 1). The predicted SOC loss in the topsoil ranged from 1.96 to 3.01 Mg ha$^{-1}$ with a potential 30% increase or reduction in biomass yield, respectively. Soil organic C stock losses were predicted even with 135 kg N ha$^{-1}$ at both soil depths in all scenarios under conventional tillage, but the losses were less severe (e.g., 0.89 vs. 1.96 with 30% yield increase) than those with CT–0N. These predictions were in agreement with reported SOC losses with N fertilization under CT in the semiarid region of the Pacific Northwest. Rasmussen and Parton (1994) reported loss of SOC in wheat–fallow rotation under CT cropping system across a range of N rates. The biological oxidation associated with tillage and erosion exacerbated SOC losses in the conventional wheat–fallow system (Collins et al., 1992; Feng et al., 2011; Rasmussen and Parton, 1994).

## Climate Change Scenarios (1.1°C MAT and 33 mm MAP from 2011 to 2020 and 1.4°C MAT and 46 mm MAP from 2021 to 2030)

There are many uncertainties in predicting the long-term effects of potential CC on SOC stock. The precipitation projections for the Pacific Northwest are for increased within-season variation and more spatial variation across the region (Littell et al., 2009). An increase of 0.2°C in MAT and 0.37 mm in MAP with changes in precipitation patterns and intensity were recorded at this site since the 1990s. Little work has been done on the effects of warming on SOC stocks in dryland agroecosystems (Dijkstra and Morgan, 2012); however, there are a few experiments in grassland (Pendall et al., 2013), grazed pasture (Ross et al., 2013), and forestry (Dawes et al., 2013) that were performed under elevated $CO_2$ and temperature that may be applicable to agroecosystems. Bhattacharyya et al. (2013) concluded that the effects of CC on belowground soil C accrual were not clear under tropical flooded soil. Climate change could alter precipitation patterns and produce higher air temperatures, resulting in warmer soil temperatures, greater soil aeration due to reduced soil water content with increased soil temperature, and higher rates of SOM and SOC decomposition (Billings et al., 1984; Oechel and Vourlitis, 1995). Elevated temperatures may modify soil water content, and stimulate plant net primary productivity. Current biomass yield with CC scenarios from 2011 to 2030 resulted in SOC losses of 0.64 and 3.81 Mg ha$^{-1}$ at rates of 0.021 and 0.127 Mg ha$^{-1}$ yr$^{-1}$ in the 0- to 30- and 30- to 60-cm soil depths, respectively, in the GP long-term experiment (Table 1). The rate of SOC loss in the top soil was 6.0 times less in this scenario than that in the 30- to 60-cm depth. However, the differences between current yield and climate vs. CC and current yield GP SOC changes were 0.41 and 0.26 Mg ha$^{-1}$ in the 0- to 30- and 30- to 60-cm depths, respectively. The immediate response of predicted SOC stocks in topsoil than the subsoil during the two decades of CC perhaps is a result of dependence of SOC decomposition in the CQESTR model on thermal units (growing degree days) and the moisture profile of the soil. Since CC will have immediate effects on topsoil temperature and water content, the response of SOC storage is expected in the top soil layer. The CC with 30% biomass yield increase scenario was the only scenario that predicted a SOC stock increase of 1.12 Mg ha$^{-1}$ by 2030, which is about 0.32 Mg ha$^{-1}$ or 22% less than the predicted SOC value with a 30% biomass yield increase but without climate change for the GP under dryland condition.

A loss of SOC stocks was predicted for the NT–0N, in both topsoil and subsoil, in all scenarios under CC except when the winter wheat–fallow crop rotation was replaced with continuous winter rotation (Table 1, Fig. 5a). Fallow in a cropping system in the intermediate precipitation zone is not conducive to SOC retention. Soil organic C stocks in the topsoil (0–30 cm) increased under all climate change scenarios for the NT–135N under dryland cropping system (Table 1, Fig. 5b). Losses of SOC occurred in 30- to 60-cm depth for all CC scenarios. This could be due to insufficient root biomass production under a winter wheat cropping system or could be due to increased biological oxidation or respiration under predicted climate change. Soil microorganisms are likely to show an immediate response to higher soil temperature by increasing soil respiration rate, provided that soil moisture does not become limiting (Giardina and Ryan, 2000). Furthermore, absolute SOC change differences between current yield and climate vs. CC and current yield were 0.53 vs. 0.30 Mg ha$^{-1}$ for the NT–135 N, and 0.79 and 0.29 Mg ha$^{-1}$ for the CT–135N in the 0- to 30- and 30- to 60-cm depths, respectively. This indicates that CC affected topsoil relatively different than subsoil. The lowest predicted SOC gain of 1.35 Mg ha$^{-1}$ at an average annual rate of 0.045 Mg ha$^{-1}$ (0.16 Mg $CO_2$ ha$^{-1}$ yr$^{-1}$) was under climate change with 30% biomass reduction during the 30-yr simulation. The highest SOC gain of 8.61 Mg ha$^{-1}$ at average

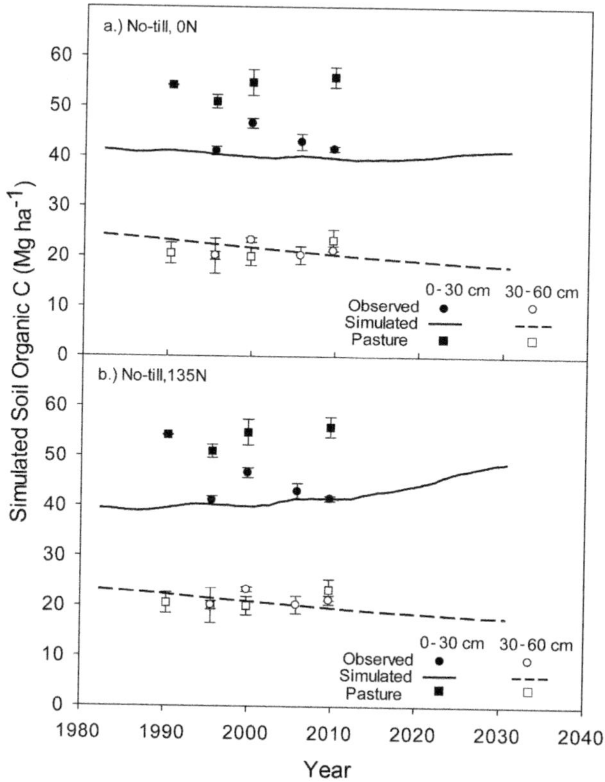


**Fig. 5. The CQESTR model simulated (lines) under climate change scenarios and observed (symbols) soil organic C stock in the 0- to 30- and 30- to 60-cm soil depths for continuous winter wheat cropping under no-tillage (NT) systems (a) without nitrogen fertilization (0N) and (b) with135 kg N ha$^{-1}$ nitrogen fertilization (135N) under climate change relative to the soil organic carbon stock for long-term grass pasture. Bars are standard errors of the mean (*n* = 4) for NT and (*n* = 2–4) for grass pasture.**

annual rate of 0.287 Mg ha$^{-1}$ (1.05 Mg $CO_2$ ha$^{-1}$ yr$^{-1}$) was under change of the crop rotation from wheat–fallow to continuous winter wheat, which was less than the current SOC stocks of topsoil in the GP (Table 1, Fig. 2). This indicates that even with crop intensification and NT cropping systems with sufficient N fertilization winter wheat was unable to produce equivalent biomass as GP did. Crop intensification with continuous spring cereal cropping under NT was proposed as a way to increase grain and crop residue production in the low precipitation zone (<300 mm) of the dryland Pacific Northwest (Young et al., 2008). Increased cropping intensity and crop rotation that includes a perennial crop combined with a reduced fallow period and tillage were suggested to increase the amount of biomass returned to the soil and increase SOC stocks (Collins et al., 2012).

Predicted CC scenarios for the CT–0N resulted in SOC losses in both topsoil and subsoil, in all scenarios under CC, with the lowest SOC lost (–1.09 Mg $ha^{-1}$) under continuous winter wheat, while the highest SOC lost (–3.64 Mg $ha^{-1}$) was under CC and 30% biomass yield reduction (Table 1). Enhanced ecosystem respiration to a greater extent than photosynthesis and greater net C loss over 4 yr in northern mixed grass prairies in Wyoming was reported by Pendall et al. (2013). Intensification of cropping system was the only CC scenario that increased SOC by 0.26 Mg $ha^{-1}$ in the top 0- to 30-cm soil under CT–135N (Table 1, Fig. 6a). Increased SOC stock, with N fertilization in continuous cropping, under CT has been reported by other researchers. Gregorich and Janzen (1996) showed a 13% increase in SOC after 32 yr of continuous corn (*Zea mays* L.) with N fertilization under CT. Huggins and Fuchs (1997) suggested that SOC changes were depen-

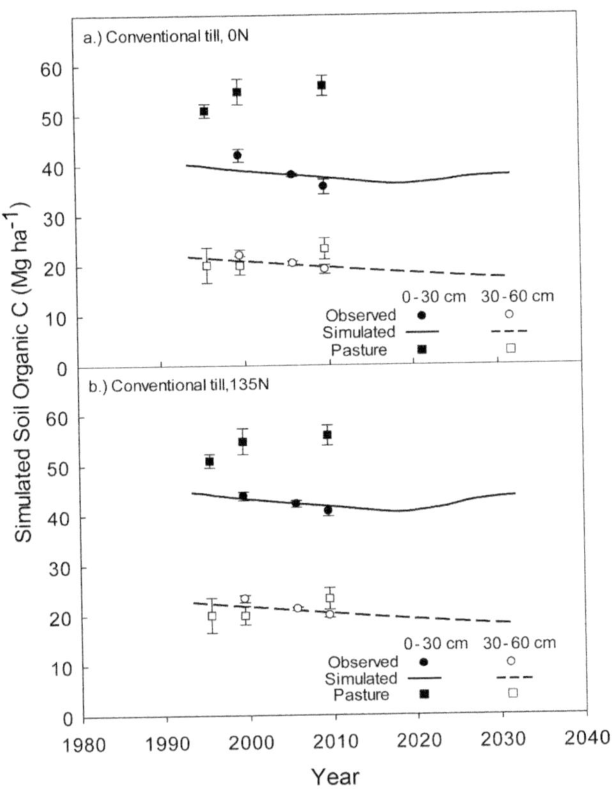


**Fig. 6. The CQESTR model simulated (lines) under climate change scenarios and observed (symbols) soil organic C) stock in the 0- to 30- and 30- to 60-cm soil depths for continuous winter wheat cropping under conventional tillage systems (a) without nitrogen fertilization (0N) and (b) with135 kg N $ha^{-1}$ nitrogen fertilization (135N) under climate change relative to the soil organic carbon stock for long-term grass. Bars are standard errors of the mean (*n* = 4) for NT and (*n* = 2–4) for grass pasture.**

dent on the aboveground and root biomass response to N. A good N fertility program in continuously cropped management systems could reverse SOC stock losses in soil under dryland semiarid conditions.

## Potential for $CO_2$ Sequestration and C Sink Capacity of Semiarid Dryland Agroecosystem

Retention and sequestration of SOC are particularly difficult in semiarid climates where crop yields and crop biomass returned to the soil are limited by low rainfall and extreme temperatures. Variation in precipitation from year to year also can cause significant differences in crop residue production (Wilhelm et al., 1987). Predicted SOC stocks in the 10 simulated scenarios indicates that there is a capacity for this soil to store C in the top 30 cm and become a sink for $CO_2$ with continuous cropping, proper N fertilization, and use of conservation tillage (e.g., an increase of 8.61 Mg $ha^{-1}$ under NT–135N) if precipitation increases in spite of increased temperature under CC (Table 1). However, winter wheat may not produce enough biomass under current management to be comparable with native grass (Fig. 5 and 6). The predicted SOC gain of 1.93 Mg $ha^{-1}$ in the NT–135N winter wheat–fallow system indicates that 0.234 Mg $CO_2$ $ha^{-1}$ $yr^{-1}$ could be sequestered from the atmosphere with a potential 30% biomass yield reduction, while with a potential increase in biomass yield of 30% a predicted SOC gain of 4.62 Mg $ha^{-1}$ or 0.562 Mg $CO_2$ $ha^{-1}$ $yr^{-1}$ could be sequestered under current dryland climatic conditions. In contrast, with the proposed CC scenarios during 2011 to 2030, the predicted SOC gain of 1.35 Mg $ha^{-1}$ in the NT–135N winter wheat–fallow system resulted in a 0.164 Mg $CO_2$ $ha^{-1}$ $yr^{-1}$ reduction with 30% biomass reduction and the highest SOC gain of 8.61 Mg $ha^{-1}$ could sequester up to 1.05 Mg $CO_2$ $ha^{-1}$ $yr^{-1}$ from the atmosphere with change of crop rotation from wheat–fallow to the continuous winter wheat at the current wheat yield. Crop intensification with continuous winter wheat or other cereal crops under NT cropping system could be used to increase C inputs to soils via crop residue, root and rhizodeposition, which in return would reduce soil erosion and the loss of SOC with each eroded particle. Furthermore, elimination of the fallow will reduce soil profile water content, which in turn would reduce biological oxidation of SOC. Reduced tillage also reduces contacts with microbial populations that degrade SOC, which are limited in NT systems where plant residues remain on the soil surface (Brussaard et al., 2007). Enhanced SOC decomposition rate and $CO_2$ emissions contribute to SOC stock losses due to increased soil profile water content during the fallow year compared to continuous dryland intensified cropping system.

The minimum C sink capacities of the NT–135, NT–0N, CT–135N, and CT–0N plots were 12.69, 14.54, 13.46, and 18.57 Mg $ha^{-1}$, respectively, relative to SOC stock values (54.25 Mg ha$^{-1}$) of the grass pasture experiment in the top 0- to 30-cm soil in 1990 (Fig. 2, 5, and 6); however, this value is itself lower than the original SOC stock values before cultivation of the land in the mid 1880s, which was estimated in 1931 at 56.83 Mg ha$^{-1}$ in the 0- to 20-cm soil depth (Rasmussen and Albrecht, 1998). Crop intensification with continuous winter wheat under NT cropping system is a more viable management system under semiarid dryland region that could improve C stock while reducing $CO_2$ in the atmosphere and will have a significant effect on biological, chemical, and physical soil properties and, consequently, yield and crop production. Increased soil organic C increases soil biodiversity, fertility, water and plant nutrient retention, improves soil structure, and prevents soil erosion and degradation (Weil and Magdoff, 2004). However, crop intensification could reduce soil nutrient stocks and soil water for the next crop, especially under dryland cropping systems. It also requires more fertilizer, herbicide and fuel usage and possibly involves more greenhouse gas emission.

## Summary and Conclusions

The potential effects of global climate change on C cycling and SOC storage or loss in agroecosystems can be assessed by process-based models. One such model is the CQESTR model, which was used to simulate the effect of soil tillage and N fertilization on soil C storage or loss under 10 biomass and CC scenarios, in three long-term experiments where a winter wheat–fallow rotation under dryland cropping system is the predominant agriculture practice. Linear regression analysis of simulated and observed SOC values were closely related ($r^2 = 0.98$, $P < 0.0001$) with slope not significantly different from one ($P < 0.0001$) and a significant Pearson correlation coefficient ($r = 0.99$), demonstrating the capacity of CQESTR to predict long-term SOC dynamics for these experiments. The model predicted SOC stock losses for the GP in 7 out of 10 scenarios at the 0- to 30-cm soil depth and in all 10 scenarios at the 30- to 60-cm depth. A biomass yield increase of at least 10% was required to observe simulated SOC gain of 0.35 Mg ha$^{-1}$ under current climate conditions. An increase of 0.2°C in MAT and 0.37 mm in MAP with changes in precipitation pattern and intensity were recorded at this site since the 1990s. Anticipated impacts of climate change on SOC due to increased MAT of 1.1°C during 2011 to 2020, and 1.4°C during 2021 to 2030 and MAP of 33 mm during 2011 to 2020, and 46 mm during 2021 to 2030 were simulated for 20 yr. With a 30% increase in biomass, predicted SOC stock in the topsoil of the GP increased by 22% under the current climate (1.44 Mg ha$^{-1}$) compared to CC scenario (1.12 Mg ha$^{-1}$), and the topsoil was more affected by CC than the subsoil. The immediate response of topsoil during the two decades of CC perhaps is a result of dependence of SOC decomposition in the CQESTR model on thermal units and moisture profile of the soil.

A loss of SOC stock was predicted at both soil depths for the CT–0N treatment in all scenarios. The same was true for CT–135N and NT–0N treatments, but SOC gains of 0.26 and 1.37 Mg $ha^{-1}$ were observed under continuous winter wheat CC scenarios, respectively. The NT–135N was the only treatment with SOC gains ranging from 1.35 Mg $ha^{-1}$ in winter wheat–fallow system with 30% biomass reduction, to 8.61 Mg $ha^{-1}$ with change of crop rotation from wheat–fallow to continuous winter wheat at current wheat yield, which could sequester up to 1.05 Mg $CO_2$ $ha^{-1}$ $yr^{-1}$ from the atmosphere. The minimum C sink capacities of the NT–135N, NT–0N, CT–135N, and CT–0N were 12.69, 14.54, 13.46, and 18.57 Mg $ha^{-1}$, respectively, relative to SOC stock values of the grass pasture experiment in 1990. The accretion of carbon in the soil was dependent on the agriculture management practice (i.e., no-till or conventional tillage, nitrogen fertilization, increased cropping intensity) and annual biomass yield. No-till management coupled with N fertilization and increased cropping intensity was predicted to be a more viable management system for SOC accretion than a winter wheat–tillage fallow cropping system for this site under potential climate change scenarios, if increased temperature and precipitation improves wheat yield, and SOC decomposition remains within the current projected rate. This cropping system can lead to increased SOC stock at this site if increased temperature and precipitation predictions and their effects on winter wheat materialize. This finding is important and could allow researchers, growers, and the public to prepare for climate change and evaluate their options for managing the amount of C sequestered in small-grain based cropping systems under dryland climatic conditions that receives 300 to 450 mm of precipitation.

## References

Albrecht, S.L., K.W. Skirvin, and D.S. Long. 2005. Winter wheat responses to nitrogen fertilization in a direct-seed, summer-fallow management system. In: S. Petrie et al., editors, Dryland agricultural research annual report. Special Rep. 1061. Agricultural experiments at Pendleton, Oregon. Oregon State University in Cooperation with Agricultural Research Service, Corvallis, OR. p. 87–912.

Angers, D.A., and N.S. Eriksen-Hamel. 2008. Full-Inversion tillage and organic carbon distribution in soil profiles: A meta-analysis. Soil Sci. Soc. Am. J. 72:1370–1374. doi:10.2136/sssaj2007.0342

Bhattacharyya, P., K.S. Roy, S. Neogi, M.C. Manna, T.K. Adhya, K.S. Rao, and A.K. Nayak. 2013. Influence of elevated carbon dioxide and temperature on belowground carbon allocation and enzyme activities in tropical flooded soil planted with rice. Environ. Monit. Assess. 185(10):8659–8671. doi:10.1007/s10661-013-3202-7

Billings, W.D., K.M. Peterson, J.O. Luken, and D.A. Mortensen. 1984. Interaction of increasing atmospheric carbon dioxide and soil nitrogen on the carbon balance of tundra ecosystems. Oecologia 65:26–29. doi:10.1007/BF00384458

Brussaard, L., P.C. de Ruiter, and G.G. Brown. 2007. Soil biodiversity for agricultural sustainability. Agric. Ecosyst. Environ. 121:233–244. doi:10.1016/j.agee.2006.12.013

Campbell, C.A., B.G. McConkey, R.P. Zentner, F.B. Dyck, F. Selles, and D. Curtin. 1996. Long-term effects of tillage and crop rotations on soil organic C and total N in a clay soil in southwestern Saskatchewan. Can. J. Soil Sci. 76:395–401. doi:10.4141/cjss96-047

Clapp, C.E., R.R. Allmaras, M.F. Layese, D.R. Linden, and R.H. Dowdy. 2000. Soil organic carbon and $^{13}C$ abundance as related to tillage, crop residue and nitrogen fertilizer under continuous corn management in Minnesota. Soil Tillage Res. 55:127–142. doi:10.1016/S0167-1987(00)00110-0

Collins, H.P., M.M. Mikha, T.T. Brow, J.L. Smith, D. Huggins, and U.M. Sainju. 2012. Agricultural management and soil carbon dynamics: Western U.S. croplands. In: M.A. Liebig, A.J. Franzluebbers, and R.F. Follett, editors, Managing agricultural greenhouse gases: Coordinated agricultural research through GRACEnet to address our changing climate. Elsevier, San Diego, CA. p. 59–77.

Collins, H.P., P.E. Rasmussen, and C.L. Douglas, Jr. 1992. Crop rotation and residue management effects on soil carbon and microbial dynamics. Soil Sci. Soc. Am. J. 56:783–788. doi:10.2136/sssaj1992.03615995005600030018x

Dawes, M.A., F. Hagedorn, I.T. Handa, K. Streit, A. Ekblad, C. Rixen, C. Körner, and S. Hättenschwiler. 2013. An alpine treeline in a carbon dioxide-rich world: Synthesis of a nine-year free-air carbon dioxide enrichment study. Oecologia 171(3):623–663. doi:10.1007/s00442-012-2576-5

Del Grosso, S.J., W.J. Parton, C.A. Keough, and M. Reyes-Fox. 2011. Special features of the DayCent modeling package and additional procedures for parameterization, calibration, validation, and applications. In: Laj R. Ahuja and Liwang Ma., editors, Methods of introducing system models into agricultural research. Advances in Agricultural System Modeling 2. p. 155–176. doi:10.2134/advagricsystmodel2.c5

Dijkstra, F.A., and J.A. Morgan. 2012. Elevated $CO_2$ and warming effects on soil carbon sequestration and greenhouse gas exchange in Agroecosystems: A review. In: M.A. Liebig, A.J. Franzluebbers, and R.F. Follett, editors, Managing agricultural greenhouse gases: Coordinated agricultural research through GRACEnet to address our changing climate. Elsevier, San Diego, CA. p. 468–486.

Feng, G., B. Sharratt, and F. Young. 2011. Soil properties governing soil erosion affected by cropping systems in the U.S. Pacific Northwest. Soil Tillage Res. 111:168–174. doi:10.1016/j.still.2010.09.008

Follett, R.F., E.A. Paul, S.W. Leavitt, A.D. Halvorson, D. Lyon, and G.A. Peterson. 1997. Carbon isotope ratios of Great Plains soils in wheat–fallow systems. Soil Sci. Soc. Am. J. 61:1068–1077. doi:10.2136/sssaj1997.03615995006100040012x

Follett, R.F., S.R. Shafer, M.D. Jawson, and A.J. Franzluebbers. 2005. Research and implementation needs to mitigate greenhouse gas emissions from agriculture in the USA. Soil Tillage Res. 83:159–166. doi:10.1016/j.still.2005.02.014

Follett, R.F., K.P. Vogel, G.E. Varvel, R.B. Mitchell, and J.M. Kimble. 2012. Soil carbon sequestration by switchgrass and no-till maize grown for bioenergy. Bioenergy Res. 5:866–875. doi:10.1007/s12155-012-9198-y

Franzluebbers, A.J. 2005. Soil organic carbon sequestration and agricultural greenhouse gas emissions in the southeastern USA. Soil Tillage Res. 83:120–147. doi:10.1016/j.still.2005.02.012

Gauch, H.G., J.T.G. Hwang, and G.W. Fick. 2003. Model evaluation by comparison of the model-based predictions and measured values. Agron. J. 95:1442–1446. doi:10.2134/agronj2003.1442

Giardina, C.P., and M.G. Ryan. 2000. Evidence that decomposition rates of organic carbon in mineral soil do not vary with temperature. Nature 404:858–861. doi:10.1038/35009076

Gollany, H.T., R.R. Allmaras, S.M. Copeland, S.L. Albrecht, and C.L. Douglas, Jr. 2005. Tillage and nitrogen fertilizer influences on carbon and soluble silica relations in a Pacific Northwest Mollisol. Soil Sci. Soc. Am. J. 69:1102–1109. doi:10.2136/sssaj2004.0284

Gollany, H.T., R.R. Allmaras, S.M. Copeland, S.L. Albrecht, and C.L. Douglas, Jr. 2006. Incorporated source carbon and nitrogen fertilization effects on carbon storage and soluble Silica in a Haploxeroll. Soil Sci. 171:585–597. doi:10.1097/01.ss.0000228042.10765.30

Gollany, H.T., R.F. Follett, and Y. Liang. 2012. CQESTR Simulations of soil organic carbon dynamics. In: M.A. Liebig, A.J. Franzluebbers, and R.F. Follett, editors, Managing agricultural greenhouse gases: Coordinated agricultural research through GRACEnet to address our changing climate. Academic Press, San Diego, CA. p. 271–292.

Gollany, H.T., A. Fortuna, M. Samuel, F.L. Young, W. Pan, and M. Pecharko. 2013. Estimating soil organic carbon accretion vs. sequestration using physicochemical fractionation and CQESTR simulation. Soil Sci. Soc. Am. J. 77:618–629. doi:10.2136/sssaj2012.0303

Gollany, H.T., J.M. Novak, Y. Liang, S.L. Albrecht, R.W. Rickman, R.F. Follett, W.W. Wilhelm, and P.G. Hunt. 2010. Simulating soil organic carbon dynamics with residue removal using the CQESTR model. Soil Sci. Soc. Am. J. 74:372–383. doi:10.2136/sssaj2009.0086

Gollany, H.T., R.W. Rickman, Y. Liang, S.L. Albrecht, S. Machado, and S. Kang. 2011. Predicting agricultural management influence on long-term soil organic carbon dynamics: Implications for biofuel production. Agron. J. 103:234–246. doi:10.2134/agronj2010.0203s

Gollany, H.T., T.E. Schumacher, M.J. Lindstrom, P. Evenson, and G.D. Lemme. 1992. Topsoil depth and desurfacing effects on properties and productivity of a Typic Argiustoll. Soil Sci. Soc. Am. J. 56:220–225. doi:10.2136/sssaj1992.03615995005600010034x

Gregorich, E.G., and M.H. Beare. 2008. Physically uncomplexed organic matter. In: M.R. Carter and E.G. Gregorich, editors, Soil sampling and methods of analysis. 2nd ed. Can. Soil Sci. Soc. Lewis Publ., Boca Raton, FL. p. 607–616. doi:10.1201/9781420005271.ch47.

Gregorich, E.G., C.F. Drury, B.H. Ellert, and B.C. Liang. 1997. Fertilization effects on physically protected light fraction organic matter. Soil Sci. Soc. Am. J. 61:482–484. doi:10.2136/sssaj1997.03615995006100020016x

Gregorich, E.G., and H.H. Janzen. 1996. Storage of soil carbon in the light fraction and macroorganic matter. In: M.R. Carter and B.A. Stewart, editors, Structure and organic matter storage in agricultural soils. Lewis Publ. CRC Press, Boca Raton, FL. p. 167–190.

Halvorson, A.D., and C.P. Jantalia. 2011. Nitrogen fertilization effects on irrigated no-till corn production and soil carbon and nitrogen. Agron. J. 103:1423–1431. doi:10.2134/agronj2011.0102

Halvorson, A.D., B.J. Wienhold, and A.L. Black. 2002. Tillage, nitrogen, and cropping system effects on soil carbon sequestration. Soil Sci. Soc. Am. J. 66:906–912. doi:10.2136/sssaj2002.9060

Huggins, D.R., G.A. Buyanovky, G.H. Wagner, J.R. Brown, R.G. Darmody, T.R. Peck, G.W. Lesoing, M.B. Vanotti, and L.G. Bundy. 1998. Soil organic C in the tallgrass prairie-derived region of the corn belt: Effects of long-term crop management. Soil Tillage Res. 47:219–234 doi:10.1016/S0167-1987(98)00108-1

Huggins, D.R., and D.J. Fuchs. 1997. Long-term N management effects on corn yield and soil C of an Aquic Haplustoll in Minnesota. In: E.A. Paul, et al., editors, Soil organic matter in temperate agroecosystems. Long-term experiments in North America. CRC Press, Boca Raton, FL. p. 122–128.

IPCC. 2001. Climate Change 2001: Synthesis report. Summary for policymakers. UNEP/GRID, Arendal, Norway. http://www.grida.no/climate/ipcc_tar/vol4/english/pdf/spm.pdf (accessed 17 Mar. 2015).

IPCC. 2014. Climate Change 2014: Synthesis report. Summary for policymakers. UNEP/GRID, Arendal, Norway. http://www.ipcc.ch/pdf/assessment-report/ar5/syr/AR5_SYR_FINAL_SPM.pdf (accessed 17 Mar. 2015).

Izaurralde, R.C., J.R. Williams, W.B. McGill, N.J. Rosenberg, and M.C. Quiroga Jakas. 2006. Simulation soil C dynamics with EPIC: Model description and testing against long-term data. Ecol. Modell. 192:362–384. doi:10.1016/j.ecolmodel.2005.07.010

Jantalia, C.P., and A.D. Halvorson. 2011. Nitrogen fertilizer effects on irrigated conventional tillage corn yields and soil carbon and nitrogen pools. Agron. J. 103:871–878. doi:10.2134/agronj2010.0455

Janzen, H.H., C.A. Campbell, R.C. Izaurralde, B.H. Ellert, N.G. Juma, W.B. McGill, and R.P. Zentner. 1998. Management effects on soil C storage on the Canadian prairies. Soil Tillage Res. 47:181–195. doi:10.1016/S0167-1987(98)00105-6

Khan, A., R.L. Mulvaney, T.R. Ellsworth, and C.W. Boast. 2007. The myth of nitrogen fertilization for soil carbon sequestration. J. Environ. Qual. 36:1821–1832. doi:10.2134/jeq2007.0099

Lal, R. 2004. Soil carbon sequestration impacts on global climate change and food security. Science 304:1623–1627. doi:10.1126/science.1097396

Lal, R., R.F. Follett, J.M. Kimble, and C.V. Cole. 1999. Managing U.S. cropland to sequester carbon in soil. J. Soil Water Conserv. 54:374–381. doi:10.2489/jswc.65.3.68A.

Lal, R., J.M. Kimble, R.F. Follett, and C.V. Cole. 1998. The potential of US cropland to sequester carbon and mitigate the greenhouse effect. Sleeping Bear Press, Ann Arbor, MI.

Leite, L.F.C., P.C. Doraiswamy, H.J. Causarano, H.T. Gollany, S. Milak, and E.S. Mendonca. 2009. Modeling organic carbon dynamics under no-tillage and plowed systems in tropical soils of Brazil using CQESTR. Soil Tillage Res. 102:118–125. doi:10.1016/j.still.2008.08.003

Liang, Y., H.T. Gollany, R.W. Rickman, S.L. Albrecht, R.F. Follett, W.W. Wilhelm, J.M. Novak, and C.L. Douglas, Jr. 2008. CQESTR Simulation of management practice effects on long-term soil organic carbon. Soil Sci. Soc. Am. J. 72:1486–1492. doi:10.2136/sssaj2007.0154

Liang, Y., H.T. Gollany, R.W. Rickman, S.L. Albrecht, R.F. Follett, W.W. Wilhelm, J.M. Novak, and C.L. Douglas, Jr. 2009. Simulating soil organic matter with CQESTR (v. 2.0): Model description and validation against long-term experiments across North America. Ecol. Modell. 220(4):568–581. doi:10.1016/j.ecolmodel.2008.11.012

Liebig, M.A., J.A. Morgan, J.D. Reeder, B.H. Ellert, H.T. Gollany, and G.E. Schumann. 2005. Greenhouse gas contributions and mitigation potential of agricultural practices in northwestern USA and western Canada. Soil Tillage Res. 83:25–52. doi:10.1016/j.still.2005.02.008

Littell, J.S., M.M. Elsner, L.C. Binder, and A.K. Snover. 2009. The Washington climate change impacts assessment: Evaluating Washington's future in a changing climate. Executive summary of the Washington Climate Change Impacts Group, University of Washington, Seattle, WA. http://cses.washington.edu/db/pdf/wacciaexecsummary638.pdf (accessed 2 Mar. 2013).

Mann, L., V. Tolbert, and J. Cushman. 2002. Potential environmental effects of corn (*Zea mays* L.) stover removal with emphasis on soil organic matter and erosion. Agric. Ecosyst. Environ. 89:149–166. doi:10.1016/S0167-8809(01)00166-9

Mertens, J., J. Vanderborght, R. Kasteel, T. Putz, R. Merckx, J. Feyen, and E. Smolders. 2007. Dissolved organic carbon fluxes under bare soil. J. Environ. Qual. 36:597–606. doi:10.2134/jeq2006.0368

Oechel, W.C., and G.L. Vourlitis. 1995. Effects of global change on carbon storage in cold soils. In: R. Lal, J. Kimble, E. Levine, and B.A. Stewart, editors, Soils and global change. CRC Press, Boca Raton, FL. p. 117–129.

Omay, A.B., C.W. Rice, L.D. Maddux, and W.B. Gordon. 1997. Changes in soil microbial and chemical properties under long-term crop rotation and fertilization. Soil Sci. Soc. Am. J. 61:1672–1678. doi:10.2136/sssaj1997.03615995006100060019x

Oorts, K., R. Merckx, E. Grehan, J. Labreuche, and B. Nicolardot. 2007. Determinants of annual fluxes of $CO_2$ and $N_2O$ in long-term no-tillage and conventional tillage systems in northern France. Soil Tillage Res. 95:133–148. doi:10.1016/j.still.2006.12.002

Parton, W.J., J.B.W. Stewart, and C.V. Cole. 1988. Dynamic of C, N, P and S in grassland soils: A model. Biogeochemistry 5:109–131. doi:10.1007/BF02180320

Paustian, K., E.T. Elliot, and K. Killian. 1998. Modeling soil carbon in relation to management and climate change in some agroecosystems in central North America. In: R. Lal, et al., editors, Soil processes and the carbon cycle. CRC Press, Boca Raton, FL. p. 459–471.

Paustian, K., W.J. Parton, and J. Persson. 1992. Modeling soil organic matter in organic-amended and nitrogen-fertilized long-term plots. Soil Sci. Soc. Am. J. 56:476–488. doi:10.2136/sssaj1992.03615995005600020023x

Pendall, E., J.L. Heisler-White, D.G. Williams, F.A. Dijkstra, Y. Carrillo, J.A. Morgan, and D.R. LeCain. 2013. Warming reduces carbon losses from grassland exposed to elevated atmospheric carbon dioxide. PLoS ONE 8(8). doi:10.1371/journal.pone.0071921

Plaza, C., H.T. Gollany, G. Baldoni, A. Polo, and C. Ciavatta. 2012. Predicting long-term organic carbon dynamics in organically-amended soils using the CQESTR model. J. Soils Sediments 12:486–493. doi:10.1007/s11368-012-0477-1

Rasmussen, P.E., and S.L. Albrecht. 1998. Crop management effects on organic carbon in semi-arid Pacific Northwest soils. In: R. Lal et al., editors, Management of carbon sequestration in soil. CRC Press, Boca Raton, FL. p. 209–219.

Rasmussen, P.E., R.R. Allmaras, C.R. Rohde, and N.C. Roager, Jr. 1980. Crop residue influences on soil carbon and nitrogen in a wheat–fallow system. Soil Sci. Soc. Am. J. 44:596–600. doi:10.2136/sssaj1980.03615995004400030033x

Rasmussen, P.E., and H.P. Collins. 1991. Long-term impacts of tillage, fertilizer, and crop residue on soil organic matter in temperate semi-arid regions. Adv. Agron. 45:93–134. doi:10.1016/S0065-2113(08)60039-5

Rasmussen, P.E., and W.J. Parton. 1994. Long-term effects of residue management in wheat–fallow. I. Inputs, yield, and soil organic matter. Soil Sci. Soc. Am. J. 58:523–530. doi:10.2136/sssaj1994.03615995005800020039x

Rasmussen, P.E., R.W. Rickman, and B.L. Kleeper. 1997. Residue and fertility effects on yield of no-till wheat. Agron. J. 89:563–567. doi:10.2134/agronj1997.00021962008900040005x

Rasmussen, P.E., and C.R. Rohde. 1988. Long-term tillage and nitrogen fertilization on organic nitrogen and carbon in a semi-arid soil. Soil Sci. Soc. Am. J. 52:1114–1117. doi:10.2136/sssaj1988.03615995005200040041x

Rasmussen, P.E., and R.W. Smiley. 1994. Long-term experiments at the Pendleton Agricultural Research Center In: Columbia Basin Agricultural Research Annual Report. Spec. Rep. 933. Agricultural Experiments at Pendleton. Oregon State University in Cooperation with Agricultural Research Service, Corvallis, OR. p. 14–20.

Rasmussen, P.E., and R.W. Smiley. 1997. Soil carbon and nitrogen change in long-term agricultural experiments at Pendleton, Oregon. In: E.A. Paul, et al., editors, Soil organic matter in temperate agroecosystems: Long-term experiments in North America. CRC Press, Boca Raton, FL. p. 353–360.

Rickman, R., C. Douglas, S. Albrecht, and J. Berc. 2002. Tillage, crop rotation, and organic amendment effect on changes in soil organic matter. Environ. Pollut. 116:405–411. doi:10.1016/S0269-7491(01)00217-2

Rickman, R.W., C.L. Douglas, S.L. Albrecht, L.G. Bundy, and J.L. Berc. 2001. CQESTR: A model to estimate carbon sequestration in agricultural soils. J. Soil Water Conserv. 56:237–242.

Ross, D.J., P.C.D. Newton, K.R. Tate, and D. Luo. 2013. Impact of a low level of $CO_2$ enrichment on soil carbon and nitrogen pools and mineralization rates over ten years in a seasonally dry, grazed pasture. Soil Biol. Biochem. 58:265–274. doi:10.1016/j.soilbio.2012.12.011

Russell, A.E., D.A. Laird, T.B. Parkin, and A.P. Mallarino. 2005. Impact of nitrogen fertilization and cropping system on carbon sequestration in Midwestern Mollisols. Soil Sci. Soc. Am. J. 69:413–422. doi:10.2136/sssaj2005.0413

SAS Institute. 2008. SAS OnlineDoc, Version 9.2. SAS Inst., Cary, NC.

Thorne, M.E., F.L. Young, W.L. Pan, R. Bafus, and J.R. Alldredge. 2003. No-till spring cereal cropping systems reduce wind erosion susceptibility in the wheat/fallow region of the Pacific Northwest. J. Soil Water Conserv. 58:250–257.

USEPA. 2014. Inventory of U.S. greenhouse gas emissions and sinks: 1990–2012. USEPA, Office of Atmospheric Programs, Washington, DC. http://www.epa.gov/climatechange/ghgemissions/usinventoryreport.html (accessed 7 July 2014).

Varvel, G.E., and W.W. Wilhelm. 2008. Soil carbon levels in irrigated Western Corn Belt rotations. Agron. J. 100:1180–1184. doi:10.2134/agronj2007.0383

Weil, R.R., and F. Magdoff. 2004. Significance of soil organic matter to soil quality and health. In: K. Magdoff and R.R. Weil, editors, Soil organic matter in sustainable agriculture. CRC Press, Boca Raton, FL. p. 1–43. doi:10.1201/9780203496374.ch1.

West, T.O., and W.M. Post. 2002. Soil organic carbon sequestration rates by tillage and crop rotation: A global data analysis. Soil Sci. Soc. Am. J. 66:1930–1946. doi:10.2136/sssaj2002.1930

Wilhelm, W.W., J.R. Hess, D.L. Karlen, J.M.F. Johnson, D.J. Muth, J.M. Baker, H.T. Gollany, J.M. Novak, D.E. Stott, and G.E. Varvel. 2010. Balancing limiting factors and economic drivers for sustainable Midwestern US agricultural residue feedstock supplies. Ind. Biotechnol. 6:271–287. doi:10.1089/ind.2010.6.271

Wilhelm, W.W., J.S. Schepers, L.N. Mielke, J.W. Doran, J.R. Elis, and W.W. Stroup. 1987. Dryland maize development and yield resulting from tillage and nitrogen fertilization practices. Soil Tillage Res. 10:167–179. doi:10.1016/0167-1987(87)90041-9

Young, F.L., L.S. Bewick, and W.L. Pan. 2008. Systems approach to crop rotation research. In: Y.U. Berklian, editor, Crop rotation. Nova Science Publishers, Inc., New York.

# Simulating Soil Organic Carbon Stock Changes in Agroecosystems using CQESTR, DayCent, and IPCC Tier 1 Methods

**Stephen J. Del Grosso,* Hero T. Gollany, and Melissa Reyes-Fox**

Abstract

Models are often used to quantify how land use change and management impact soil organic carbon (SOC) stocks because it is often not feasible to use direct measuring methods. Because models are simplifications of reality, it is essential to compare model outputs with measured values to evaluate model performance. In this chapter we compare the abilities of two process-based models (CQESTR and DayCent) and one empirical model (IPCC Tier 1 methodology) to represent how tillage intensity and nitrogen fertilizer inputs affect SOC stocks in a dryland wheat (*Triticum aestivum* L.)–fallow system in Oregon and an irrigated maize (*Zea mays* L.) system in Colorado. All three models correctly predicted SOC losses in Oregon, but the models (particularly the Tier 1 method) tended to underpredict the magnitude of losses. In Colorado, all three models underestimated SOC losses observed with conventional tillage and no N fertilizer addition. At both sites, observations showed higher losses (or smaller gains) compared to the models as tillage intensity increased and N fertilizer additions decreased. The IPCC Tier 1 method is meant to be used for large-scale applications and does not account for all of the processes that influence SOC stocks, so it is not surprising that this method does not perform particularly well when compared with field level observations. The process-based models performed better than the Tier 1 methodology, but could be improved by better accounting of erosion, the vertical distribution of SOC in the soil profile, and the impacts of nitrogen on plant production and carbon allocation.

Abbreviations: CT, conventional moldboard tillage; IPCC, Intergovernmental Panel on Climate Change; NT, no-till; SOC, soil organic carbon; SOM, soil organic matter.

Stephen J. Del Grosso and Melissa Reyes-Fox (Melissa.Reyes@ars.usda.gov), USDA, 2150 Centre Ave., Bldg. D, Suite 100, Fort Collins, CO 80526. Hero T. Gollany, USDA-ARS-CPCRC, 48037 Tubbs Ranch Rd., Adams, OR 97810 (hero.gollany@ars.usda.gov). *Corresponding author (steve.delgrosso@ars.usda.gov).

doi:10.2134/advagricsystmodel6.2013.0001.5


Synthesis and Modeling of Greenhouse Gas Emissions and Carbon Storage in Agricultural and Forest Systems to Guide Mitigation and Adaptation
S. Del Grosso, L. Ahuja, and W. Parton, Editors
Advances in Agricultural Systems Modeling, Volume 6. Lajpat R. Ahuja, Series Editor

Cropped soils are a substantial anthropogenic source of greenhouse gas emissions, particularly nitrous oxide, but can be a source or sink of carbon dioxide (Del Grosso and Grant, 2011). Although soil $CO_2$ fluxes can be measured directly using ground-based gas sampling chambers or micrometeorological techniques, $CO_2$ fluxes are often inferred based on changes in measured carbon stocks (Mosier et al., 2006). In this chapter we focus on SOC stock changes because large amounts of C can be lost or sequestered in soils (Necpálová et al., 2014), and SOC levels influence important soil properties such as nutrient and water retention (Milne et al., 2015; Powlson et al., 2011). Soil organic C stock changes in cropped systems are determined by the balance between C inputs from unharvested plant residues (above and below ground), root exudates, amendments, and deposition and C losses from decomposition of litter and soil organic matter (SOM), erosion (Rasmussen and Parton, 1994), and leaching (Mertens et al., 2007). Inputs are influenced by crop variety, rotations, nutrient additions, and amendments (i.e., manure, biosolid, or biochar), harvesting practices, and residue management; Gollany et al., 2012), while C losses are influenced by tillage practices (Campbell et al., 1996). Both inputs and outputs are strongly influenced by weather and soil properties such as texture and clay mineral content (Horwath, 2015; Page et al., 2013). In addition to C inputs and outputs from plant material, amendments (Williams, 2008), and decomposition (Rasmussen and Parton, 1994; Williams, 2004), movement of soil across the landscape from deposition and erosion (Gollany et al., 1992) and translocation or leaching of dissolved organic C (Mertens et al., 2007) also influence SOC stocks. For example, Gollany et al. (1992) reported 1 to 4 g C $kg^{-1}$ soil loss in the top 24- to 25-cm soil depth of moderately and severely eroded backslope position on a Beadle clay loam soil (fine, montmorillonitic, mesic Typic Argiustoll) under moldboard plowed tillage in a continuous corn cropping system in eastern South Dakota. In general, SOC stocks are thought to be positively correlated with plant residue inputs and negatively correlated with tillage intensity (Kumar et al., 2014; Stockmann et al., 2013), although there are exceptions (e.g., Dimassi et al., 2014; Powlson et al., 2014). Systems with low residue inputs, such as wheat–fallow rotations, typically lose SOC, while practices that increase plant growth and biomass inputs, such as fertilizer and water additions, tend to increase SOC stocks. Intensive tillage tends to break aggregates and increase decomposition rates (Six et al., 1999), as well as increase vulnerability to erosion by moving protective surface litter into the soil profile.

The global SOC pool is large but uncertain, with most estimates being close to 1500 Pg C (Pg = $10^{15}$ g) but ranging from 500 to 3000 Pg C (Scharlemann et al., 2014). For comparison, the atmospheric pool is ~835 Pg C (Ciais et al., 2014). Similarly, the impact of different land management practices on soil C stock changes is also uncertain, although large changes in SOC stocks have been documented (e.g., Necpálová et al., 2014). One reason for the uncertainty is that SOC varies at relatively fine spatial scales, so direct measurement of SOC at scales larger than experimental plots is prohibitively expensive. Even at the plot level, measurements often have high coefficients of variation. For these reasons, models of varying complexity are often used to predict the impacts of different land management options on SOC changes. However, model results are also uncertain and must be compared with field observations to establish credibility. Furthermore, models are often calibrated for specific site(s), landscapes, cropping systems, and biomes, which can compromise their general use when applied at regional and greater scales. Our objectives were to: (i) compare the abilities of two process based models (i.e., CQESTR and DayCent) and an empirical model (PCC Tier 1 methodology) to represent the impacts of crop system, fertilization rate, and tillage intensity for long-term research sites in eastern Oregon (dryland) and northern Colorado (irrigated) and (ii) identify model limitation(s) or situations of inapplicability and suggest improvements. We first describe the field experiments, then describe the models and explain how simulations were conducted, summarize the model predictions, and lastly compare model outputs with observations from the two sites and suggest model improvements.

## Oregon Field Study

The Pendleton study was conducted at the Columbia Plateau Conservation Research Center, which is located 5 km northeast of Pendleton, OR, USA (45°72′ N, 118°62′ W, elevation 458 m). The climate is temperate semiarid with a 30-yr mean annual temperature of 10.4°C and mean annual precipitation of 437 mm $yr^{-1}$, 90% of which occurs mainly as winter rain during the period of 1 October to 30 June. Average air temperature ranges from −0.6°C in January to 21.2°C in July (Rasmussen and Smiley, 1997). The site is in the intermediate precipitation zone (300–450 mm) of the US Pacific Northwest (Gollany et al., 2013). The soil is a Walla Walla silt loam (coarse-silty, mixed, superactive, mesic Typic Haploxeroll) developed in loess. The Ap horizon contained 72% silt, 19% sand, and 9% clay. Previous research at Columbia Plateau Conservation Research Center, on long-term crop residue plots at Oregon State University Columbia Basin Agricultural Research Center (<1 km north of these plots) indicated that measured soil erosion ranged from 0.08 Mg $ha^{-1}$ $yr^{-1}$ in plots with standing stubble to 0.85 Mg $ha^{-1}$ $yr^{-1}$ in the plots with fertilizer (90 kg N $ha^{-1}$) or manure (111 kg N $ha^{-1}$) added with or without crop residue burned and 3.3 Mg $ha^{-1}$ $yr^{-1}$ in plots with no fertilizer addition and residue burned (Williams, 2008). The data reported here are from two long-term winter wheat experiments, a long-term no-till experiment (NT) and a conventional moldboard tillage experiment (CT). The field was put into agriculture production in the mid 1880s and was previously under CT winter wheat–fallow rotation before initiation of this NT experiment in 1982 (Albrecht et al., 2005). The NT is a randomized complete block design experiment of four replicates of five N fertilization rates. We used data collected from 0 and 180 kg N $ha^{-1}$ fertilization levels for the NT. The CT experiment is also a randomized complete block design consisting of four replications of two N fertilization (0 and 135 kg N $ha^{-1}$) rates and lies to the east of the NT. The winter wheat NT (direct-seed) fallow rotation was seeded (~4 cm deep) in 25-cm rows to soft-white winter wheat in September or early October using a John Deere (HZ 616) deep furrow drill, and a Noble no-till drill to place a fertilizer (Uran 32, Polyphos, and Thiosul) solution 5 cm below the seed. Nitrogen fertilizer was applied as urea-ammonium nitrate, P as ammonium polyphosphate, and S as ammonium thiosulfate. In the CT experiment primary tillage (moldboard plowed ~22 cm deep) was used in April, followed by secondary tillage (field cultivator ~13 cm deep, tine harrow ~13 cm deep, and three or four rod-weeder passes depending on weather, with a set of three 3-m rod-weeders with 1.9-cm square rods set to a depth of 5 cm to kill weeds), and fertilized at seeding in September the following year. Four soil cores were collected following the fall harvests in 2002 and 2009 from each tillage and N treatment using a Giddings hydraulic probe 3.2 cm in diameter and 1.8 m in length (Giddings Machine Co.). Soil cores were sectioned at 0 to 5, 5 to 10, 10 to 20, 20 to 30, 30 to 60, 60 to 100, and 100 cm to a depth of 150 cm unless a restricted layer was reached. Only the top 30-cm increments were used in the simulation comparison. Two cores from each plot were dried at 105°C and weighed to determine soil bulk density. The air-dried mass was converted to oven-dry basis, and sample volume was used to calculate soil bulk density. Total C was analyzed using a combustion analyzer (Thermo Finnigan FlashEA 1112 Elemental Analyzer). Total inorganic C was determined by dry combustion (Nelson and Sommers, 1996) using a Skalar Primacs TOC Analyzer (Skalar Corp.). Total organic carbon was determined by subtracting inorganic C from total C for each sample.

## Colorado Field Study

The Colorado study was conducted at the Agricultural Research Development and Education Center in northeastern Colorado near Fort Collins (40°39′ N, 104°59′ W). The region has a semiarid temperate climate with typical mean annual temperature of 8.9°C and precipitation of 383 mm $yr^{-1}$. Growing-season precipitation averages 254 mm, and irrigation additions are approximately 455 mm $yr^{-1}$. The soil is a Fort Collins clay loam classified as fine-loamy, mixed, superactive, mesic Aridic Haplustalfs (34% clay, 41% sand). We used data collected from continuous irrigated corn cropping under CT and NT production systems at high (200–225 kg N $ha^{-1}$) and 0 N fertilization levels. The amount of irrigation water added to the CT and NT systems was identical for the plots considered here, except that an additional 2.5 cm was added to CT plots before emergence in 2005 due to very dry soil conditions. For both CT and NT, nitrogen fertilizer was applied as urea-ammonium nitrate in a single application near planting time from 1999 to 2004. In 2005, half the N fertilizer rate was applied as urea-ammonium nitrate at corn planting, and the second half of the N rate was applied as polymer-coated urea in mid June. In 2006, half the N rate was applied as a surface banded polymer-coated urea at corn emergence and the second half of the N rate as dry granular urea fertilizer in June. From 2007 to 2009, a polymer-coated urea was surface band applied near the corn row at emergence in May. More detail is provided by Halvorson et al. (2006, 2008). The CT system implemented the following: shredding of corn stalks, tandem disking, and moldboard plowing to a depth of about 25 to 30 cm in fall and roller harrowing (two operations) and landplaning (two operations) and cultivation, if needed, to reduce potential for wind erosion during spring (Jantalia and Halvorson, 2011). Soil carbon was sampled once per year for both systems beginning in 1999. One soil core (5-cm diam.) sample was collected from near the same location within each plot (3 replications) each fall after harvest in depth increments of 0 to 7.6, 7.6 to 15.2, and 15.2 to 30.4 cm to determine SOC. For details on the experimental design and data collection, see Mosier et al. (2006), Halvorson et al. (2008), and (Jantalia and Halvorson, 2011).

## CQESTR Model Description and Simulations

CQESTR is a process-based C model that simulates organic matter decomposition and predicts soil C stock changes based on: crop yields (above- and belowground biomass), N content of residues and organic amendments, soil properties (soil texture and drainage class), weather variables (temperature, precipitation), and thermal time (growing degree-days) (Gollany et al., 2012; Liang et al., 2009). An overview of the CQESTR model can be found in Gollany (2016, this volume). Briefly, CQESTR simulates the effects of management practices or climate on SOC stocks in a soil with up to five depth increments (Liang et al., 2009; Rickman et al., 2002). It operates on a daily time step and can perform long-term simulations. The C pools are depicted as a continuum. Crop residues and organic amendments are categorized by their placement in the soil and their identities are maintained throughout the decomposition phases. After each organic material placement in the soil, decomposition occurs in two phases: a rapid decomposition phase covers the first 1000 units of thermal time (cumulative degree-days), approximating the oxidation of readily metabolized substrate, and a slow decomposition phase representing oxidization of more recalcitrant organic materials (Fig. 1). Each organic residue addition is tracked separately according to its placement in soil horizons (layers). In the final phase (Phase III), when organic matter decomposition is completed, the composted residue is transferred to the recalcitrant SOM pool. Decomposition of crop biomass, amendments and SOM are functions of amount of substrate availability, substrate quality (N, C/N ratio), mechanical disturbance such as tillage, location of residue or substrate, soil water, and thermal time.

Primary model inputs are: (i) number and thickness of soil horizons; (ii) soil texture, drainage class/hydraulic properties, soil bulk density, and SOM/SOC of each horizon; (iii) average monthly mean air temperature and precipitation (daily maximum and minimum air temperature and precipitation is calculated by CQESTR); (iv) aboveground and belowground crop biomass added; (v) root distribution characteristics of crops (belowground crop biomass and root distribution

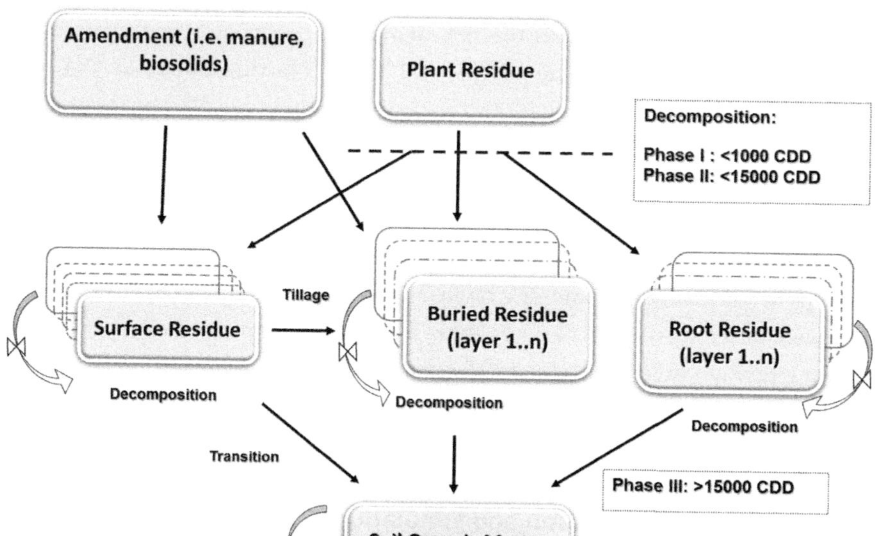


**Fig. 1. Flow diagram for the CQESTR model. The solid arrows depict biomass additions to the residue pools or transfers from the residue pools to the stable soil organic matter pool. The empty arrows represent the computed change in mass on a daily basis. Cumulative growing degree-days (CDD) is the thermal time (adapted after Liang et al., 2009).**

characteristics can be obtained from the CQESTR database; (vi) percentages of crop residues and organic amendments buried and left on the soil surface after tillage; (vii) dates of amendment applications and tillage operations; (viii) initial N contents of crop residues and organic amendments; (ix) depth of tillage, type of tillage and number of operations and timing of each field operations; and (x) crop rotation information (planting schedules and management operation and timing—these are compiled into RUSLE, Version 1, C-factor crop management files; Renard et al., 1996).

Model SOC outputs in any given year are dependent on amount of biomass or amendment input, N content of the crop residue or amendment added, soil edaphic conditions, and tillage, which, in turn, are influenced by previous land use and agricultural management practices. Consequently, CQESTR simulations start with a spin-up procedure typically ranging from 3 to 15 yr before the time period of interest, so that soil organic C pools are properly established and initialized. For the Pendleton field study used in this exercise, we simulated SOC stocks for three incremental depths (0–10, 10–20, and 20–30 cm), as well as for the 0- to 30-cm depth. Conventional moldboard tillage and NT winter wheat–fallow rotations were simulated from 1995 and NT wheat from 1996. The CQESTR outputs from 1995 to 2015 and 1996 to 2015 were saved and used as initial conditions for the four sets of simulations to represent four of the experimental treatments: no N fertilization under conventional tillage (CT 0 N), no N fertilization under NT (NT 0 N), high N under CT (CT high N) and under no-till (NT high N). The treatments were simulated from 1999 to 2015. The simulated SOC stocks for respective years were compared to observed values. For the Colorado field study used in this exercise, we simulated irrigated CT corn from 1995 and irrigated NT corn from 1996. The CQESTR files that schedule management events (day of year when fertilizer was applied and amount, and day of year fields were tilled and intensity of tillage) were based on records of when these events were implemented. For the day of year irrigation water was added and amount, we added the amount of irrigated water to the precipitation values for the day of event occurred. Model outputs for SOC were saved and compared with measured values.

## DayCent Model Description and Simulations

DayCent (Fig. 2) is a biogeochemical model that simulates fluxes of C and N among the atmosphere, vegetation, and soil (Parton et al., 1998; Del Grosso et al., 2001). Flows of C and N between the different pools are controlled by the size of the pools, C/N ratio and lignin content of material, and abiotic water and temperature controls (Parton et al., 1994). Soil water content, temperature, and nitrate are simulated for each horizon throughout the defined depth of the soil profile

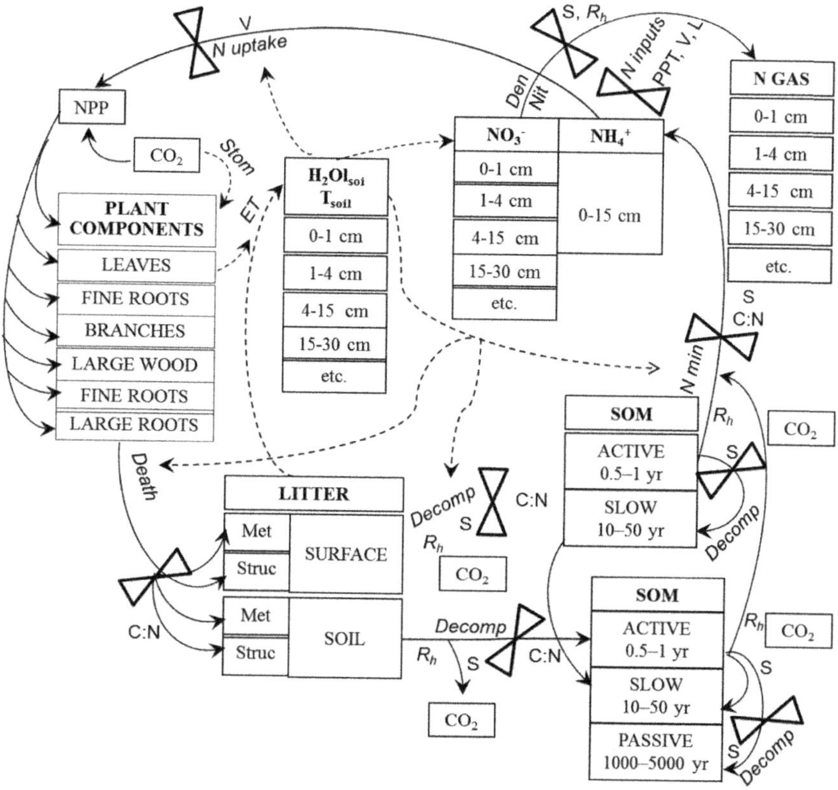


**Fig. 2. Flow diagram for the DayCent model. The solid arrows depict flows of carbon and nutrients (N, P) between live biomass, litter, soil organic matter, and soil mineral pools (bowties represent controls on flow rates). The dashed arrows represent feedbacks and information flows. V is vegetation, S is soil texture, $R_h$ is heterotrophic respiration, Stom is stomatal conductance, and Decomp is decomposition (adapted after Del Grosso et al., 2011).**

(Parton et al., 1998). Typical layering structure going down from the surface is 0- to 2-, 2- to 5-, 5- to 10-, 10- to 20-, 20- to 30-, and 15-cm layers down to 150 cm. The model simulates saturated water flow on days that receive rainfall, irrigation, or snow melt and unsaturated flow every day on a subdaily time step. Plant production is a function of genetic potential, phenology, nutrient availability, water and temperature stress, and solar radiation. Net primary productivity is allocated to plant components (e.g., roots vs. shoots) based on vegetation type, phenology, and water and nutrient stress. Nutrient concentrations of plant components vary within specified limits depending on vegetation type, phenology, and nutrient availability relative to plant demand. Crop germination date can be specified or made a function of soil temperature. Similarly, harvest date can be specified or made a function of accumulated growing degree days since germination.

Litter and SOM decomposition and nutrient mineralization are functions of substrate availability, substrate quality (lignin content, C/N ratio), water and temperature stress, and tillage intensity. The model contains two surface SOM pools: active, which represents microbial biomass and the highly labile byproducts of microbial activity, and slow, which represents more recalcitrant material. In addition to active and slow, the soil contains a third pool designated passive SOM, which represents the highly resistant products of decomposition that can remain for hundreds to thousands of years. Soil organic matter is simulated only for the 0- to 20- or 30-cm layer, depending on the value used for the constant controlling the maximum decomposition rate for the passive SOM pool. Decomposition of organic matter releases ammonium, which is assumed to be immobile and confined to the top 15-cm layer of soil. Ammonium can be taken up by plants, or nitrified to nitrite, which also results in N gas ($N_2O$, $NO_x$) emissions. Unlike ammonium, nitrate is mobile and susceptible to leaching below the rooting zone and also provides substrate for denitrification and associated N gas ($N_2O$, $N_2$) emissions. Nitrification and denitrification are driven by concentrations of ammonium and nitrate in soil, soil water content and temperature, an index of gas diffusivity, and labile C availability (Parton et al., 2001).

Primary model inputs are: daily maximum–minimum air temperature and precipitation, soil texture and hydraulic properties by horizon, and land cover and land use data (e.g., vegetation type, cultivation and planting schedules, amount and timing of nutrient amendments). Outputs include: daily N-gas flux ($N_2O$, $NO_x$, $N_2$), $CH_4$ uptake, SOM (top 20-cm layer), $CO_2$ flux from heterotrophic soil respiration, actual evapotranspiration, soil $NO_3$, water content, and temperature by horizon, soil $NH_4$ in top 15 cm, $NO_3$ leaching, live biomass, plant growth, surface litter, standing dead litter, and other ecosystem parameters. The ability of DayCent to simulate plant growth, SOC, $N_2O$ emissions, $NO_3$ leaching, and $CH_4$ oxidation has been tested with data from various native and agricultural systems (Del Grosso et al., 2000, 2001, 2002, 2005), and the model is used to estimate SOC stock changes and $N_2O$ emissions from cropped and grazed soil in the United States for the National Greenhouse Gas Inventory compiled annually by the USEPA (2015).

Model outputs during any given year are influenced by SOC and mineral N levels, which, in turn, are influenced by previous land use. Consequently, DayCent simulations typically begin at least 1000 yr before the time period of interest so that soil C and N pools are properly initialized. For the Colorado field study used in this exercise, we simulated shortgrass steppe until 1949, irrigated, conventional tillage rotations of corn, dry bean (*Phaseolus vulgaris* L.), sugarbeet (*Beta vulgaris* L. subsp. *vulgaris*), and barley (*Hordeum vulgare* L.) from 1950 to 1992, and irrigated, conventional tillage corn from 1993 to 1998. Model outputs from 1998 were saved and used as initial conditions for the four sets of simulations to represent four of the experimental treatments: no N fertilization under conventional tillage (CT 0 N), no N fertilization under NT (NT 0 N), high N under CT (CT high N) and under no-till (NT high N). The treatments were simulated from 1999 to 2009. The DayCent files that schedule management events (day of year when fertilizer was applied and amount, day of year irrigation water was added and amount, and day of year fields were tilled and intensity of tillage) were based on records of when these events were implemented. Model outputs for SOC were saved and compared with measured values of these variables.

## IPCC Tier 1 Methodology

The Intergovernmental Panel on Climate Change (IPCC) has developed guidelines to estimate GHG emissions for different economic sectors (IPCC, 2006). For the agricultural sector, the guidelines include detailed methods to estimate soil $N_2O$ and $CH_4$ fluxes, as well as biomass and SOC stock changes (see Ogle et al., 2016, this volume, for details). Explicit in the guidelines are tiers reflecting complexity, with Tier 1 methods based on emission factors being the simplest, Tier 3 methods employing the most complex process-based models, and Tier 2 methods being intermediate. The Tier 1 method for SOC stock change provides a reference SOC stock based on climate and soil type and then applies stock change factors based on practices such as land use change to cropping, tillage intensity, and residue input amount. For example, the Tier 1 method estimates initial SOC stock of 5000 g C $m^{-2}$ for high activity clay soils in cold temperate, dry climate. The factor for land use change to long-term cropping is 0.82 for temperate dry climate, implying that 18% of the initial SOC stock would be lost after long-term (>20 yr) cropping. Factors can be greater than 1 and predict lower loss rates or even SOC gains if, for example, cultivated land is converted from CT to NT, if low residue crops are replaced with high residue crops, or if cover crops are included in the rotation.

## Results of Model Testing

### Pendleton Site

The models correctly represented some, but not all, of the observed SOC changes in response to tillage and fertilization rates. All three models and the observations showed losses of soil C (0–30 cm) for all treatments (Fig. 3). However, the Tier 1 method underestimated losses for all treatments, CQESTR underestimated losses for the CT 0 N treatment, DayCent underestimated losses for the CT high N treatment by more than 50%, and both process-based models underestimated losses for the NT high N treatment. The predicted results indicate that CQESTR

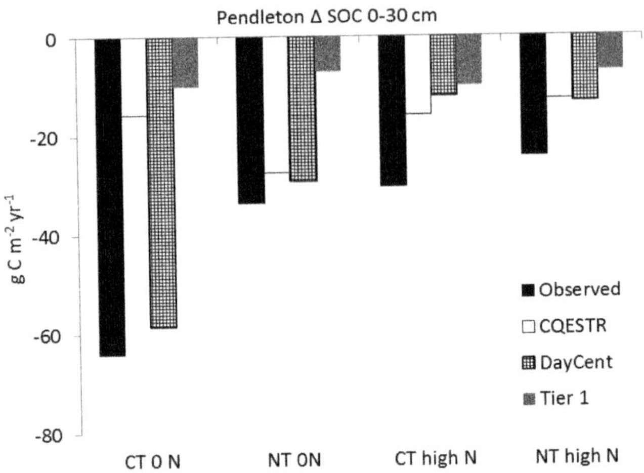


**Fig. 3. Mean annual observed and modeled changes in soil organic carbon (SOC) stocks for conventional tillage (CT) during 2000–2009 and no-till (NT) during 1991–2009 for winter wheat–fallow cropping at Pendleton, OR. High N was 135 kg N ha$^{-1}$ for CT and 180 kg N ha$^{-1}$ for NT.**

underestimated SOC losses by more than 60% for the CT 0 N treatment, which is most likely due to soil erosion and limited residue cover on these plots (Williams, 2008) and the lack of soil erosion algorithms in the CQESTR model to account for soil erosion losses. Not accounting for the impacts of erosion also at least partially explains why DayCent tended to underestimate SOC losses at this site. The CQESTR model SOC predictions tended to most closely match observed SOC losses in the 20- to 30-cm layer but underestimated losses in the 0- to 10- and 10- to 20-cm layers, except for the NT 0 N treatment (Fig. 4). CQESTR underestimation of SOC stock losses in the 0- to 10- and 10- to 20-cm layers can be explained by depth of moldboard plow tillage (0 to ~22 cm) for the CT and translocation of SOM associated soil particles in these plots. Note that DayCent does not explicitly model SOC by soil layer. Although DayCent tended to represent mean annual SOC changes fairly well (Fig. 3), the DayCent model overestimated SOC stocks, whereas CQESTR did a good job predicting SOC stocks for each sampling time at the Pendleton site (Fig. 5).

## Colorado Site

As with the Pendleton site, model performance was mixed for the Colorado site. Both observations and estimations showed that tillage and fertilizer affect SOC changes, but the estimated stocks did not always match the observed magnitude of changes. All models severely (>75% error) underestimated SOC losses under CT without N fertilizer (Fig. 6), DayCent underestimated the relatively small

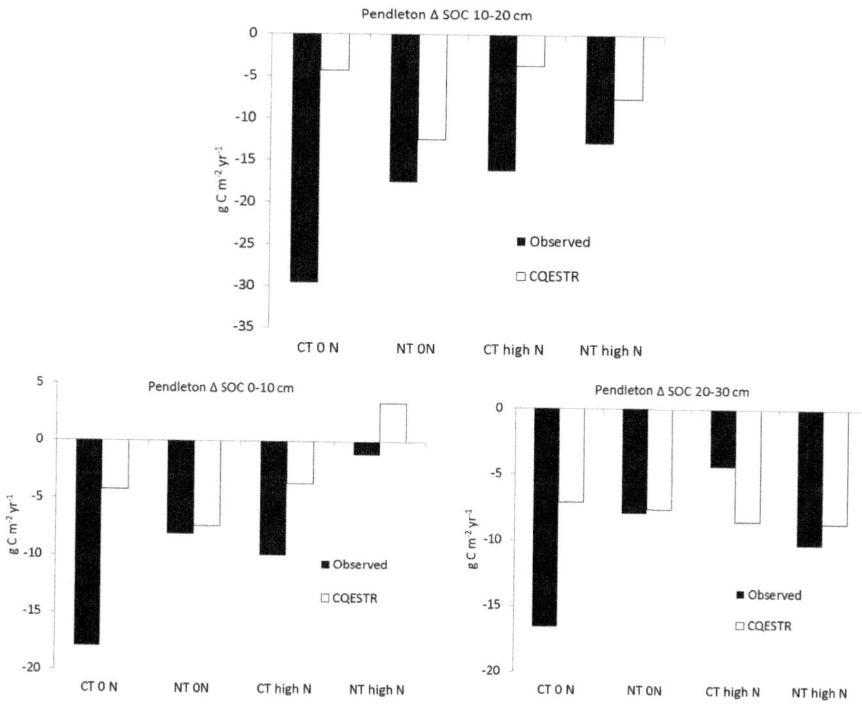


**Fig. 4. Mean annual observed and modeled changes in soil organic carbon (SOC) stocks by soil horizon for conventional tillage (CT) during 2000–2009 and no-till (NT) during 1991–2009 for winter wheat–fallow cropping at Pendleton, OR. High N was 135 kg N ha$^{-1}$ for CT and 180 kg N ha$^{-1}$ for NT.**

observed SOC gains with NT and 0 N and predicted substantial gain under conventional tillage and high N, whereas observations showed very small loss, but matched well the observed gains with NT high N. On the other hand, CQESTR underestimated SOC gains observed for NT high N by about 75%, but correctly predicted the close to neutral SOC changes observed for CT high N (Fig. 6). The Tier 1 method predicted the small SOC losses observed for the conventional till high N treatment.

CQESTR predicted SOC stocks matched observations well in the 0- to 15-cm layer for the 0 N treatments, but underestimated observed SOC gains for the high N treatments (Fig. 7). For the 15- to 30-cm layer, CQESTR underestimated observed SOC losses for the CT treatments, but overestimated losses for the NT treatments (Fig. 7). DayCent results for SOC stocks agreed with observations, but CQESTR underestimated SOC stocks for this site (Fig. 8). CQESTR lacks mechanisms for simulating root biomass and root exudate contributions to SOC stocks and uses measurements or estimates of these as model inputs. Root biomass inputs may have been underestimated due to limited information available for

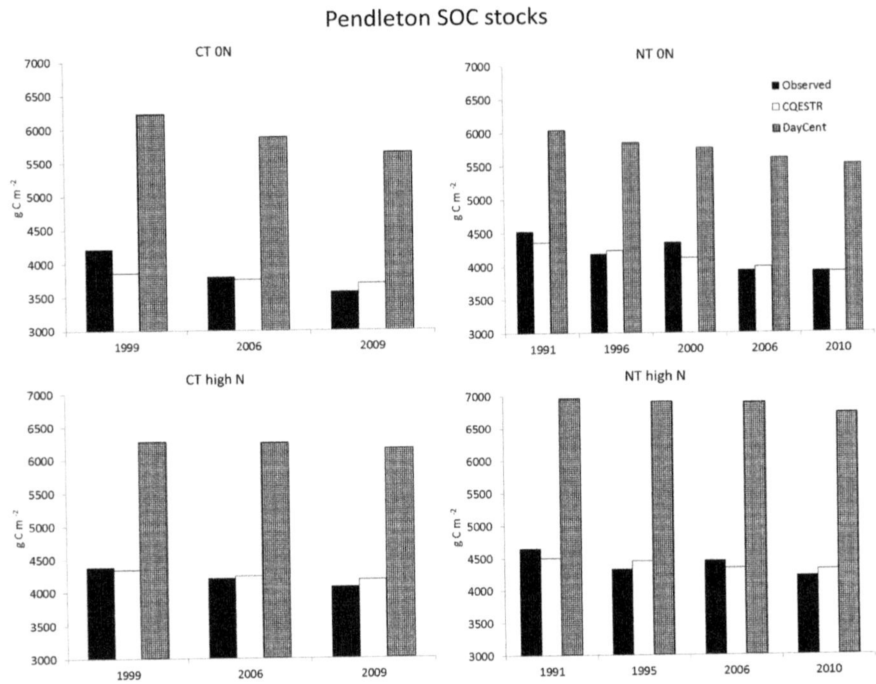


Fig. 5. Observed and modeled soil organic carbon (SOC) stocks for conventional tillage (CT) and no-till (NT) winter wheat–fallow cropping at Pendleton, OR. High N was 135 kg N ha$^{-1}$ for CT and 180 kg N ha$^{-1}$ for NT.

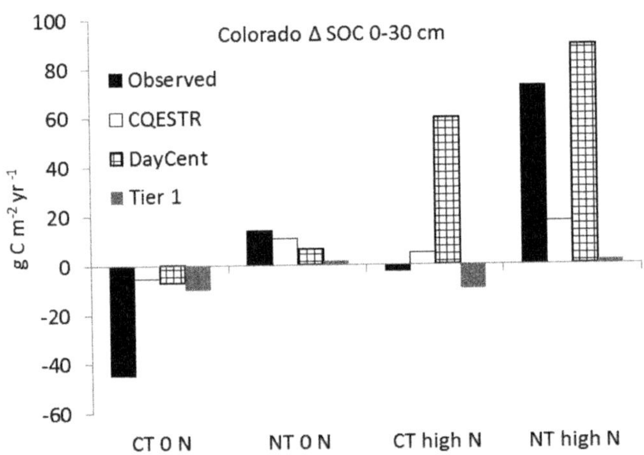


Fig. 6. Mean annual observed and modeled changes in soil organic carbon (SOC) stocks for conventional tillage (CT) during 1999–2008 and no-till (NT) during 1999–2009 for irrigated maize cropping in northeastern Colorado. High N was 200–225 kg N ha$^{-1}$ for both CT and NT.

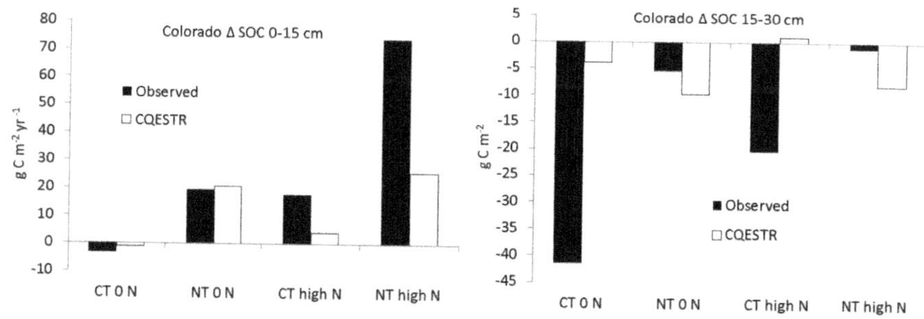


**Fig. 7. Mean annual observed and modeled changes in soil organic carbon (SOC) stocks by soil horizon for conventional tillage (CT) during 1999–2008 and no-till (NT) during 1999–2009 for irrigated maize cropping in northeastern Colorado. High N was 200–225 kg N ha$^{-1}$ for both CT and NT.**

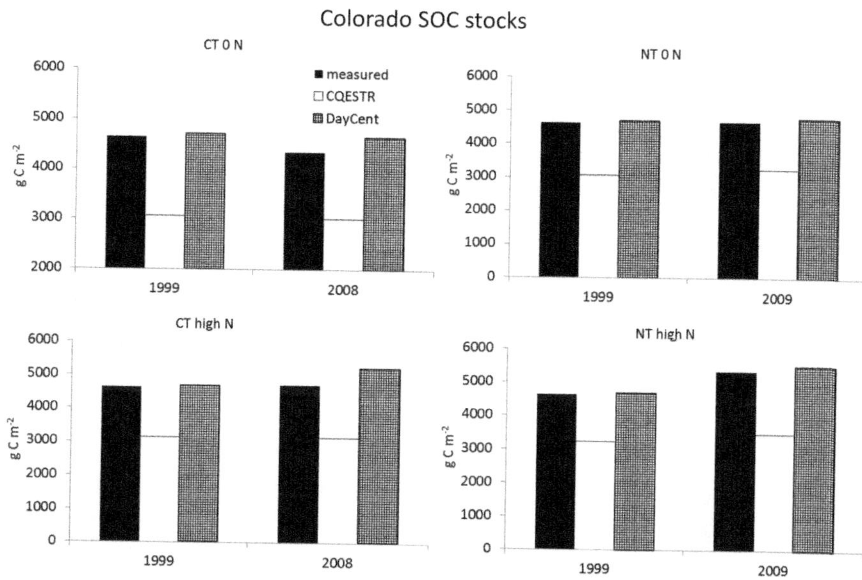


**Fig. 8. Observed and modeled soil organic carbon (SOC) stocks for conventional tillage (CT) during 1999–2008 and no-till (NT) during 1999–2009 for irrigated maize cropping in northeastern Colorado. High N was 200–225 kg N ha$^{-1}$ for both CT and NT.**

root biomass and root distribution under irrigated and rainfed corn, although CQESTR predicted SOC stocks very well ($r^2$ = 0.82) for an irrigated dryland site in Nebraska (Wienhold et al., 2015). Root biomass used in these simulations was adopted from the RUSLE vegetation database file without verification under actual growing conditions at this irrigated dryland site. A more appropriate root/shoot ratio, if known, can be inserted into the RUSLE c-factor files to improve CQESTR predicted SOC values for this irrigated site.

## Discussion and Suggestions for Model Improvement

When comparing methods it is essential to consider the purposes for which the different methods were designed. The IPCC Tier 1 method is meant to represent average SOC changes over large land areas that integrate many different cropping systems and practices that are not all explicitly represented in the method (Smith et al., 2012a). The method is sufficiently simple, making it relatively easy to apply for purposes such as national GHG inventories. Consequently, one should not expect this method to do particularly well when applied to plot-level observations. Process-based models, on the other hand, include many of the factors (e.g., weather, cropping history, soil properties) not represented in the Tier 1 method and are expected to compare more favorably with plot-level measurements. Comparisons of regressions for changes in SOC at the two sites represented by the three methods with observations generally support these expectations. All three methods had relatively high (0.67–0.79) correlation coefficients (Fig. 9). However, the DayCent regression had a slope of 1, and the CQESTR regression had a slope of 0.30, while the Tier 1 regression had a slope of 0.10. The CQESTR regression had an intercept close to 0 (−1.23), while the DayCent intercept was close to 19, and the Tier 1 intercept was close to −5 (Fig. 9). In aggregate, these statistics demonstrate that the process based models are more accurate than the Tier 1 method.

Observations and all models indicated SOC losses at Pendleton for all treatments, but the models (particularly Tier 1) tended to underestimate the magnitude of losses, and CQESTR underestimated the magnitude of losses particularly for the CT 0 N treatment. One reason for this is that the model simulations did not account for erosion impacts on SOC, but it is also possible that the models do not fully account for the impacts of the wheat–fallow rotation on SOC dynamics. Erosion rates in the vicinity of the long-term plots ranged from 0.08 to 3.3 Mg $ha^{-1}$ $yr^{-1}$ depending on the agricultural management practice (Williams, 2008). In an experiment 10 km east of the current plots Douglas et al. (1998) reported total N losses in surface runoff from a wheat–pea (*Pisum sativum* L.) rotation ranged from 5 to 70 kg N $ha^{-1}$ $yr^{-1}$ and that from nonfertilized continuous fallow plots ranged from 20 to 100 kg N $ha^{-1}$ $yr^{-1}$. The C/N ratio of the soil at this site was 1.50. If we consider the middle of these estimates (e.g., an erosion rate of 1.65 Mg $ha^{-1}$ $yr^{-1}$ and an N loss rate of 0.035 Mg N $ha^{-1}$ $yr^{-1}$) and calculate the rate of C loss due to erosion in the vicinity of these plots, the estimated SOC loss would be 0.087 Mg $ha^{-1}$ $yr^{-1}$. This is within the uncertainty of estimated SOC loses of 0.11 Mg $ha^{-1}$ $yr^{-1}$ reported by Williams (2008) for the fertilized (90 kg N $ha^{-1}$) plots in wheat–fallow rotation under CT at Pendleton. Most of the remaining C stocks lost by mineralization can be accounted for in the models.

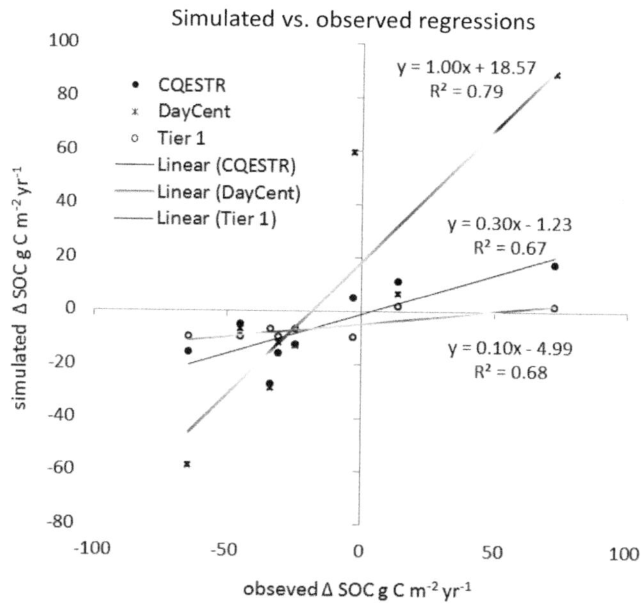


**Fig. 9. Regressions of simulated vs. observed changes in soil organic carbon (SOC) for the Colorado and Oregon sites.**

At the Colorado site, the models (particularly DayCent) also underestimated SOC losses for the conventionally tilled treatments. This suggests that the parameters controlling the impacts of tillage on decomposition rates may need adjustment or that the models do not fully represent the impacts of tillage on soil properties, such as aggregate stability, that influence organic matter protection and decomposition rates (Six et al., 1999), nor the impacts of wind erosion during spring (Jantalia and Halvorson, 2011). For the no-till treatments, CQESTR underestimated SOC gains for the high N treatment, whereas DayCent underestimated gains for the low N treatment. This implies that DayCent is perhaps oversensitive to the impact of N limitation on plant production and accompanying C inputs to soil, while CQESTR underestimated this control on SOC.

Some caveats should be kept in mind when evaluating model performance using field observations. Spatial and temporal variation in SOC stocks as well as handling and processing the collected soil samples can be sources of error in measured SOC stocks (Gollany et al., 2013). Measurements of SOC stocks vary horizontally across plots as well as vertically within a given surface area. Most carbon turnover and cycling occurs in the upper 20- to 30-cm layer (Stockmann et al., 2013); however, recent work has shown that deeper layers are also vulnerable to SOC loss and gain (e.g., Follett et al., 2013). It is difficult, if not impossible, for models intended to be applied at the plot level and greater scales to account for the observed horizontal and vertical variability in SOC stocks. Consequently, most models represent SOC dynamics only in the upper 30-cm layer (Stockmann et al., 2013) and do not account for the impacts of factors such as microtopography and heterogeneity of soil properties, such as texture within the plot, and model simulations typically represent the average conditions across the plot. For example, mean values for soil texture, bulk density, and hydraulic properties are used as model inputs. This should not be a major limitation for processes that respond linearly to changes in drivers, but unfortunately, many processes that influence SOC dynamics respond nonlinearly to key drivers (e.g., the response of decomposition to soil water content is sigmoidal or hyperbolic). Process-based models are continually being improved to account for some of these complications, but lack of sufficiently resolved model input and testing data limit the potential for model improvement.

The analyses presented here suggest that the ability of process-based models to predict SOC stocks is not highly correlated with their ability to predict SOC changes. DayCent overestimated SOC stocks for the Pendleton site (Fig. 5) but still predicted SOC changes reasonably well (Fig. 3). Similarly, CQESTR underestimated SOC stock for the Colorado site (Fig. 8) but predicted SOC changes comparatively well (Fig. 6). From a GHG emission–C sequestration perspective, it can be argued that it is more important for models to accurately predict SOC changes than it is to accurately predict SOC stocks. Model performance reported here is similar to other studies in that, although the models tended to show the observed trends in SOC changes, in some cases model error can be large (Smith et al., 2012b). The regression statistics in Fig. 9 showing mixed results regarding the values for the slopes, intercepts, and correlation coefficients are also similar to those reported by Wu et al. (2015). We conclude that it is unlikely that any single model will be adequate for all applications, as suggested by Hillier et al. (2016, this volume). Users may consider using multiple models, depending on the goals, error tolerances, and resources available to conduct and analyze simulations. Because model errors for at least some land management–location combinations are likely to remain high, model developers should make clear to stakeholders the model limitations and improve model documentation (Tonitto et al., 2016, this volume).

## References

Albrecht, S.L., K.W. Skirvin, and D.S. Long. 2005. Winter wheat responses to nitrogen fertilization in a direct-seed, summer-fallow management system. In: S. Petrie et al., editors, Dryland agricultural research annual report. Spec. Rep. 1061. Agricultural experiments at Pendleton, Oregon. Oregon State Univ., USDA-ARS, Pendleton, OR. p. 87–92.

Campbell, C.A., B.G. McConkey, R.P. Zentner, F.B. Dyck, F. Selles, and D. Curtin. 1996. Long-term effects of tillage and crop rotations on soil organic C and total N in a clay soil in southwestern Saskatchewan. Can. J. Soil Sci. 76:395–401. doi:10.4141/cjss96-047

Ciais, P., C. Sabine, G. Bala, L. Bopp, V. Brovkin, J. Canadell, and P. Thornton. 2014. Carbon and other biogeochemical cycles. In: Climate change 2013: The physical science basis. Contribution of Working Group I to the Fifth Assessment Report of the Intergovernmental Panel on Climate Change. Cambridge Univ. Press, Cambridge, UK. p. 465–570.

Del Grosso, S.J., A.R. Mosier, W.J. Parton, and D.S. Ojima. 2005. DAYCENT model analysis of past and contemporary soil $N_2O$ and net greenhouse gas flux for major crops in the USA. Soil Tillage Res. 83:9–24. doi:10.1016/j.still.2005.02.007

Del Grosso, S.J., D.S. Ojima, W.J. Parton, A.R. Mosier, G.A. Peterson, and D.S. Schimel. 2002. Simulated effects of dryland cropping intensification on soil organic matter and greenhouse gas exchanges using the DAYCENT ecosystem model. Environ. Pollut. 116:S75–S83. doi:10.1016/S0269-7491(01)00260-3

Del Grosso, S.J., W.J. Parton, A.R. Mosier, M.D. Hartman, C.A. Keough, G.A. Peterson, D.S. Ojima, and D.S. Schimel. 2001. Simulated effects of land use, soil texture, and precipitation on N gas emissions using DAYCENT. In: R.F. Follett and J.L. Hatfield, editors, Nitrogen in the environment: Sources, problems, and management. Elsevier, Amsterdam, the Netherlands. p. 413–431.

Del Grosso, S.J., and D.W. Grant. 2011. Reducing agricultural greenhouse gas emissions: Role of biotechnology, organic systems, and consumer behavior. Carbon Manage. 2(5):505–508. doi:10.4155/cmt.11.47

Del Grosso, S.J., W.J. Parton, C.A. Keough, and M. Reyes-Fox. 2011. Special features of the DayCent modeling package and additional procedures for parameterization, calibration, validation, and applications. In: L.R. Ahuja and Liwang Ma, editors, Methods of introducing system models into agricultural research. Advances in Agricultural Systems Modeling 2. ASA, CSSA, and SSSA, Madison, WI. p. 155–176. doi:10.2134/advagricsystmodel2.c5

Del Grosso, S.J., W.J. Parton, A.R. Mosier, D.S. Ojima, C.S. Potter, W. Borken, R. Brumme, K. Butterbach-Bahl, P.M. Crill, K. Dobbie, and K.A. Smith. 2000. General $CH_4$ oxidation model and comparisons of $CH_4$ oxidation in natural and managed systems. Global Biogeochem. Cycles 14:999–1019. doi:10.1029/1999GB001226

Dimassi, B., B. Mary, R. Wylleman, J. Labreuche, D. Couture, F. Piraux, and J.P. Cohan. 2014. Long-term effect of contrasted tillage and crop management on soil carbon dynamics during 41 years. Agric. Ecosyst. Environ. 188:134–146. doi:10.1016/j.agee.2014.02.014

Douglas, C.L., K.A. King, and J.F. Zuzel. 1998. Nitrogen and phosphorous in surface runoff and sediment from a what–pea rotation in northeastern Oregon. J. Environ. Qual. 37:1337–1344.

Follett, R.F., C.P. Jantalia, and A.D. Halvorson. 2013. Soil carbon dynamics for irrigated corn under two tillage systems. Soil Sci. Soc. Am. J. 77(3):951–963. doi:10.2136/sssaj2012.0413

Gollany, H. 2016. CQESTR simulation of dryland agroecosystem soil organic carbon changes under climate change scenarios. In: S.J. Del Grosso, L.R. Ahuja, and W.J. Parton, editors, Synthesis and modeling of greenhouse gas emissions and carbon storage in agricultural and forest systems to guide mitigation and adaptation. Advances in Agricultural Systems Modeling 6. ASA, CSSA, and SSSA, Madison, WI. doi:10.2134/advagricsystmodel6.2013.0004

Gollany, H.T., R.F. Follett, and Y. Liang. 2012. CQESTR Simulations of Soil Organic Carbon Dynamics. In: M.A. Liebig, A.J. Franzluebbers, and R.F. Follett, editors, Managing agricultural greenhouse gases: Coordinated agricultural research through GRACEnet to address our changing climate. Academic Press, San Diego, CA. p. 271–292.

Gollany, H.T., A. Fortuna, M. Samuel, F.L. Young, W. Pan, and M. Pecharko. 2013. Estimating soil organic carbon accretion vs. sequestration using physicochemical fractionation and CQESTR simulation. Soil Sci. Soc. Am. J. 77:618–629. doi:10.2136/sssaj2012.0303

Gollany, H.T., T.E. Schumacher, M.J. Lindstrom, P. Evenson, and G.D. Lemme. 1992. Topsoil depth and desurfacing effects on properties and productivity of a Typic Argiustoll. Soil Sci. Soc. Am. J. 56:220–225. doi:10.2136/sssaj1992.03615995005600010034x

Halvorson, A.D., S.J. Del Grosso, and C.A. Reule. 2008. Nitrogen, tillage, and crop rotation effects on nitrous oxide emissions from irrigated cropping systems. J. Environ. Qual. 37:1337–1344. doi:10.2134/jeq2007.0268

Halvorson, A.D., A.R. Mosier, C.A. Reule, and W.C. Bausch. 2006. Nitrogen and tillage effects on irrigated continuous corn yields. Agron. J. 98:63–71. doi:10.2134/agronj2005.0174

Hillier, J., M. Abdalla, J. Bellarby, F. Albanito, A. Datta, M. Dondini, N. Fitton, P. Hallett, A. Hastings, E. Jones, M. Kuhnert, D. Nayak, M. Pogson, M. Richards, J. Smith, S. Vetter, J. Yeluripati, and P. Smith. 2016. Mathematical modelling of greenhouse gas emissions from agriculture for different end users. In: S.J. Del Grosso, L.R. Ahuja, and W.J. Parton, editors, Synthesis and modeling of greenhouse gas emissions and carbon storage in agricultural and forest systems to guide mitigation and adaptation. Advances in Agricultural Systems Modeling 6. ASA, CSSA, and SSSA, Madison, WI. doi:10.2134/advagricsystmodel6.2013.0038

Horwath, W. 2015. Carbon Cycling: The dynamics and formation of organic matter. In: E.A. Paul, editor, Soil microbiology, ecology, and biochemistry. 4th ed. Academic Press, New York. p. 339–382.

IPCC. 2006. 2006 IPCC guidelines for national greenhouse gas inventories. Vol. 4. Agriculture, forestry and other land use. http://www.ipcc-nggip.iges.or.jp/public/2006gl/vol4.html (accessed 22 Sept. 2015).

Jantalia, C.P., and A.D. Halvorson. 2011. Nitrogen fertilizer effects on irrigated conventional tillage corn yields and soil carbon and nitrogen pools. Agron. J. 103(3):871–878. doi:10.2134/agronj2010.0455

Kumar, S., T. Nakajima, E.G. Mbonimpa, S. Gautam, U.R. Somireddy, A. Kadono, and N. Fausey. 2014. Long-term tillage and drainage influences on soil organic carbon dynamics, aggregate stability and corn yield. Soil Sci. Plant Nutr. 60(1):108–118. doi:10.1080/00380768.2013.878643

Liang, Y., H.T. Gollany, R.W. Rickman, S.L. Albrecht, R.F. Follett, W.W. Wilhelm, J.M. Novak, and C.L. Douglas, Jr. 2009. Simulating soil organic matter with CQESTR (v. 2.0): Model description and validation against long-term experiments across North America. Ecol. Model. 220(4):568–581. doi:10.1016/j.ecolmodel.2008.11.012

Mertens, J., J. Vanderborght, R. Kasteel, T. Pütz, R. Merckx, J. Feyen, and E. Smolders. 2007. Dissolved organic carbon fluxes under bare soil. J. Environ. Qual. 36:597–606. doi:10.2134/jeq2006.0368

Milne, E., S.A. Banwart, E. Noellemeyer, D.J. Abson, C. Ballabio, F. Bampa, and K. Paustian. 2015. Soil carbon, multiple benefits. Environ. Develop. 13:33–38. doi:10.1016/j.envdev.2014.11.005

Mosier, A.R., A.D. Halvorson, C.A. Reule, and X.J. Liu. 2006. Net global warming potential and greenhouse gas intensity in irrigated cropping systems in northeastern Colorado. J. Environ. Qual. 35:1584–1598. doi:10.2134/jeq2005.0232

Necpálová, M., R.P. Anex, A.N. Kravchenko, L.J. Abendroth, S.J. Del Grosso, W.A. Dick, M.J. Helmers, D. Herzmann, E.J. Kladivko, J.G. Lauer, E.D. Nafziger, J.E. Sawyer, P.C. Scharf, J.S. Strock, and M.B. Villamil. 2014. What does it take to detect a change in soil carbon stock? A regional comparison of minimum detectable difference and experiment duration in the North-Central United States. J. Soil Water Conserv. 69:517–531. doi:10.2489/jswc.69.6.517

Nelson, D.W., and L.E. Sommers. 1996. Total carbon, organic carbon, and organic matter. In: D.L. Sparks et al., editors, Methods of soil analysis. Part 3. Chemical methods. SSSA Book Ser. 5. SSSA and ASA. p. 961–1010. doi:10.2136/sssabookser5.3.c34

Ogle, S.M., S. Spencer, M. Hartman, L. Buendia, L. Stevens, L. du Toit, and J. Witi. 2016. Developing national baseline GHG emissions and analyzing mitigation potentials for agriculture and forestry using an advanced national GHG inventory software system. In: S.J. Del Grosso, L.R. Ahuja, and W.J. Parton, editors, Synthesis and modeling of greenhouse gas emissions and carbon storage in agricultural and forest systems to guide mitigation and adaptation. Advances in Agricultural Systems Modeling 6. ASA, CSSA, and SSSA, Madison, WI. doi:10.2134/advagricsystmodel6.2013.0009

Page, K.L., R.C. Dalal, M.J. Pringle, M. Bell, Y.P. Dang, B. Radford, and K. Bailey. 2013. Organic carbon stocks in cropping soils of Queensland, Australia, as affected by tillage management, climate, and soil characteristics. Soil Res. 51:596–607. doi:10.1071/SR12225

Parton, W.J., M.D. Hartman, D.S. Ojima, and D.S. Schimel. 1998. DAYCENT: Its land surface submodel: Description and testing. Global Planet. Change 19:35–48. doi:10.1016/S0921-8181(98)00040-X

Parton, W.J., E.A. Holland, S.J. Del Grosso, M.D. Hartman, R.E. Martin, A.R. Mosier, D.S. Ojima, and D.S. Schimel. 2001. Generalized model for $NO_x$ and $N_2O$ emissions from soils. J. Geophys. Res. 106(D15):17403–17420. doi:10.1029/2001JD900101

Parton, W.J., D.S. Ojima, C.V. Cole, and D.S. Schimel. 1994. A general model for soil organic matter dynamics: Sensitivity to litter chemistry, texture, and management. In: R.B. Bryant and R.W. Arnold, editors, Quantitative modeling of soil forming processes. SSSA Spec. Publ. 39. SSSA, Madison, WI. p. 147–167. doi:10.2136/sssaspecpub39.c9

Powlson, D.S., A.P. Whitmore, and K.W.T. Goulding. 2011. Soil carbon sequestration to mitigate climate change: A critical re-examination to identify the true and the false. Eur. J. Soil Sci. 62(1):42–55. doi:10.1111/j.1365-2389.2010.01342.x

Powlson, D.S., C.M. Stirling, M.L. Jat, B.G. Gerard, C.A. Palm, P.A. Sanchez, and K.G. Cassman. 2014. Limited potential of no-till agriculture for climate change mitigation. Nature Climate Change 4(8):678–683.

Rasmussen, P.E., and W.J. Parton. 1994. Long-term effects of residue management in wheat–fallow. I. Inputs, yield, and soil organic matter. Soil Sci. Soc. Am. J. 58(2):523–530. doi:10.2136/sssaj1994.03615995005800020039x

Rasmussen, P.E., and R.W. Smiley. 1997. Soil carbon and nitrogen change in long-term agricultural experiments at Pendleton, Oregon. In: E.A. Paul, K. Paustian, E.T. Elliot, and C.V. Cole, editors, Soil Organic Matter in Temperate Agroecosystems: Long-Term Experiments in North America. CRC Press, Boca Raton, FL. p. 353–360.

Renard, K.G., G.R. Foster, G.A. Weesies, D.K. McCool, and D.C. Yoder. 1996. Predicting soil erosion by water: A guide to conservation planning with the Revised Universal Soil Loss Equation (RUSLE). USDA Agric. Handb. 703. USDA, Washington, DC.

Rickman, R., C. Douglas, S. Albrecht, and J. Berc. 2002. Tillage, crop rotation, and organic amendment effect on changes in soil organic matter. Environ. Pollut. 116:405–411. doi:10.1016/S0269-7491(01)00217-2

Scharlemann, J.P., E.V. Tanner, R. Hiederer, and V. Kapos. 2014. Global soil carbon: Understanding and managing the largest terrestrial carbon pool. Carbon Manage. 5(1):81–91. doi:10.4155/cmt.13.77

Six, J., E.T. Elliot, and K. Paustian. 1999. Aggregate and soil organic matter dynamics under conventional and no-tillage systems. Soil Sci. Soc. Am. J. 63:1350–1358. doi:10.2136/sssaj1999.6351350x

Smith, P., C.A. Davies, S. Ogle, G. Zanchi, J. Bellarby, N. Bird, and A.K. Braimoh. 2012a. Towards an integrated global framework to assess the impacts of land use and management change on soil carbon: Current capability and future vision. Glob. Change Biol. 18(7):2089–2101. doi:10.1111/j.1365-2486.2012.02689.x

Smith, W.N., B.B. Grant, C.A. Campbell, B.G. McConkey, R.L. Desjardins, R. Kröbel, and S.S. Malhi. 2012b. Crop residue removal effects on soil carbon: Measured and inter-model comparisons. Agric. Ecosyst. Environ. 161:27–38. doi:10.1016/j.agee.2012.07.024

Stockmann, U., M.A. Adams, J.W. Crawford, D.J. Field, N. Henakaarchchi, M. Jenkins, and M. Zimmermann. 2013. The knowns, known unknowns and unknowns of sequestration of soil organic carbon. Agric. Ecosyst. Environ. 164:80–99. doi:10.1016/j.agee.2012.10.001

Tonitto, C., N.P. Gurwick, and P.B. Woodbury. 2016. Quantifying greenhouse gas emissions from agricultural and forest landscapes for policy development and verification. In: S.J. Del Grosso, L.R. Ahuja, and W.J. Parton, editors, Synthesis and modeling of greenhouse gas emissions and carbon storage in agricultural and forest systems to guide mitigation and adaptation. Advances in Agricultural Systems Modeling 6. ASA, CSSA, and SSSA, Madison, WI. doi:10.2134/advagricsystmodel6.2013.0007

USEPA. 2015. Inventory of U.S. greenhouse gas emissions and sinks: 1990–2013. http://www.epa.gov/climatechange/emissions/usinventoryreport.html (accessed 22 Sept. 2015).

Williams, J.D. 2004. Effects of long-term winter wheat, summer fallow residue and nutrient management on field hydrology for a silt loam in north-central Oregon. Soil Tillage Res. 75:109–119. doi:10.1016/j.still.2003.08.010

Williams, J.D. 2008. Soil erosion from dryland winter wheat–fallow in a long-term residue and nutrient management experiment in north-central Oregon. J. Soil Water Conserv. 63:53–59. doi:10.2489/jswc.63.2.53

Wienhold, B.J., M.R. Schmer, V.L. Jin, G.E. Varvel, and H.T. Gollany. 2015. CQESTR simulated changes in soil organic carbon under residue management practices in continuous corn systems. Bioenerg. Res. doi:10.1007/s12155-015-9654-6.

Wu, X., A. Akujärvi, N. Lu, J. Liski, G. Liu, Y. Wang, M. Holmberg, F. Li, Y. Zeng, and B. Fu. 2015. Dynamics of soil organic carbon stock in a typical catchment of the Loess Plateau: Comparison of model simulations with measurements. Landscape Ecol. 30:381–397. doi:10.1007/s10980-014-0110-3

# Measuring and Modeling Carbon Stock Change Estimates for US Forests and Uncertainties from Apparent Inter-annual Variability

**James E. Smith* and Linda S. Heath**

## Abstract

Our approach is based on a collection of models that convert or augment the USDA Forest Inventory and Analysis program survey data to estimate all forest carbon component stocks, including live and standing dead tree aboveground and belowground biomass, forest floor (litter), down deadwood, and soil organic carbon, for each inventory plot. The data, which include estimates of forest area, can then be used in calculations for total stocks or change. We describe our approach, which has been used to estimate forest carbon stocks and stock change for the annual US greenhouse gas inventory compiled by the USEPA for reporting to the United Nations Framework Convention on Climate Change. Results in the national stock change trends can exhibit relatively large annual changes in these change trends, which are called inter-annual variability. We examine the inter-annual variability and underlying data for the possible causes, and discuss implications.

Abbreviations: FIA, Forest Inventory and Analysis; FIADB, Forest Inventory and Analysis Database; IAV, inter-annual variability; NGHGI, national greenhouse gas inventory.

James E. Smith and Linda S. Heath (Lheath@fs.fed.us), USDA Forest Service, Northern Research Station, 271 Mast Road, Durham, NH 03824. *Corresponding author (Jsmith11@fs.fed.us).

doi:10.2134/advagricsystmodel6.2013.0005


Synthesis and Modeling of Greenhouse Gas Emissions and Carbon Storage in Agricultural and Forest Systems to Guide Mitigation and Adaptation
S. Del Grosso, L. Ahuja, and W. Parton, Editors
Advances in Agricultural Systems Modeling, Volume 6. Lajpat R. Ahuja, Series Editor

Sequestration by US forests could offset about 10% of emissions from fossil fuel annually.

The forest carbon estimates for the United States that are part of the nation's participation in the United Nations Framework Convention on Climate Change (UNFCCC) are succinctly summarized as net annual stock change as in Fig. 1. These estimates are included in the annual US national greenhouse gas inventory (NGHGI) compiled by USEPA (2011). Net uptake, or sequestration, by US forests averaged over 160 Tg C/yr over 20 yr, an amount that could offset on average about 10% of emissions from fossil fuel annually over the period. The approach that produced these estimates (Woodbury et al., 2007; Smith et al., 2010; Heath et al., 2011a) has also been used for region-, state-, and county-level estimates (e.g., Mickler et al., 2004; McKinley et al., 2011; also see Heath, 2012), and the models that produce the plot-level estimates have been adopted by the USDA Forest Service, Forest Inventory and Analysis (FIA) program and included in their national dataset called the FIADB (Woudenberg et al., 2010). A number of models have used the estimates for calibration, model validation, as input data, or for a mixture of these activities (e.g., Hayes et al., 2012; Heath et al., 2010; Liu et al., 2008; Potter et al., 2008).

As indicated in Fig. 1, the results are characterized by sometimes relatively large annual changes in this rate and trends such as the extreme around the year 2000 or the apparently steady multi-year changes before and after that time. Both the year-to-year variability and the trends can affect perception of the forest carbon estimates. For example, the variability can indicate rapidly changing conditions of US forest lands or alternatively be a reflection of the precision in the estimates. Similarly, any trend could suggest some sustained influence on forest land across the United States.

The year-to-year variability in NGHGI emissions or removals estimates is also known as inter-annual variability (IAV) in UNFCCC documentation, and its significance is related to how the change estimates are developed (Eggleston et al., 2006, in particular see volume 1, chapter 3, Uncertainties). Estimates developed around multi-year averages will use IAV to describe precision in the underlying data. However, where empirically based estimates are developed as specific to data of a given year, the expectation is that the IAVs serve to indicate change in activity data as influenced by specific, temporally linked human activity or biophysical processes (Richards, 2011; Eggleston et al., 2006). That is, the value and underlying cause of IAV is unique to each year's estimate. This single-year specificity of emission/removals estimates is considered good

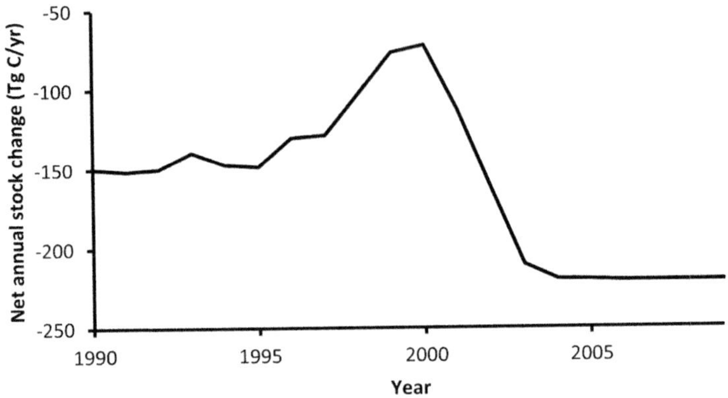


**Fig. 1. Net annual carbon stock change (USEPA, 2011).**

practice by some, especially for some reporting such as to the Kyoto Protocol. However, others recognize the limitations and costs of estimating change in forests for a single year.

Our approach is a collection of models that convert or augment the FIA forest survey data to estimate all forest carbon component stocks, including live and standing dead tree aboveground and belowground biomass, forest floor (litter), down deadwood, and soil organic carbon for each inventory plot. The plot data can then be used in calculations for total stocks or change. We first present a summary of the FIA data and briefly describe the models to calculate carbon and the approach used to derive the stock change estimates. We then use the results from an application that produced a series of the estimates reported in USEPA (2011) to examine and diagnose apparent year-to-year variability in the estimate, and discuss implications. Although carbon in harvested wood products is a notable contribution to sequestration related to the forest sector, we focus only on carbon in forests in this study. More information on the modeling approach for carbon in harvested wood products can be found in Skog (2008).

The stock change estimates of carbon on US forest lands are primarily based on annualized forest inventory surveys. Briefly, forest inventory surveys are each defined in terms of tons of carbon at the mean date of field data collection—carbon stocks at a specific point in time. The difference between two such successive stocks provides change over an interval, which is expressed as tons of carbon per year. This inventory-to-carbon stock and stock change methodology (Penman et al., 2003) is based on whole-state estimates or sub-state disaggregation where necessary for consistency (USEPA, 2011; Smith et al., 2010). Additionally, the same methodology can be applied to less aggregate classifications, including individual counties (McKinley et al., 2011) or major ownerships such as National Forest lands by region (Heath et al., 2011b). Continuing development of the stock change method is applied for forest carbon estimates of the United States. The NGHGI continues with future modifications focused on increased availability of annual inventories and adaptation to separate land use and land use change classifications. This study represents a step in the process of refining and improving estimates.

We explicitly identify influences on the IAV of Fig. 1 through sensitivity analysis of the stock change method to the inventory-with-carbon data. The purpose is to examine the link between the year of the stock change estimate and the corresponding influence on IAV (Richards, 2011; Eggleston et al., 2006). The analysis is specific to the forest carbon estimates of USEPA 2011 because the IAV issue was highlighted in the review of that year's NGHGI (United Nations Framework Convention on Climate Change, 2012), which noted that "the signals of sudden change may not be registered in the estimates at the proper point in time." The signals of large sudden change will tend to be averaged over time using this approach, but the time period will bracket the time the change occurred. We identify influences on IAV, explain the mechanism, and discuss future approaches to address this issue.

## Methods

Interpretation or understanding of an assessment of forest carbon change such as shown in Fig. 1 depends on an understanding the data and methods applied to produce the estimate. In this approach, the underlying data are the FIA forest inventories and the carbon conversion factors/models. These data are then applied using a stock change method (Penman et al., 2003). Commonly, "stock change" within the IPCC/UNFCCC publications is used to refer to relatively short time steps such as the difference between two successive stocks so that stock change reflects a rate unique to a particular year or an average over a short period of years. In this scenario, IAV would be the change between two successive short-interval stock changes, and the IAV would reflect activity or emissions events within that short interval.

Our inventory-based forest carbon stock change differs from this "standard" interpretation of stock change due to two important characteristics of the forest inventories. Specifically, the periodic intervals allocate an estimate of change to multi-year intervals, yet change-in-change, or IAV, is allocated to infrequent instances. Second, the total for the United States is a composite of many separate state-based stock change estimates at independent (i.e., out of phase) intervals. The following methods and analysis explain this distinction and provide a basis for interpreting the IAV of the estimates examined in this study.

## Forest Inventory Data

Forest inventories are the basis for the carbon estimates, and these inventory data are obtained by continuous systematic sampling of US forest lands. The data are compiled and made available by the FIA program of the US Forest Service (USDA Forest Service, 2013a), and these publicly available surveys are the starting point for the carbon estimates. Some elements of the structure of FIA forest inventories are relevant to understanding the methods and analysis presented below. Forest land is defined for purposes of the inventory according to cover and land use (Smith et al., 2009). This includes information on a large system of permanent plots over forest land (currently representing about 270 million hectares) in the conterminous United States (i.e., 48 states), but generally excludes wooded agricultural or urban land. Inventory methods are established according to a national standard so that data are consistent, but surveys are organized, and conducted within individual states. Surveys from the last 10 to 15 yr are known as "annual inventory" where a portion of the survey data is collected each year on a continuous cycle, whereas older surveys are known as "periodic inventory" with all statewide data collected in 1 or 2 yr followed by a 5 to 15 yr interval before a state was resurveyed. The publically available data, the FIADB (Woudenberg et al., 2010), are organized as separate surveys by state and by years of data collection. Statewide inventory totals are summarized within the FIADB on completion of each survey cycle, that is, one pass through a state's inventory plots. Each cycle is identified by a nominal inventory year, which is usually the year the cycle is completed, but the data of the survey can also be summarized according to an average date for collection of all field data (i.e., average of all site visits; Smith et al., 2013). Thus, the basic inventory data are summarized as statewide totals, each associated with an average date associated with field data collection. Over time, series of these inventory year paired values accumulate at intervals of five or more years apart, depending on the state. Cycle length can vary, and whole-state forest inventories that focus on standing stocks rather than change within the FIADB are typically based on data collected over the number of years associated with each cycle, where possible (Roesch et al., 2002; Woudenberg et al., 2010). Our results focus on change, so the data we use for our calculations are based on different years than those data used for stock estimates.

## Carbon Conversion Factors

Carbon factors and models were developed to provide plot-level estimates of carbon density (e.g., tons of carbon per hectare) for the following distinct, non-overlapping forest ecosystem carbon pools: live trees, understory vegetation, standing dead trees, down deadwood, forest floor, and soil organic carbon.

Expansion of plot estimates to whole-state total carbon stocks for each survey is according to the population estimates defined for the FIADB (also identified as evaluations within the FIADB; USDA Forest Service, 2013b; also see Woudenberg et al., 2010; Bechtold and Patterson, 2005). For additional details on the carbon conversion factors, their application to inventories, expansion to population totals, or approaches to determining stock change, see the Carbon Calculation Tool model (Smith et al., 2010), USEPA (2011), FIADB documentation (Woudenberg et al., 2010), early soil carbon methods (Heath et al., 2002), or Smith et al. (2013).

## Forest Carbon Stock Change Calculations

The approach to developing the national estimates of carbon change in forest ecosystems using survey data following Smith et al. (2010) can be described by the steps outlined in Fig. 2. Resolution of forest inventory data as plot-level carbon stocks that are then expanded to statewide or population totals at specified times (i.e., the paired values of tons and year) are discussed above and represented in the upper dashed box of Fig. 2. The next step involves summarizing the series of carbon stocks for a state by interpolating or extrapolating to the annualized stock and stock change for the reporting interval starting with 1990 (Smith et al., 2010). This step is represented in the lower dashed box of Fig. 2.

To define the within-state series of stocks consistent over time, which is essential for stock change calculations, forest land of some states is subdivided so that the entire series of stock and stock change calculations were recast as sub-state populations (Smith et al., 2010). For example, in the past, National For-

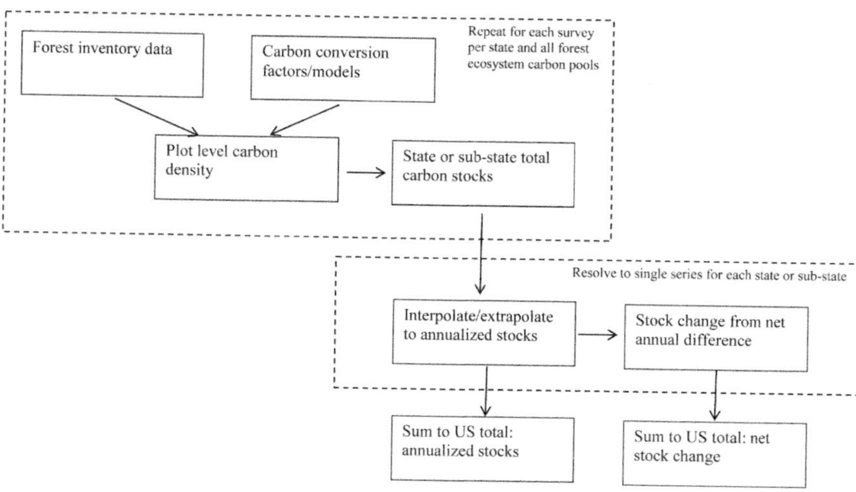


**Fig. 2. Outline of basic approach to develop forest carbon stock change estimates based on forest inventory data.**

est lands on western states were measured separately and at different intervals than most other forest land in those states; therefore "National Forest" versus "not National Forest" is the basis for many of the sub-state divisions. Once all surveys are defined as carbon stocks within consistent series, linear interpolation or extrapolation defines a series of annualized stocks for each state or sub-state series, for example, stocks defined for each year, 1990 through 2009 (USEPA, 2011). By convention, carbon stock for a given year is considered the stock at the beginning of the year so that when summed with stock change during that year the total is the next year's (beginning) stock value. Net annual stock change is the difference between annualized stocks, with negative values indicating sequestration, by convention. From this, the calculation for 2002, for example, is defined as the annualized stock for 2002 minus the annualized stock for 2003.

The annualized estimates for sub-state classification are then summed to the whole-state estimates, and all annualized net change estimates are summed to the US total (Fig. 1, and the lowest two boxes of Fig. 2). See USEPA (2011) for an exact list of inventory sources, sub-state classifications, and surveys used per state associated with the estimates presented here. In general, the years used to identify the surveys are the nominal years, which are associated with the state's data within the forest inventories.

Estimated uncertainties are calculated using Monte Carlo simulation, following IPCC guidance (Penman et al., 2003; Eggleston et al., 2006). Probability density functions are defined for each of the plot-level conversion factors. The sampling error is also determined for each carbon pool and included in the uncertainty estimates using Monte Carlo sampling. More details of the methods to estimate uncertainty are given in Heath and Smith (2000), Smith and Heath (2001), and Smith et al. (2013). The 95% confidence interval for the 2009 stock change estimates for all forest ecosystem pools is –261.6 to –180.5 Tg C/yr, which would often be written in the format –220.6 ± 18% Tg C/yr.

## Examining Inter-annual Variability

The set of forest inventory data available and applicable to USEPA (2011) was classified as 88 separate state and sub-state series (i.e., as in the lower dashed box of Fig. 2). The number of surveys available for each series varied and was generally 2 to 4, with a total of 236 surveys (i.e., as in the last box within the upper dashed box of Fig. 2). To illustrate the basic process described above, we use the forest inventories for Tennessee, which was one of the 88 series and had 4 surveys available, all from the FIADB (Fig. 3). The four surveys resolved to carbon stocks at year 1989.0, 1998.2, 2002.7, and 2005.8 (which are the nominal inventory years

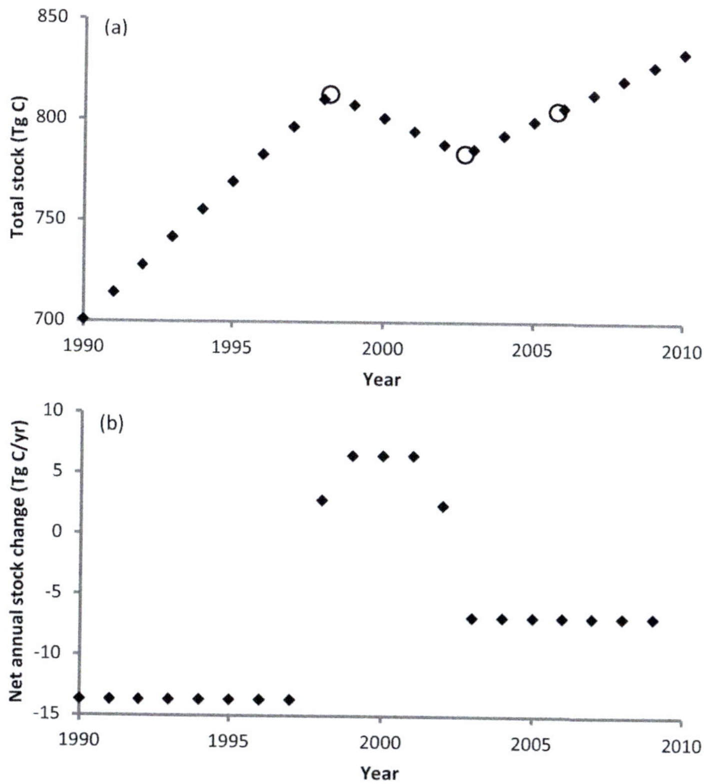


**Fig. 3. Example of individual state forest (a) stock and (b) stock change (b) for Tennessee. Circles represent carbon stock summaries from forest surveys, and other symbols represent annualized (a) carbon stock and (b) carbon stock change. Note in Fig. 3a that the stock estimate for the first survey is 687 Tg C in the year 1989. It is not shown here to keep the year axis the same throughout.**

1989, 1999, 2004, and 2007; Fig. 3a, noted by circle symbols). Note that the stock for the first survey is not shown because it is for year 1989, at 687 Tg C. These four carbon stocks (three visible in Fig. 3a) represent the effect of the first three steps in the approach described above and in Fig. 2. The annualized stocks (diamond symbols) are from linear interpolation/extrapolation. The values for net annual stock change are determined as described above (Fig. 3b); for example, net stock change for 1998 is based on the difference between the 1999 and 1998 stocks (Fig. 3a). The annualized series from 1990 to 2009 (Fig. 3a and 3b) are the result of step four (above and Fig. 2).

The inter-annual variability is defined as the change in annualized estimate for 1 yr relative to the previous years; that is, the annual steps in Fig. 1. While change in net change is expected, our interest is in the patterns or extremes and our ability to link each state's survey-with-carbon to the values of Fig. 1.

As illustrated in Fig. 3, interpolation between successive periodic stocks (Fig. 3a) produces constant values for net stock change over the interval (Fig. 3b) so that all of the change-in-change (or IAV) is confined to the two short 2-yr intervals 1998 to 1999 and 2002 to 2003. The plotted point indicating stock change for 1998 is the first positive value in the sequence, and the IAV for 1998 is that value minus the estimate for 1997, a large positive IAV. The stock change graph for Tennessee (Fig. 3b) is typical of the state-level stock change pattern in that IAV occurs over very brief intervals, with constant net stock change over most of the reporting years 1990 to 2009. This is a different pattern than for the summed stock change for the United States (Fig. 1), with IAV occurring between most years.

### Inter-annual Variability Sensitivity to Forest Surveys

Sensitivity of net annual stock change to the underlying forest inventory surveys is identified by repeated stock change calculations while systematically removing individual surveys. The point of this process is to look at each national level annual step change—or IAV of stock change—to identify (i) specific surveys that have an influence and (ii) the relative level, or percent, effect of each. Again, we define IAV for a particular year as the difference, or change, relative to the previous year; for the Fig. 1 example, the change in net change for 1999 is 26 Tg C (a positive value because the 1999 stock change is less negative than that for 1998).

Sensitivity to the individual surveys is based on the base stock change (Fig. 1, as in USEPA, 2011) and an additional 236 similar annualized totals for 1990 to 2009, which are each based on deleting a single survey. Pairwise comparisons were made for each year between each partial model and the base (complete) model. Each pairwise non-zero contribution of one survey for a particular year is divided by the summed absolute values of all contributions for that particular year to attribute a portion of change to each influential survey. This same process can be applied to identify sensitivity of IAV to each of the 88 state or sub-state series of surveys as well.

## Results

Summed net annual stock change for US forests and the pattern in Fig. 1 are from the sum of the 88 separate stock change sequences such as the one illustrated in Fig. 3b. Each yearly estimate for each state or sub-state sequence contributes to the magnitude of net stock change for the country. However, only a few of the annualized estimates of stock change contribute to the year-to-year change, or IAV of the total. For example, from Fig. 3b for Tennessee forests the annualized estimates for 1998, 1999, 2002, and 2003 are the only values that affect the IAV of Fig. 1.

Identifying sensitivities of IAV, as in Fig. 1, is based on the effect of the selective deletion of components. This process as applied to the 88 series of state or sub-state stocks is illustrated for the sensitivity of the 1999 estimate (Fig. 4). The solid line represents the base estimates (same as Fig. 1), and the dashed line represents the estimates without the Georgia inventory, which has the greatest effect on the magnitude of net annual stock change for 1999 but no effect on IAV for 1999 (result not shown). The dotted line represents the estimates without the Oregon non–National Forest Eastside sequence, which has the greatest influence/effect on IAV for 1999. That is, the net change in 1999 relative to that in 1998 where the addition of those particular Oregon surveys changes the annual interval from the dotted to the solid line has the greatest effect on the summed slope in Fig. 1 for those years. The surveys (the set of 236) with the greatest influence on each of the yearly IAVs is shown in Table 1, which includes the 3 most influential surveys for

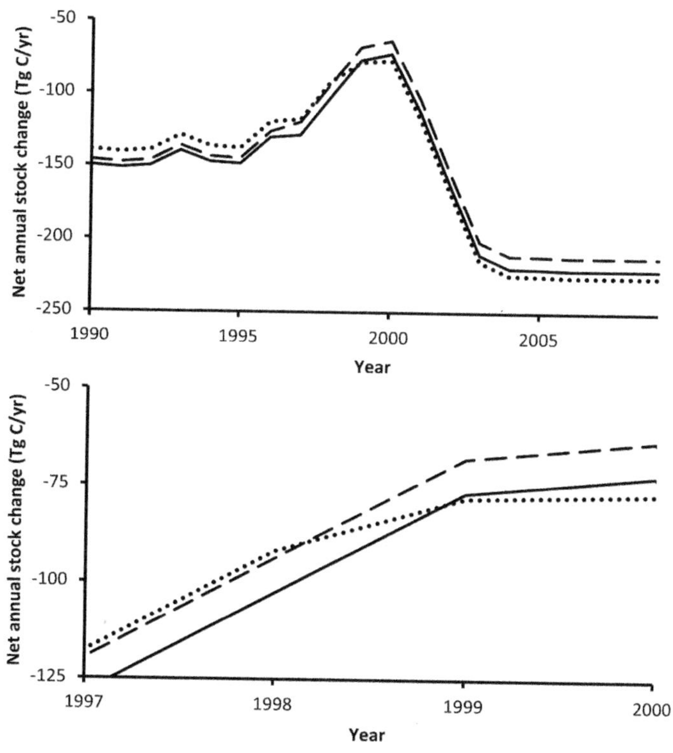


**Fig. 4. Example of influence on net forest stock change, or inter-annual variability (IAV), for 1999, with the second figure providing detail about the 1999 estimates. The solid line represents the base stock change totals as in Fig. 1. The dashed line represents the totals without the Georgia inventories, which have the greatest effect on the magnitude of net annual change for 1999 but not the effect on IAV for 1999 (i.e., the change between 1998 and 1999). The dotted line represents the totals without the Oregon non–National Forest Eastside forest survey, which has the greatest effect on total IAV for 1999 (i.e., the change between 1998 and 1999).**

**Table 1. The three individual surveys of state or sub-state populations that have the greatest influence on each inter-annual change in net stock change, that is, change in change or inter-annual variability (IAV). The year represents the nominal year associated with the survey. The proportion is the summed relative influence on IAV of the three listed surveys. Surveys listed in italics are influential, but in the opposite direction of the summed IAV as shown in Fig. 1.**

| Year | Influence on change in net change | Proportion |
|---|---|---|
| 1991 | Louisiana 1991, 2005; Alabama 1990 | 0.47 |
| 1992 | Michigan 1993, 1980; Texas (East) 1986 | 0.47 |
| 1993 | Michigan 1993, 1980, 2004 | 0.69 |
| 1994 | Florida 1995, 2007, 1987 | 0.76 |
| 1995 | Florida 1995, 2007; *Wisconsin 1995* | 0.45 |
| 1996 | Montana National Forest 1989, 2009; Arkansas 1995 | 0.48 |
| 1997 | *Georgia 1997*; Indiana 1986, 1998 | 0.30 |
| 1998 | Tennessee 1999; Indiana 1986, 1998 | 0.48 |
| 1999 | Oregon non–National Forest Eastside IDB† (1999), 2009; Alabama 2000 | 0.43 |
| 2000 | Idaho other National Forest 1991, 2009; *North Carolina 2002* | 0.36 |
| 2001 | North Carolina 2002, 2007; Minnesota 2003 | 0.43 |
| 2002 | Wisconsin 2004, 2008; Minnesota 2003 | 0.26 |
| 2003 | Michigan 2004, 2008; Tennessee 1999 | 0.36 |
| 2004 | Illinois 2005, 2008; Arkansas 2005 | 0.59 |

† IDB is a periodic inventory dataset for California, Oregon, and Washington. See http://www.fs.fed.us/pnw/fia/publications/data/data.shtml or Smith et al. (2010) for more information.

each year and the summed relative effect of the 3 in determining IAV. Note that sensitivities to the series of surveys (the set of 88) are not separately provided, but the most influential state or sub-state series is that listed first in Table 1 (i.e., the most influential series includes the most influential survey).

The Tennessee 1999 survey appears in Table 1 as influential for 1998 and 2003 IAV; the effect of that one survey on national totals is shown in Fig. 5. Additionally, the effect of a few highly influential surveys is shown in Fig. 6, which is based on removing only 5 of the 236 surveys: those most influential in each of the 5 highest IAV years—1998, 1999, 2001, 2002, and 2003. The surveys are as follows: Tennessee 1999, Oregon non–National Forest Eastside (1995/IDB), North Carolina 2002, Wisconsin 2004, and Michigan 2004 (Table 1).

## Discussion

A relatively small number of the 236 surveys underlying Fig. 1 have a significant influence on IAV in any 1 yr; frequently, the 3 surveys listed in Table 1 account for a large proportion of the influence. In addition, sensitivity to surveys changes continually so that the influence is generally only for one or 2 yr. The pattern of annualized stock change estimates in Fig. 3b is consistent with these observations; 4 of the 20 estimates affect IAV with 2-yr intervals each.

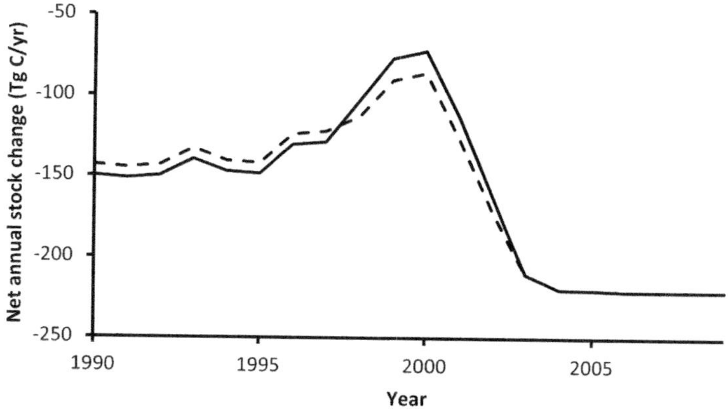


**Fig. 5. Effect of removing only the 1999 Tennessee survey (dashed line) relative to the total of Fig. 1 (solid line).**

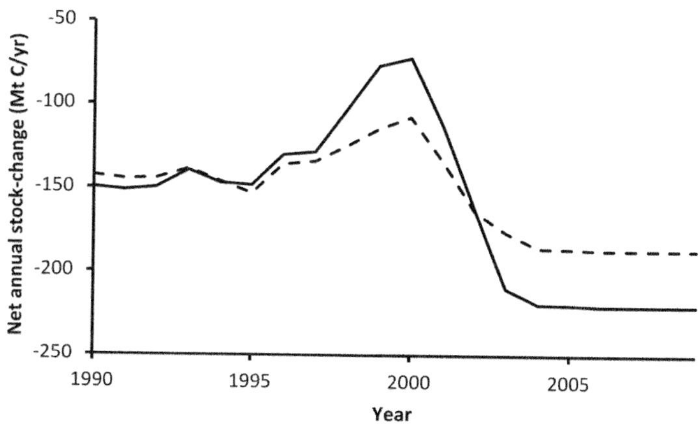


**Fig. 6. Effect of removing 5 (of the 236) surveys: those most influential in each of the 5 highest IAV years—1998, 1999, 2001, 2002, and 2003. The surveys are as follows: Tennessee 1999, Oregon non–National Forest Eastside (1995/IDB), North Carolina 2002, Wisconsin 2004, and Michigan 2004. The dashed line is the reduced model (231 surveys) relative to the total of Fig. 1 (solid line).**

The pattern of the most influential surveys (Table 1) where consecutive pairs of surveys within a state are frequently listed is readily explained because sensitivity is related to the slope over a 1-yr interval as illustrated in Fig. 4. If either of the points defining a line segment (as in Fig. 3a) is modified, then the change in net change at each node is also modified. The Tennessee 1999 survey represents an example where paired stocks do not have equal influence. This survey is most influential for overall IAV in 1998, but the 1989 survey would be fourth on the Table 1 list for 1998. Similarly, Tennessee 1999 is listed as influential for the 2003 IAV; the Tennessee 2004 survey would be fourth on this list. These examples demonstrate that sensitivity can be separated from the inventory field data collection by varying numbers of years.

The apparent disconnect between sensitivity of IAV and what are most likely the actual in-the-forest changes in net annual changes in carbon stocks can be interpreted from Fig. 3. In the interval between the carbon stocks at the approximate average years of 1989 and 2003 (Fig. 3a), the stock change construct places all the IAV in 1998 and 1999 (Fig. 3b). In reality, change is continuous, probably not linear, and abrupt step changes are unlikely. This is repeated for the intervals on either side of the stock plotted for 2003 as well. The important point here is that this combination of data and methods means that the IAV apparent for a single year is not the product of change—or activity data—exclusive to that year as is expected of IAV (Richards, 2011; Eggleston et al., 2006). Figures 5 and 6 also provide examples of sensitivity of IAV to one or a few surveys, where the two lines are not parallel.

In contrast to our use of periodic surveys, stock change developed on the basis of true annual stock estimates would, by definition, place IAV and activity as jointly occurring in the same year. The difference of USEPA (2011) (Fig. 1) with the presumed definition of IAV of Eggleston et al. (2006) is the multi-year intervals of the stocks from FIA data. These data tend toward the step changes such as in Fig. 3b. A mechanism to smooth the stock-to-stock series, and thus the stock change, would be one step toward better linking cause and IAV. If done properly, the sensitivities would be more dispersed and Fig. 1 and the respective IAVs would be smoothed, yet accurate. The ideal approach to clearly link activity and resulting removals/emissions would likely be to transform the entire series of periodic surveys to annual estimates where each year is based on an estimate of inventory characteristic of that year rather than the current interpolated-between-estimates approach. However, in many forest types, the change within a year is so small that if one tried to take measurements every year for a validation dataset, the change would likely be within the measurement error. It may be difficult to prove the attribution to a specific year.

## Conclusions

Carbon stocks and stock change estimation in forests often employ a more data-driven approach. The major carbon changes in forests are typically in trees; changes in tree carbon are relatively more rapid and responsive to influences on change, such as forest management, relative to soil organic carbon. Trees are relatively easy to measure than carbon change in soil. Soil carbon can appear to change rapidly depending on how the accounting framework accounts for the changing forest land base. Trees in forests have been measured and statistically sampled for volume estimation for decades, if not centuries (Scott and Gove, 2002). These same measurements can be used for estimating carbon, although additional measurements may be needed to improve carbon estimates with this type of approach. Note that carbon is not directly measured with this approach; models must be used with the measurements although the models may be fairly basic, direct, well-documented, and considered accurate. The estimates produced with the approach discussed here are often used as the standard with which process models and remote-sensing-based models are calibrated or validated.

Because the forest surveys in the United States are conducted at the state level, because the survey data were often traditionally associated with ending survey dates rather than the year of survey, and because the survey design has changed during the period of interest, calculations using the plot data can be complicated. These changes have contributed to what appears to be inter-annual variability at the national level. However, in the example we investigated, the IAV appears to be due to new data from individual states or sub-state areas representing estimates for a short, but multi-year, time period that nonetheless was noticeably influential when summed with all the other forested areas of the United States. The change is a true change, and the inventory-based estimates of stock change are specific to each of the reported years, but the IAV cannot be attributed to specific activity data for each of those years.

More research is needed to continue to reduce perceived uncertainties of the estimates and ensure the estimates correspond to the true years the sequestration or emissions occurred.

Population totals as presented for the FIADB are not currently provided as true annual estimates because fewer data can be used for evaluating annual estimates according to single inventory years and this results in a lower precision that has been considered unacceptable (Roesch et al., 2002; Bechtold and Patterson, 2005). However, techniques are under development that may provide estimates closely matched to year that the activity occurred with only a small loss of precision (e.g., see Eskelson et al., 2009). Development of optional approaches and improved estimates are expected over time as more years of annual data become available.

## References

Bechtold, W.A., and P.L. Patterson, editors. 2005. The enhanced forest inventory and analysis program- national sampling design and estimation procedures. Gen. Tech. Rep. SRS-80. USDA Forest Service, Asheville, NC.

Eggleston, S., L. Buendia, K. Miwa, T. Ngra, and K. Tanabe, editors. 2006. 2006 IPCC guidelines for national greenhouse gas inventories. Inst. for Global Environ. Strategies for the Intergovernmental Panel on Clim. Change, Kanagawa, Japan.

Eskelson, B.N.I., H. Temesgen, and T.M. Barrett. 2009. Estimating current forest attributes from paneled inventory data using plot-level imputation: A study from the Pacific Northwest. For. Sci. 55(1):64–71.

Hayes, D.J., D.P. Turner, G. Stinson, A.D. McGuire, Y. Wei, T.O. West, L.S. Heath, B. Jong, B.G. McConkey, R.A. Birdsey, W.A. Kurz, A.R. Jacobson, D.N. Huntzinger, Y. Pan, M.W. Post, and R.B. Cook. 2012. Reconciling estimates of the contemporary North American carbon balance among terrestrial biosphere models, atmospheric inversions, and a new approach for estimating net ecosystem exchange from inventory-based data. Global Change Biol. 18:1282–1299. doi:10.1111/j.1365-2486.2011

Heath, L.S. 2012. Using FIA data to inform United States forest carbon national-level accounting needs: 1990–2010. In: A.E. Camp, L.C. Irland, and C.J.W. Carroll, editors, Long-term silvicultural & ecological studies: Results for science and management, Vol II. GISF Res. Pap. 013. Yale Univ. School of For. and Environ. Stud., New Haven, CT. p. 149–160. http://www.nrs.fs.fed.us/pubs/43402 (accessed 5 May 2012).

Heath, L.S., R.A. Birdsey, and D.W. Williams. 2002. Methodology for estimating soil carbon for the carbon budget model of the United States, 2001. Environ. Pollut. 116:373–380. doi:10.1016/S0269-7491(01)00213-5

Heath, L.S., M.C. Nichols, J.E. Smith, and J.R. Mills. 2010. FORCARB2: An updated version of the US Forest Carbon Budget Model. [CD-ROM]. Gen. Tech. Rep. NRS-67. USDA Forest Service, Newtown Square, PA.

Heath, L.S., and J.E. Smith. 2000. An assessment of uncertainty in forest carbon budget projections. Environ. Sci. Policy 3:73–82. doi:10.1016/S1462-9011(00)00075-7

Heath, L.S., J.E. Smith, K. Skog, D. Nowak, and C. Woodall. 2011a. Managed forest carbon estimates for the US Greenhouse Gas Inventory, 1990–2008. J. For. 109:167–173.

Heath, L.S., J.E. Smith, C.W. Woodall, D.L. Azuma, and K.L. Waddell. 2011b. Carbon stocks on forestland of the United States, with emphasis on USDA Forest Service ownership. Ecosphere. 2(1):1–21.

Liu, J., S. Liu, T.R. Loveland, and L.L. Tieszen. 2008. Integrating remotely sensed land cover observations and a biogeochemical model for estimating forest ecosystem carbon dynamics. Ecol. Modell. 219:361–372. doi:10.1016/j.ecolmodel.2008.04.019

McKinley, D.C., M.G. Ryan, R.A. Birdsey, C.P. Giardina, M.E. Harmon, L.S. Heath, R.A. Houghton, R.B. Jackson, J.F. Morrison, B.C. Murray, D.E. Pataki, and K.E. Skog 2011. A synthesis of current knowledge on forests and carbon storage in the United States. Ecol. Appl. 21(6):1902–1924. doi:10.1890/10-0697.1

Mickler, R.A., J.E. Smith, and L.S. Heath. 2004. Forest carbon trends in the Southern United States. In: H.M. Rauscher and K. Johnsen, editors, Southern forest science: Past, present, and future. Gen. Tech. Rep. SRS-75. USDA Forest Service, Asheville, NC. p. 383–394.

Penman, J., M. Gytarsky, T. Hiraishi, T. Krug, D. Kruger, R. Pipatti, L. Buendia, K. Miwa, T. Ngara, K. Tanabe, and F. Wagner, editors. 2003. Good practice guidance for land use, land use change, and forestry. Inst. for Global Environ. Strategies for the Intergovernmental Panel on Climate Change, Kanagawa, Japan.

Potter, C., P. Gross, S. Klooster, M. Fladeland, and V. Genovese. 2008. Storage of carbon in US forests predicted from satellite data, ecosystem modeling, and inventory summaries. Clim. Change 90:269–282. doi:10.1007/s10584-008-9462-5

Richards, G. 2011. Issues arising from the use of Tier 3 models in AFOLU and inter-annual variability. In: H.S. Eggleston, N. Srivastava, K. Tanabe, J. Baasansuren, and M. Fukuda, editors, Use of models and facility-level data in greenhouse gas inventories. Report of the IPCC Expert Meeting on Use of Models and Measurements in GHG Inventories, 9–11 Aug. 2010, Sydney, Australia. Inst. for Global Environ. Strategies for the Intergovernmental Panel on Climate Change, Kanagawa, Japan. p. 34–38.

Roesch, F.A., J.R. Steinman, and M.T. Thompson. 2002. Annual forest inventory estimates based on the moving average. In: R.E. McRoberts, G.A. Reams, P.C. Van Deusen, and J.W. Moser, editors, Proceedings of the Third Annual Forest Inventory and Analysis Symposium. Gen. Tech. Rep. NC-230. USDA Forest Service, St. Paul, MN. p. 21–30.

Scott, C.T., and J. Gove. 2002. Forest inventory. In: A.H. El-Shaaraw and W.W. Piegorsch, editors, Encyclopedia of environmetrics. Vol. 2. John Wiley & Sons, Chichester, UK. p. 814–820.

Skog, K.E. 2008. Sequestration of carbon in harvested wood products for the United States. For. Prod. J. 58:56–72.

Smith, J.E., and L.S. Heath. 2001. Identifying influences on model uncertainty: An application using a forest carbon budget model. Environ. Manage. 27:253–267. doi:10.1007/s002670010147

Smith, J.E., L.S. Heath, and C.M. Hoover. 2013. Carbon factors and models for forest carbon estimates for the 2005– 2011 National Greenhouse Gas Inventories of the United States. For. Ecol. Manage. 307:7–19. doi:10.1016/j.foreco.2013.06.061

Smith, J.E., L.S. Heath, and M.C. Nichols. 2010. US forest carbon calculation tool: Forest-land carbon stocks and net annual stock change (revised for use with FIADB 4.0). Gen. Tech. Rep. NRS-13. USDA Forest Service, Newtown Square, PA.

Smith, W.B., P.D. Miles, C.H. Perry, and S.A. Pugh. 2009. Forest resources of the United States, 2007. Gen. Tech. Rep. WO-78. USDA Forest Service, Washington, DC.

United Nations Framework Convention on Climate Change. 2012. Report of the individual review of the inventory submission of the United States of America submitted in 2011. FCCC/ARR/2011/USA. http://unfccc.int/documentation/documents/advanced_search/items/6911.php?priref=600007068 (accessed 3 June 2013).

USDA Forest Service. 2013a. FIA datamart. USDA Forest Service, Washington, DC. http://apps.fs.fed.us/fiadb-downloads/datamart.html (accessed 3 June 2013).

USDA Forest Service. 2013b. Forest inventory and analysis national program: FIA library, database documentation. USDA Forest Service, Washington, DC. http://www.fia.fs.fed.us/library/database-documentation/ (accessed 3 June 2013).

USEPA. 2011. Inventory of US greenhouse gas emissions and sinks: 1990–2010. EPA 430-R-11-001. USEPA Office of Atmospheric Programs, Washington, DC. http://www.epa.gov/climatechange/ghgemissions/usinventoryreport/archive.html (accessed 22 Apr. 2013).

Woodbury, P.B., J.E. Smith, and L.S. Heath. 2007. Carbon sequestration in the US forest sector from 1990 to 2010. For. Ecol. Manage. 241:14–27. doi:10.1016/j.foreco.2006.12.008

Woudenberg, S.W., B.L. Conkling, B.M. O'Connell, E.B. LaPoint, J.A. Turner, and K.L. Waddell. 2010. The Forest Inventory and Analysis Database: Database description and user's manual, version 4.0 for Phase 2. Gen. Tech. Rep. RMRS-GTR-245. USDA Forest Service, Fort Collins, CO.

# Developing National Baseline GHG Emissions and Analyzing Mitigation Potentials for Agriculture and Forestry Using an Advanced National GHG Inventory Software System

**Stephen M. Ogle,* Shannon Spencer, Melannie Hartman, Leandro Buendia, Luanne Stevens, Linde du Toit, and Jongikhaya Witi**

## Abstract

National greenhouse gas (GHG) inventories are compiled by governments for all sectors, including agriculture and forestry, and provide historical emissions data for assessing Nationally Appropriate Mitigation Actions (NAMAs). National inventory guidelines have been developed by the Intergovernmental Panel on Climate Change (IPCC), and the Agriculture and Land Use National Greenhouse Gas Inventory (ALU) software is an advanced system applying the IPCC inventory methods, along with functionality to assess GHG mitigation potentials. Two GHG mitigation case studies are provided for rice (*Oryza sativa* L.) management in the Philippines and livestock and manure management in South Africa. The Philippines example demonstrates that managing rice straw by burning the residues will decrease GHG emissions, relative to the current practice of leaving some rice straw on the field. Rice straw is a substrate that enhances methane emissions when the soil is flooded, and it is a source of nitrogen leading to soil nitrous oxide emissions. However, this option may not be a NAMA because retaining some residue in the field is important for longer-term sustainability of the soil for crop production, as well as the negative effects of burning residues on air quality and human health. For South Africa, enteric methane emissions can be reduced with feed additives in ruminant diets that are managed in confined operations. In addition, GHG emissions from manure management can be reduced with the more widespread use of digesters. The ALU software provides a framework to quantitatively assess mitigation options in a country and inform policy decisions about NAMAs.

Abbreviations: ALU, Agriculture and Land Use National Greenhouse Gas Inventory (software); GHG, greenhouse gas; IPCC, Intergovernmental Panel on Climate Change; NAMA, Nationally Appropriate Mitigation Action; TACCC, transparency, accuracy, completeness, consistency, and comparability; TMR, total mixed ration.

Stephen M. Ogle, Shannon Spencer (Shannon.Spencer@colostate.edu), and Melannie Hartman (Melannie.Hartman@colostate.edu), Natural Resource Ecology Laboratory, Campus Delivery 1499, Colorado State Univ., Fort Collins, CO 80523. *Corresponding author (stephen.ogle@colostate.edu). S.M. Ogle, also Dep. of Ecosystem Science and Sustainability, Fort Collins, CO 80523. Leandro Buendia, 2113-A Pula Street, College Ville, Los Banos, Laguna 4030, Philippines (leandro.buendia@gmail.com). Luanne Stevens, School of Geo and Spatial Science, Unit for Environmental Sciences and Management, Bldg. E4, Room G3, North-West Univ., Potchefstroom campus, Hoffman Street, Potchefstroom 2531, South Africa (luanne@jacali.net). Linde du Toit, Dep. of Animal Science, Faculty of Science, Tshwane University of Technology, Bldg. 5 Room 627, Private Bag x680, Pretoria 0001, South Africa (DuToitCJL@tut.ac.za). Jongikhaya Witi, Dep. of Environmental Affairs, Environment House, 473 Steve Biko Street, Arcadia, Pretoria 0001, South Africa (Jwiti@environment.gov.za).

doi:10.2134/advagricsystmodel6.2013.0009


Synthesis and Modeling of Greenhouse Gas Emissions and Carbon Storage in Agricultural and Forest Systems to Guide Mitigation and Adaptation
S. Del Grosso, L. Ahuja, and W. Parton, Editors
Advances in Agricultural Systems Modeling, Volume 6. Lajpat R. Ahuja, Series Editor

National GHG inventories provide information on GHG emissions for various sectors in a country, including agriculture and forestry. Inventories can form the basis for analyzing GHG mitigation potentials by governments to determine historical emissions and evaluate mitigation options that are considered candidates for NAMAs. Moreover, the GHG emissions data are needed to assess synergies between mitigation options and related policy issues such as food security (Smith et al., 2013) and adaptation to climate change (Godfray et al., 2011; Beddington et al., 2012; Ogle et al., 2014).

The IPCC has developed a series of guidelines for countries to estimate and report anthropogenically driven GHG emissions to the United Nations Framework Convention on Climate Change (UNFCCC) (IPCC, 1997, 2000, 2003, 2006). Greenhouse gas emission sources from agriculture and forestry include changes in carbon stocks that result in the release or removal of carbon dioxide ($CO_2$) associated with changes in biomass, litter (or dead biomass as described in the IPCC Guidelines), and soil carbon stocks; release of soil nitrous oxide ($N_2O$) from nitrogen management practices; enteric methane ($CH_4$) emissions from ruminant livestock; release of $N_2O$ and $CH_4$ due to livestock manure management; release of non-$CO_2$ GHGs (CO, $N_2O$, $NO_x$, $CH_4$) due to biomass burning; $CO_2$ emissions from urea and lime applications; $CH_4$ emissions from wetland rice production systems; $CO_2$ and $N_2O$ emissions due to peatland management; and storage of C in durable harvested wood products (IPCC, 2006).

Our objective was to present an overview of the emission sources for agriculture and forestry, the data requirements for estimating those emissions, and an advanced inventory software system that has been developed to estimate historical national GHG emissions and evaluate mitigation potentials. We highlighted some of the key principles and good practices for compiling a national GHG inventory (IPCC, 2000, 2006), ensuring that the resulting estimates are reliable for policy planning and development (Ogle et al., 2013). We provided an overview of an advanced software system, the ALU software, highlighting key features that facilitate application of the IPCC good practice. We then provided two applications of the ALU software to estimate GHG mitigation potentials for management options that could become part of a portfolio of NAMAs in the Philippines and South Africa.

## IPCC National Greenhouse Gas Inventory Guidelines

The general approach for conducting an inventory of GHG emissions involves combining activity data based on human management with emission factors that quantify an amount of GHG emission or carbon stock change per unit of anthropogenic activity (IPCC, 1997, 2000, 2003, 2006). The calculation of emissions or removals of GHGs involves using a set of simple algebraic equations relating an amount of activity, emission factors, unit conversions, and in some cases empirical constants. A fundamental objective of the IPCC Guidelines is to provide "good practice" promoting high-quality inventories "that do not over- or underestimate emissions or removals, as far as can be judged, and where uncertainties are reduced as much as is practical" (IPCC, 2000). To meet this objective, the IPCC recommends that GHG emissions are estimated with the principles of transparency, accuracy, completeness, consistency, and comparability (TACCC). Transparency means an inventory has sufficient documentation so that inventory compilers and users of the inventory can understand the basis for the results. Accuracy, the agreement between the true underlying value of emissions and removals and their estimates, implies that emissions should be neither overestimated nor underestimated (i.e., no bias). Bias can occur because of a failure to capture all relevant processes influencing emissions or because of a lack of representative data. Completeness means that inventory estimates have been prepared for all inventory sources and sink categories that are nationally relevant and that all geographic areas within the country have been included. Consistency means that estimates for different inventory years and categories are made in such a way that trends reflect real differences in emissions and are not subject to changes resulting from methodological differences. Comparability means the inventory is reported in a way that allows it to be compared with national GHG inventories from other countries. This comparability is reflected in the appropriate choice of key emission source categories, the use of reporting guidance and tables, the use of classification and definition of categories, and the use of SI units (International System of Units). The IPCC Guidelines provide guidance on ensuring quality for all steps of the inventory compilation, from data collection to reporting (IPCC, 2006).

The IPCC Guidelines define three levels of estimation methods from a Tier 1 level, which represents a default method, to a Tier 3 level, which is the most detailed and complex estimation method (IPCC, 2006). Tier 1 methods are simple so that any country can estimate emissions with limited data and information. Though a country implementing a Tier 1 analysis must collect aggregate level activity data, the IPCC provides default stock and emission factors for most equations. Additionally, some activity data can be collected from international organizations (IPCC, 2010; Tubiello et al., 2013) or through the use of expert knowledge (e.g., Maia et al., 2010). A Tier 2 method uses the same methodological approach and equations as Tier 1, but with country-specific emission factors instead of global or continental default values provided by the IPCC. In some cases, such as enhanced livestock characterization, the Tier 2 method provides a separate set of detailed equations to estimate country-specific emission factors. The Tier 3 level uses higher order estimation methods that typically include complex models, national inventory measurement systems, and highly disaggregated activity data (e.g., Ogle et al., 2010; Spencer et al., 2011). In general, it is good practice for a country to move to higher tier analyses for larger emissions sources because higher tier methods can produce more accurate and precise inventory estimates (e.g., Del Grosso et al., 2011). However, many countries do rely on the Tier 1 methods and default emission factors, even for their largest emission sources (Ogle et al., 2014), which is likely due to a lack of information and software to organize data and conduct higher tier inventories (Ogle et al., 2013).

Estimating GHG emissions for agriculture and forestry management activity is inherently complex because of the variability in management options and because of the interaction of management with environmental variables such as weather patterns and soil characteristics. Thus, agriculture and forestry arguably require more information than other sectors in a national inventory, and thus can lead to more error, particularly if IPCC good practice is not applied in the inventory compilation.

## Agriculture and Land Use National Greenhouse Gas Inventory Software

The ALU software was developed to assist inventory compilers with compilation of Tier 1 or 2 inventories using IPCC good practice (IPCC, 2006) (http://www.nrel.colostate.edu/projects/ALUsoftware/). In turn, the resulting inventories can form the basis to develop a robust baseline for estimating GHG mitigation potentials and evaluating NAMAs associated with agriculture and forestry management.

The ALU software guides inventory compilers through the steps of activity data entry, emission factor assignment, and emissions calculations, while the inventory data are managed and stored in a relational database. In addition, the ALU software maintains data integrity and provides feedback based on the status of data entry for each emission source category. Uncertainty with activity data and emission factors can be entered to derive confidence intervals for GHG emission estimates. The ALU software provides detailed reports that include all calculations, activity data, and emission factors.

The software includes most emission source categories for agriculture and forestry, including livestock enteric $CH_4$ emissions, $CH_4$ and $N_2O$ emissions from manure management systems, biomass burning non-$CO_2$ GHGs, $N_2O$ emissions from soils, $CH_4$ emissions from rice management, and biomass C and soil C stock changes. The ALU software also provides a method for estimating non-$CO_2$ GHG emissions due to peatland burning.

There are three basic steps to completing an inventory in the ALU software, including (i) activity data entry, (ii) emission factor assignments, and (iii) emission computations. All of these steps are accomplished within a session that the compiler creates when initiating activity data entry. A session represents a specific spatial unit and year. Spatial units could be as large as the entire country or could be created for sub-regions or politically defined areas within the country, such as provinces or states. After initial data entry, the ALU software guides the compiler through the basic steps of the inventory, which comprise three modules. The compiler enters activity data for agriculture and forestry in Module I and assigns emission factors in Module II, which can be default IPCC factors or country-specific values, and then the program calculates and displays the emissions in Module III.

The ALU software applies the IPCC good practice standards to ensure transparency, accuracy, completeness, time series consistency, and comparability (i.e., TACCC principles). Transparency and comparability of results are facilitated through detailed emission reports that show all activity data and emission factors used in each equation. The ALU software also allows the compiler to insert documentation about activity data and also the sources that were used to develop country-specific emission factors, and this information can be exported for reporting purposes. Accuracy can be evaluated by estimating uncertainty for emissions sources. Uncertainty analysis is a critical component of the inventory process, and uncertainty information is integrated with data entry in the ALU software (Colomb et al., 2013). At each level of data entry, the compiler may enter the lower and upper values of 95% confidence intervals expressed as a percentage of the mean data values. The ALU software also promotes accuracy with an integrated quality assurance/quality control (QA/QC) process at each stage of data entry. Completeness of data is encouraged through data validation at each data entry step to ensure the entire land base and livestock population is tracked through the inventory process from activity data entry to the emission calculations. An example of data validation is that the compiler enters activity data in a hierarchically framework in Module I, starting with larger scale data such as land use areas, which are then disaggregated into management systems. The program ensures that data from the larger scales in the hierarchy are included in more disaggregated levels, such as the area of crop management systems compared with the total land use area for croplands.

A primary tenet of IPCC good practice is to encourage the inventory compilers to apply higher tier methods for large emission sources (IPCC, 2006). The ALU software allows the compiler to apply country-specific emission factors and advance to a Tier 2 method. Advancing to higher tier methods has been a challenge in many countries (Lokupitiya and Paustian, 2006; Ogle et al., 2014), but nonetheless it is a good practice because emission factors based on country-specific emissions data can more accurately represent emission rates in a country than IPCC default values.

Greenhouse gas mitigation potentials can be estimated using the inventory data to inform baseline emission rates for the country, along with a specific set of management alternatives for one or more emission source categories. The first step is to determine the baseline emissions, which are the expected future emissions with no mitigation activities (Bernoux et al., 2010). In the ALU software, baseline emissions can be set according to the historical emissions in the last year of the inventory, or the baseline can be estimated according to a linear trend in the historical emissions data. The second step is to estimate emissions with one or more mitigation activities. The last step is to subtract the estimated emissions with mitigation activities from the baseline to determine the potential for reducing emissions.

The ALU software supports two types of mitigation analyses, including a "practice-based" approach and a "whole-session" approach. The practice-based approach is an analysis for a specific set of management options or activity data. The focus of this approach is to investigate the impact of technological advances or conservation management on GHG mitigation potentials. Examples of practice-based approaches in the ALU software include reducing deforestation; altering crop residue management; reducing grassland burning; changing water and residue management for rice production; altering manure management systems to reduce $CH_4$ and $N_2O$ emissions; the use of livestock feed additives to reduce enteric $CH_4$ emissions; and reducing soil $N_2O$ emissions through reduced N fertilizer use, application of nitrification inhibitors, or slow-release fertilizers. In contrast, the "whole-session" approach allows the compiler to modify any activity data or emission factors associated with an inventory session. This approach has more options and could be combined with economic analyses to evaluate a broader and more complex suite of mitigation drivers, including population growth, economic growth, technological advancements, and conservation management.

## ALU Software: Applications Assessing National GHG Mitigation Potentials

Although there are numerous constraints to implementing GHG mitigation options in forestry and agriculture, there is also a significant potential for mitigation of GHG emissions (Smith et al., 2007). Two applications of the ALU software are presented for the agricultural sector, demonstrating how the software can be used to conduct an initial analysis of GHG mitigation potential to facilitate discussion and policy development. The first application is an analysis of GHG mitigation potential associated with rice straw management in the Philippines. Rice is one of the key crops for the agricultural sector in the Philippines, and alternative straw management practices could reduce those emissions in the future. The second application is an analysis of GHG mitigation potential associated with changing diets or manure management in South Africa. Livestock production is a large industry in South Africa, and this analysis provides information on GHG mitigation potentials that are possible with alternative management practices.

## Philippines: GHG Mitigation Potentials for Rice Management

Rice cultivation is a major industry in the Philippines, and rice straw or residues, which are produced in an abundant quantity (almost 50% of the total rice aboveground biomass), are piled or spread in the field, burned in situ, or in some cases removed from the field for horticulture usage or as construction materials. Retaining rice straw in the field is considered the best practice for longer-term sustainability of rice production because it enhances soil carbon and improves soil fertility (Dobermann and Fairhurst, 2002). However, this practice also increases $CH_4$ emission from rice fields because rice straw becomes the carbon source for $CH_4$ production through the microbial process of methanogensis (Takai, 1970; Neue and Sass, 1994; Smith et al., 2008). Also, mineralization of nitrogen from rice straw contributes to $N_2O$ emissions. In contrast, on-site burning of rice straw, particularly during the dry season, is an effective practice for eliminating the rice straw and preparing the field to plant the next crop. Burning residues also emits GHGs including $CH_4$ and $N_2O$ in addition to precursors that can form GHGs, carbon monoxide, and oxides of nitrogen (IPCC, 2006).

For the national policy in the Philippines, several questions arise in relation to reducing GHG emissions from the agricultural sector. In particular, which residue management practice has the lowest GHG emissions, and thus is it possible to reduce emissions with a management change from the current practice? If it is possible to reduce emissions, could this practice serve as a NAMA? The analysis examined the effect of rice straw management on GHG emissions using the practice-based approach in the ALU software. In year 2000, the Philippines had a total harvested rice area of 4 million hectares, 57% of which is planted during the wet season, and about 67% of this total area is irrigated.

On the basis of the IPCC methodology (IPCC, 2000), annual $CH_4$ emissions were estimated as the product of a seasonally integrated emission factor (EFi) and annual area harvested (A) in the ALU software. The EFi is a function of the $CH_4$ emission factor for continuously flooded fields without organic amendment (EFc), combined with scaling factors for water management during cultivation period (SFw) and amount of organic amendment applied (SFo). Country-specific $CH_4$ emission factors (EFc) for dry and wet seasons were developed for the Philippines using data from $CH_4$ measurement experiments conducted by the International Rice Research Institute (IRRI) between 1993 and 1999 (Wassmann et al., 2000). The SFo factor was set on the basis of the reported 4 to 5 t ha$^{-1}$ of dry matter rice straw that are produced each cropping season in irrigated rice areas, and a reduced amount for rice straw produced in rainfed areas (Philippine Rice Research Institute, 1997).

The current practice for residue management in most rice production regions of the Philippines is to retain 25% of the total harvested rice straw in the fields and to burn the remaining 75%, particularly after dry season harvest when conditions are more favorable for burning. A full range of residue management options were considered from 100% retained in the field to 100% burned (Table 1). The amounts of residue retained and burned on-site were estimated as well as the resulting impacts on GHG emissions for $CH_4$ emissions from rice fields, $N_2O$ emissions from crop residue in rice fields, and non-$CO_2$ emissions from residue burning.

Greenhouse gas emissions in the inventory for 2000 were estimated at 15,296 Gg $CO_2$ eq. yr$^{-1}$ (Table 2). Of this total amount, 93% is in the form of $CH_4$ and the remainder is $N_2O$. In terms of mitigation options, GHG emission would be 25,302 Gg $CO_2$ eq. yr$^{-1}$ if all harvested rice straw was returned to fields, which is a 65% increase over the emissions from the current practice (Table 2). $CH_4$ emissions would be 98.4% of the total, while $N_2O$ emission would be 1.6%. In contrast, emissions are lowest when all residues are burned, largely because rice straw enhances GHG emissions from methanogenesis more than the effect of burning on a $CO_2$ equivalent basis. Although not assessed in this study, burning 100% of the residue could also decrease carbon inputs to the soil and the amount of soil C (Paustian et al., 1997). Furthermore, burning residues may create health problems

**Table 1. Scaling factor for organic amendment SFo based on different management options associated with residue management in rice production systems of the Philippines.†**

| Residue management practice scenarios | | | | |
|---|---|---|---|---|
| Retained | Burned | ROA | CFOA | SFo |
| % | | t ha$^{-1}$ of dry matter | dimensionless | |
| 100 | 0 | 4.50 | 1 | 2.73 |
| 75 | 25 | 3.38 | 1 | 2.39 |
| 50 | 50 | 2.25 | 1 | 2.00 |
| 25 | 75 | 1.13 | 1 | 1.56 |
| 0 | 100 | 0.00 | 1 | 1.0 |

† ROA is the rate of application of straw in dry matter; CFOA is a conversion factor for the type of organic amendment (i.e., straw in this analysis). SFo values apply to both dry and wet cropping seasons and to both irrigated and rainfed rice ecosystems.

**Table 2. Effects of changing residue management on greenhouse gas (GHG) emissions from rice cultivation in the Philippines, based on data for inventory year 2000. The current case is 75% burned and 25% retained in the field. The largest mitigation potential is to burn 100% of the residues according to this analysis.**

| Residue management | | Total residue retained | Total residue burned | Rice methane emission | Soil $N_2O$ emission from crop residue | GHG from residue burning | Total GHG emission |
|---|---|---|---|---|---|---|---|
| Retained | Burned | | | | | | |
| % | | t wet matter | | Gg $CO_2$ eq. | | | |
| 100 | 0 | 21,482,612 | 0 | 24,902 | 400 | 0 | 25,302 |
| 75 | 25 | 16,111,959 | 5370,653 | 21,804 | 300 | 322 | 22,426 |
| 50 | 50 | 10,741,306 | 10,741,306 | 18,243 | 200 | 644 | 19,086 |
| 25 | 75 | 5370,653 | 16,111,959 | 14,231 | 100 | 966 | 15,296 |
| 0 | 100 | 0 | 21,482,612 | 9124 | 0 | 1288 | 10,412 |

for local populations and black carbon has a direct radiative forcing effect in the atmosphere (Bond et al., 2013), which is not addressed in GHG inventories. These additional effects would need to be considered when evaluating a NAMA associated with rice residue management.

## South Africa: GHG Mitigation Potentials for Livestock Enteric Methane Emissions and Manure Management

Agricultural emissions, which include livestock and manure management emissions, were estimated to contribute 11.6% to national GHG emissions in South Africa during 1990 (van der Merwe and Scholes, 1998), but have declined to 9.3% in 1994 and 8.4% in 2000 (Department of Environmental Affairs and Tourism, 2009). Inventories with more detailed methodologies have been used to estimate enteric fermentation emissions, and this has led to increased emission estimates, changing from 916 Gg $CH_4$ in 1998 (van der Merwe and Scholes, 1998) to 1225 Gg $CH_4$ in 2004 (Department of Agriculture, Forestry and Fisheries, 2010) and 1171 Gg $CH_4$ (excluding game) in 2010 (du Toit et al., 2013b, 2013c, 2013d). Regardless of the trend, all national inventories have shown enteric fermentation to be one of the top five key emission sources in South Africa. For this reason, South Africa is investigating possible mitigation options for emissions from cattle production. The ALU software was applied in this study to evaluate two options. The first option was the effect of various feed additives on the total enteric fermentation emissions, and the second option was the effect of changing manure management systems. This is not intended to be an exhaustive analysis of options for mitigation of livestock emissions, but rather an initial case study of two common mitigation options.

Livestock are raised throughout South Africa, with numbers and species varying according to environmental conditions and production systems, which include commercial, emerging, and communal. Livestock production occurs on an estimated 38,500 commercial farms and a further 2 million emerging and communal farms, occupying 80% of the agricultural land (Meissner et al., 2013; see du Toit et al., 2013a, 2013b, 2013c, 2013d for regional estimates). Intensive production systems tend to be near metropolitan markets and feed suppliers, but the majority of farms are operated on extensive production systems. Manure management is practiced in intensive production systems and mainly with non-ruminants such as swine and poultry.

Livestock population data can be compiled using a basic or enhanced characterization (IPCC, 2000, 2006). The basic characterization includes total population estimates for the livestock categories, while the enhanced characterization further disaggregates livestock populations by production groups and phases, such as lactating or non-lactating phases; breeder and grower systems; and wool or meat production systems, in addition to gender and age structure to better capture country-specific livestock management conditions. Consequently, the enhanced characterization facilitates the application of Tier 2 methods. A basic characterization was used for activity data on buffalo (which represent privately owned game), goats, horses, mules and asses, and sheep (du Toit et al., 2013a, 2013c, 2013d; Department of Agriculture, Forestry and Fisheries, 2012). These livestock categories are not major contributors to GHG emissions in South Africa. An enhanced characterization was developed for the dairy cows, non-dairy cattle, and swine (Department of Agriculture, Forestry and Fisheries, 2012; du Toit et al., 2013a, 2013c, 2013d). Dairy cows constitute < 2% of the population, while non-dairy cattle constitute about 17%. All non-dairy cattle graze in large areas in South Africa, while only the total mixed ration (TMR) (i.e., lactating cows and lactating heifers) and feedlot cattle use stalls.

The IPCC default emission factors (i.e., Tier 1) were used for the livestock categories with a basic categorization, and the factors were automatically assigned in Module II of the ALU software. For dairy cows and non-dairy cattle, which had an enhanced characterization, the emission factors were calculated in the ALU software from country-specific data on live weights, weight gain, milk production, digestible energy, lactation, and other characteristics (IPCC, 2000, 2006; data from du Toit et al., 2013b).

## Enteric Methane Emissions

Extensive beef production is the major contributor to enteric $CH_4$ emissions in South Africa. The extensive nature of South African production systems make the application of short-term mitigation options challenging. Nutritional management strategies have the greatest short-term impact, particularly methods for inhibiting methanogens, including feed additives such as fats and oils, organic acids, essential oils, and secondary plant metabolites such as tannins, saponins, and ionophores; altering the concentrate to roughage ratio; and increasing the amount of rumen resistant starch in the diet (Smith et al., 2008). The additives reduce the methane conversion rate in the livestock rumen, which is designated as Ym in the IPCC method (IPCC, 2006).

Using the practice-based approach, the ALU software provides functionality to modify the percentage of the livestock population receiving feed additives (namely monensin, which is an antibiotic, edible oils, or both additives). Several mitigation scenarios were selected and evaluated in the software (Table 3). Most of the options that were investigated addressed additives given to feedlot cattle in intensive production systems. Incorporating feed additives into the diet of livestock is unlikely to be feasible in the extensive grazing systems that rely mostly on forage grasses. The scenarios for options 1 through to 7 are based on increasing feedlot cattle numbers over the last 10 yr that were projected into the future using the ALU software. The increasing numbers of livestock may not occur or at least may decline in the future, as feedlot cattle numbers stabilize under market forces. Consequently, additional mitigation scenarios (options 8 to 10, Table 2) were considered under the assumption that feedlot cattle populations had reached their maximum by the end of the inventory in 2010.

The largest GHG mitigation potential among the scenarios is about 230 Gg $CO_2$ eq. $yr^{-1}$ for the small population of livestock that can be feasibly given the additives in their rations (Fig. 1). The ALU software indicates that the $CH_4$ mitigation potential increases as more additives are provided to the livestock, and the mitigation potential over the long-term period (to 2050) increases most sharply when all feedlot cattle receive both monensin and edible oils (option 3). However, the software shows that mitigation potential is not changing much over the time series for some options, such as providing half of the commercial lactating dairy cow and heifer population with edible oils (option 4).

If it is assumed that feedlot cattle population numbers have stabilized by 2010, providing monensin and edible oils as additives on all feedlots (option 9) will lead to a mitigation potential of 86.2 Gg $CO_2$ eq. $yr^{-1}$. In addition, including additives in the diets of 50% of TMR and 25% of growing commercial beef cattle (option 10) provides the highest mitigation potential, which increases slightly from 180.7 to 182.9 Gg $CO_2$ eq. $yr^{-1}$ from 2025 to 2050, respectively.

**Table 3. Selected enteric fermentation mitigation options for South Africa that were analyzed with the Agriculture and Land Use National Greenhouse Gas Inventory software.†**

| Mitigation scenario | Feed additives | Reason for selecting mitigation option |
|---|---|---|
| Option 1 | 100% of feedlot cattle receive monensin | All feedlot cattle already receive monensin in their diet; however, this has not been taken into account in any of the previous inventories. |
| Option 2 | 50% of feedlot cattle receive monensin;<br>50% receive both monensin and edible oils | It is feasible to include edible oils into the diet of feedlot cattle. |
| Option 3 | 100% of feedlot cattle receive both monensin and edible oils | |
| Option 4 | 50% of TMR lactating cows and heifers receive edible oils | Edible oils can also be used in the TMR ration of lactating dairy cattle. There is some concern that the oils may affect the quality of the milk, so this option was limited to <50% of the population. |
| Option 5 | 25% of growing commercial beef cattle receive monensin | Farmers could include monensin in supplements provided to grazing beef cattle. This is, however, not widely practiced in South Africa, so this option was limited to <25% of the growing population. |
| Option 6 = option 2 + option 4 | 50% of feedlot cattle receive monensin;<br>50% receive both monensin and edible oils;<br>50% of TMR lactating cows and heifers receive edible oils | See options 2 and 4 above. |
| Option 7 = option 2 + option 4 + option 6 | 50% of feedlot cattle receive monensin;<br>50% receive both monensin and edible oils;<br>50% of TMR lactating cows and heifers receive edible oils;<br>25% of growing commercial beef cattle receive monensin | See options 2, 4, and 6. |
| Option 8 | Option 2 but with a constant feedlot population | In these options it was assumed that there was no increase in the feedlot population number since 2001. |
| Option 9 | Option 3 but with a constant feedlot population | |
| Option 10 | Option 7 but with a constant feedlot population | |

† TMR, total mixed ration.

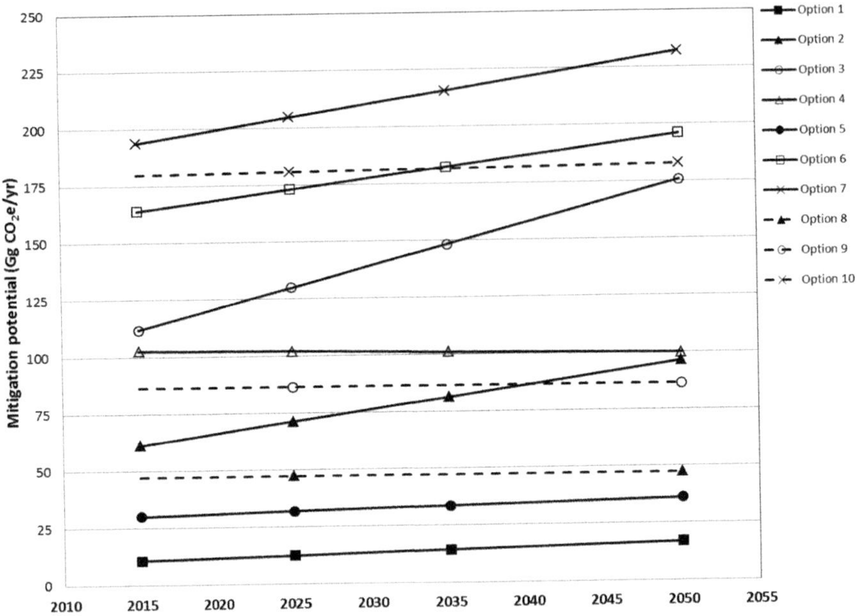


**Fig. 1. Greenhouse gas mitigation potential (Gg $CO_2$ eq. $yr^{-1}$) for reducing $CH_4$ emissions from enteric fermentation between 2015 and 2050 using the mitigation scenarios listed in Table 3.**

## Manure Management

All manure from non-dairy cattle, buffalo (privately owned game), sheep, goats, horses, mules, and asses is left on the field or in the paddock (i.e., pasture/range/paddock) and therefore is not managed (du Toit et al., 2013a, 2013b, 2013c). We assumed that there were limited opportunities for mitigating GHG emissions for this portion of the manure. Similarly, most of the manure from lactating dairy cows is deposited directly onto pasture/range/paddock and is not managed (90% for pasture cows and 88.5% for TMR cows). However, a small percentage of the manure from dairy cows is collected in milking facilities and managed in anaerobic lagoons (3%) or by daily spread (7%). For TMR cows, 10% of manure is managed in anaerobic lagoons and 1.5% as daily spread and liquid/slurry. In contrast, about 92% of swine manure is managed in anaerobic lagoons, 1.5% in daily spread and also liquid/slurry, and 5% as dry lot. Poultry manure is managed in dry lot systems (80%) and intensive compost (20%).

The amount of GHG emissions from manure varies considerably and depends on the management system used as well as the climate, technical and infrastructure resources, and manure management practices (Smith et al., 2007). Because of the South African climate and the nature of extensive production systems, only a relatively small amount of $CH_4$ and $N_2O$ are emitted from production systems where the manure is managed. Active manure management is limited to intensive production systems such as dairies, piggeries, and poultry production systems. The goal of strategies to reduce $CH_4$ and $N_2O$ from manure is to change manure properties or the conditions under which the manure is stored or treated (Sejian and Naqvi, 2012).

Biogas technology is limited in South Africa, but it does have some potential as a mitigation option (Boyd, 2012). Small-scale farm digesters can be utilized in South African emerging and communal farming systems for electricity as well as heat production and are an effective strategy for utilizing and thus reducing $CH_4$ emissions from manure (Sejian and Naqvi, 2012). Commercial dairy and pork producers both utilize liquid-based systems in which $CH_4$ is recovered in the digester. Methane recovery technologies include covered anaerobic digesters, complete mix digesters, and plug flow digesters.

The effectiveness of digesters for reducing GHG emissions from manure was investigated using the practice-based approach in the ALU software, with several scenarios that are considered feasible for South Africa (Table 4). Assuming no population increase from current livestock numbers, the mitigation potential for manure management could be as large as 234 Gg $CO_2$ eq. $yr^{-1}$ (Fig. 2) for the manure from South African livestock production systems that have liquid manure management. The highest mitigation potential would be achieved where 75% of anaerobic lagoon manure went to a digester that had an efficiency of 85% (option 4). Furthermore, data show that a 10% increase in digester efficiency leads to a 13% increase in the mitigation potential, while a 10% increase in the amount

**Table 4. Selected manure management mitigation options in South Africa that were analyzed with the Agriculture and Land Use National Greenhouse Gas Inventory software. †**

| Mitigation scenario | Manure management | Reason for selection |
|---|---|---|
| Option 1 | 50% of the manure in anaerobic lagoons (TMR cattle, feedlot cattle, swine) goes to a digester; digester has an efficiency of 75% | These two options test the impact of installing digesters. |
| Option 2 | 75% of the manure in anaerobic lagoons (TMR cattle, feedlot cattle, swine) goes to a digester; digester has an efficiency of 75% | |
| Option 3 | 50% of the manure in anaerobic lagoons (TMR cattle, feedlot cattle, swine) goes to a digester; digester has an efficiency of 85% | These two options test the impact of improved digester efficiency. |
| Option 4 | 75% of the manure in anaerobic lagoons (TMR cattle, feedlot cattle, swine) goes to a digester; digester has an efficiency of 85% | |

† TMR, total mixed ration.

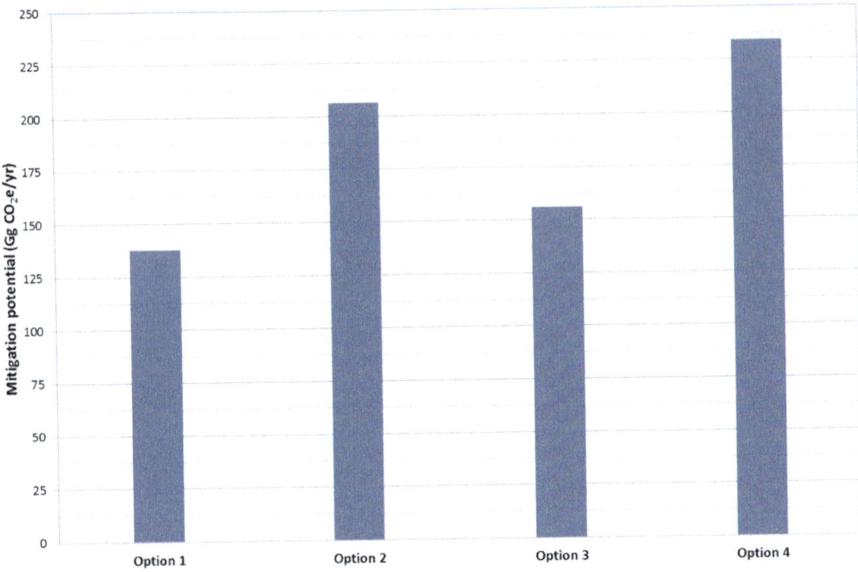


**Fig. 2. Annual greenhouse gas mitigation potential (Gg $CO_2$ eq. $yr^{-1}$) from 2015 to 2050 for manure management mitigation options listed in Table 4.**

of anaerobic lagoon manure that is directed through a digester leads to a 20% increase in the mitigation potential. Consequently, more effort should be spent on increasing the amount of manure going to a digester, rather than improving digester efficiency if resources are limited for implementing this potential practice in a mitigation program.

## Conclusions

The ALU software implements international standards for estimating GHG emissions associated with agriculture and forestry, which have been developed by the IPCC (1997, 2000, 2003, 2006). The software includes functionality to facilitate application of IPCC good practice and TACCC principles, in addition to functionality for estimating GHG mitigation potentials using the historical inventory data as the baseline. Assessing mitigation options and their potential for reducing GHG emissions provides essential information to determine which management practices could serve as NAMAs.

For the Philippines, an analysis of rice straw with the ALU software revealed that burning all rice residue may be the best option for maintaining low GHG emissions. While burning all rice straw is the scenario with the lowest GHG emissions, this practice may not be a NAMA because retaining some residue in the field is important for longer-term sustainability of soil, improving soil fertility and crop production. Burning residues may also create health problems for local populations; reduce soil C stocks (Paustian et al., 1997); and result in more black C in the atmosphere, which has a direct radiative forcing effect on the climate (Bond et al., 2013). All of these effects would need to be considered to determine if the practice is a NAMA for the Philippines. Other management options may also be explored in the future to further evaluate possible NAMAs for rice production. For example, some studies suggest that composting rice straw and returning the compost to the field could minimize $CH_4$ emissions from rice fields with co-benefits of increased soil fertility and crop productivity (Khosa et al., 2010; Wassmann et al., 2000; Shin et al., 1996).

In South Africa, the mitigation potential for reducing enteric $CH_4$ emissions with feed additives in addition to $CH_4$ from manure management with digesters is relatively modest, but could be part of a larger portfolio of NAMAs. Moreover, this analysis highlights the need to further investigate strategies for reducing emissions from enteric fermentation and possibly manure management to meet GHG mitigation goals. Other strategies for reducing emissions from enteric fermentation can be investigated, particularly in extensive production systems, such as direct selection for more adapted animals, residual or net feed intake, selection and breeding of improved forages, incorporating legumes, improvement of natural rangeland and soil fertility, and the implementation of improved production systems (Scholtz et al., 2012).

Future planning and policy development can build on these analyses using the ALU software system. Other scenarios and mitigation options can be tested in the ALU software to inform policy as decision makers develop NAMAs. In addition, uncertainty will need to be quantified in the emission estimates and projections, but this information was not available for the activity data associated with the case studies. Unfortunately, it is too often the case that governments compile statistics without providing information on the uncertainty for the data. To meet the reporting requirements for GHG inventories, governments will need to advance their surveys and other data collection programs so that the uncertainty is provided. Regardless, the ongoing assessments highlight the GHG mitigation potentials, which when combined with other information such as sustainability of a practice, economic viability, and its implications for food security can provide the necessary information for governments to further develop climate change policy programs in their countries.

## Acknowledgments

We appreciate the funding support provided by the US Environmental Protection Agency (contract EP-W-13-005) in preparation of the manuscript and development of the Agriculture and Land Use National Greenhouse Gas Inventory Software.

# References

Beddington, J.R., M. Asaduzzaman, M.E. Clark, A. Fernández Bremauntz, M.D. Guillou, D.J.B. Howlett, M.M. Jahn, E. Lin, T. Mamo, C. Negra, C.A. Nobre, R.J. Scholes, N. Van Bo, and J. Wakhungu. 2012. What next for agriculture after Durban? Science 335:289–290. doi:10.1126/science.1217941

Bernoux, M., G. Branca, A. Carro, L. Lipper, G. Smith, and L. Bockel. 2010. Ex-ante greenhouse gas balance of agriculture and forestry development programs. Sci. Agric. 67:31–40. doi:10.1590/S0103-90162010000100005

Bond, T.C., S.J. Doherty, D.W. Fahey, P.M. Forster, T. Berntsen, B.J. DeAngelo, M.G. Flanner, S. Ghan, B. Kärcher, D. Koch, S. Kinne, Y. Kondo, P.K. Quinn, M.C. Sarofim, M.G. Schultz, M. Schulz, C. Venkataraman, H. Zhang, S. Zhang, N. Bellouin, S.K. Guttikunda, P.K. Hopke, M.Z. Jacobson, J.W. Kaiser, Z. Klimont, U. Lohmann, J.P. Schwarz, D. Shindell, T. Storelvmo, S.G. Warren, and C.S. Zender. 2013. Bounding the role of black carbon in the climate system: A scientific assessment. J. Geophys. Res. Atmos. 118:5380–5552. doi:10.1002/jgrd.50171

Boyd, A. 2012. Informing international UNFCCC technology mechanisms from the ground up: Using biogas technology in South Africa as a case study to evaluate the usefulness of potential elements of an international technology agreement in the UNFCCC negotiations process. Energy Policy 51:301–311. doi:10.1016/j.enpol.2012.08.020

Colomb, V., L. Bockel, J.-L. Chotte, S. Martin, O. Touchemoulin, M. Tinlot, and M. Bernoux. 2013. Selection of appropriate calculators for landscape-scale greenhouse gas assessment for agriculture and forestry. Environ. Res. Lett. 8:015029. doi:10.1088/1748-9326/8/1/015029

Del Grosso, S.J., S.M. Ogle, and W.J. Parton. 2011. Soil organic matter cycling and greenhouse gas accounting methodologies. In: L. Guo, A. Gunasekara, and L. McConnell, editors, Understanding greenhouse gas emissions from agricultural management. Chap 1. American Chemical Society, Washington, DC. p. 3–13.

Department of Agriculture, Forestry and Fisheries. 2010. The South African agricultural GHG inventory for 2004. Department of Agriculture, Forestry and Fisheries, Pretoria, South Africa.

Department of Agriculture, Forestry and Fisheries. 2012. Abstract of agricultural statistics. 2012. Department of Agriculture, Forestry and Fisheries, Pretoria, South Africa. http://www.nda.agric.za/docs/statsinfo/Ab2012.pdf (accessed 20 Aug. 2013).

Department of Environmental Affairs and Tourism. 2009. Greenhouse gas inventory South Africa: 1990–2000. National inventory report. Department of Environmental Affairs and Tourism, Pretoria, South Africa.

Dobermann, A., and T.H. Fairhurst. 2002. Rice straw management. Better Crops Int. 16(spec. suppl.)7–11.

du Toit, C.J.L., H.H. Meissner, and W.A. van Niekerk. 2013b. Direct greenhouse gas emissions of the game industry in South Africa. S. Afr. J. Anim. Sci. 43:367–393.

du Toit, C.J.L., H.H. Meissner, and W.A. van Niekerk. 2013c. Direct methane and nitrous oxide emissions of South African dairy and beef cattle. S. Afr. J. Anim. Sci. 43:320–339. doi:10.4314/sajas.v43i3.7

du Toit, C.J.L., W.A. van Niekerk, and H.H. Meissner. 2013a. Direct methane and nitrous oxide emissions of monogastric livestock in South Africa. S. Afr. J. Anim. Sci. 43:362–375. doi:10.4314/sajas.v43i3.9

du Toit, C.J.L., W.A. van Niekerk, and H.H. Meissner. 2013d. Direct greenhouse gas emissions of South African small stock sectors. S. Afr. J. Anim. Sci. 43:340–361. doi:10.4314/sajas.v43i3.8

Godfray, H.C.J., J. Pretty, S.M. Thomas, E.J. Warham, and J.R. Beddington. 2011. Linking policy on climate and food. Science 331:1013–1014. doi:10.1126/science.1202899

IPCC. 1997. Revised 1996 IPCC guidelines for national greenhouse gas inventories: Reference manual. Vol. 3. Intergovernmental Panel on Climate Change, Bracknell, UK.

IPCC. 2000. IPCC good practice guidance and uncertainty management in national greenhouse gas inventories. Institute for Global Environmental Strategies, Hayama, Japan.

IPCC. 2003. Good practice guidance for land use, land-use change and forestry. Institute for Global Environmental Strategies, Hayama, Kanagawa, Japan.

IPCC. 2006. 2006 IPCC guidelines for national greenhouse gas inventories. Institute for Global Environmental Strategies, Hayama, Japan.

IPCC. 2010. Datasets for use in the IPCC Guidelines. In: H.S. Eggleston, N. Srivastava, K. Tanabe, and J. Baasansuren, editors, Meeting Report of the IPCC–FAO–IFAD Expert Meeting on FAO Data for LULUCF/AFOLU Rome, Italy, 20–22 October, 2009. Institute for Global Environmental Strategies, Hayama, Japan.

Khosa, M.K., B.S. Sidhu, and D.K. Benbi. 2010. Effect of organic materials and rice cultivars on methane emission from rice field. J. Environ. Biol. 31:277–280.

Lokupitiya, E., and K. Paustian. 2006. Agricultural soil greenhouse gas emissions: A review of national inventory methods. J. Environ. Qual. 35(4):1413–1427.

Maia, S.M.F., S.M. Ogle, C.E.P. Cerri, and C.C. Cerri. 2010. Soil organic carbon stock change due to land use activity along the agricultural frontier of the southwestern Amazon, Brazil between 1970 and 2002. Global Change Biol. 16:2775–2788. doi:10.1111/j.1365-2486.2009.02105.x

Meissner, H.H., M.M. Scholtz, and H.C. Schonfeldt. 2013. The status, socio-economic and environmental impact, and challenges of livestock agriculture in South Africa. http://www.rmrdsa.co.za/ProjectDownloads/FocusArea8Climateeffectadaptation.aspx (accessed 20 Aug. 2013).

Neue, H.U., and R.L. Sass. 1994. Rice cultivation and trace gas exchange. In: R.G. Prinn, editor, Global atmospheric-biospheric chemistry. Plenum Press, New York. p. 119–147.

Ogle, S.M., F.J. Breidt, M. Easter, S. Williams, K. Killian, and K. Paustian. 2010. Scale and uncertainty in modeled soil organic carbon stock changes for US croplands using a process-based model. Global Change Biol. 16:810–820. doi:10.1111/j.1365-2486.2009.01951.x

Ogle, S.M., L. Buendia, K. Butterbach-Bahl, F.J. Breidt, M. Hartman, K. Yagi, R. Nayamuth, S. Spencer, T. Wirth, and P. Smith. 2013. Advancing national greenhouse gas inventories for agriculture in developing countries: Improving activity data, emission factors and software technology. Environ. Res. Lett. 8:015030. doi:10.1088/1748-9326/8/1/015030

Ogle, S.M., L. Olander, L. Wollenberg, T. Rosenstock, F. Tubiello, K. Paustian, L. Buendia, A. Nihart, and P. Smith. 2014. Reducing agricultural greenhouse gas emissions in developing countries: Providing the basis for action. Global Change Biol. 20:1–6. doi:10.1111/gcb.12361

Paustian, K., H.P. Collins, and E.A. Paul. 1997. Management controls on soil carbon. In: E.A. Paul, K. Paustian, and C.V. Cole, editors, Soil organic matter in temperate agroecosystems: Long-term experiments in North America. CRC Press, Boca Raton, FL. p. 15–49.

Philippine Rice Research Institute. 1997. Characteristics of popular Philippine rice varieties. Rice Technol. Bull. 19. Department of Agriculture, Philippine Rice Research Institute, Muñoz, Phillippines.

Scholtz, M.M., Y. Steyn, E. van Marle-Koster, and H.E. Theron. 2012. Improved production efficiency in cattle to reduce their carbon footprint for beef production. S. Afr. J. Anim. Sci. 42:450–453. doi:10.4314/sajas.v42i3.9

Sejian, V., and S.M.K. Naqvi. 2012. Livestock and climate change: Mitigation strategies to reduce methane production. In: G. Liu, editor, Greenhouse gases—capturing, utilizing and reduction. In Tech, Rijeka, Croatia, p. 255–276.

Shin, Y.-K., S.-H. Yun, M.-E. Park, and B.-L. Lee. 1996. Mitigation options for methane emission from rice fields in Korea. Ambio 25:289–291.

Smith, P., H. Haberl, A. Popp, K. Erb, C. Lauk, R. Harper, F.N. Tubiello, A. de Siqueira Pinto, M. Jafari, S. Sohi, O. Masera, H. Böttcher, G. Berndes, M. Bustamante, H. Ahammad, H. Clark, H. Dong, E.A. Elsiddig, C. Mbow, N.H. Ravindranath, C.W. Rice, C. Robledo Abad, A. Romanovskaya, F. Sperling, M. Herrero, J.I. House, and S. Rose. 2013. How much land-based greenhouse gas mitigation can be achieved without compromising food security and environmental goals? Global Change Biol. 19(8):2285–2302.

Smith, P., D. Martino, Z. Cai, D. Gwary, H. Janzen, P. Kumar, B. Mccarl, S. Ogle, F. O'Mara, C. Rice, B. Scholes, O. Sirotenko, M. Howden, T. Mcallister, G. Pan, V. Romanenkov, U. Schneider, and S. Towprayoon. 2007. Policy and technological constraints to implementation of greenhouse gas mitigation options in agriculture. Agric. Ecosyst. Environ. 118:6–28. doi:10.1016/j.agee.2006.06.006

Smith, P., D. Martino, Z. Cai, D. Gwary, H. Janzen, P. Kumar, B. McCarl, S. Ogle, F. O'Mara, C. Rice, B. Scholes, O. Sirotenko, M. Howden, T. McAllister, G. Pan, V. Romanenkov, U. Schneider, S. Towprayoon, M. Wattenbach, and J. Smith. 2008. Greenhouse gas mitigation in agriculture. Philos. Trans. R. Soc. Lond. B Biol. Sci. 363:789–813. doi:10.1098/rstb.2007.2184

Spencer, S., S.M. Ogle, F.J. Breidt, J. Goebel, and K. Paustian. 2011. Designing a national soil carbon monitoring network to support climate change policy: A case example for US agricultural lands. Greenhouse Gas Meas. Manage. 1:167–178. doi:10.1080/20430779.2011.637696

Takai, Y. 1970. The mechanism of methane fermentation in flooded paddy soil. Soil Sci. Plant Nutr. 16:238–244.

Tubiello, F.N., M. Salvatore, S. Rossi, A. Ferrara, N. Fitton, and P. Smith. 2013. The FAOSTAT GHG Database. Environ. Res. Lett. 8:015009. doi:10.1088/1748-9326/8/1/015009

van der Merwe, M., and R.J. Scholes. 1998. Greenhouse gas emissions inventory of 1990 for South Africa, Phase 1. Council for Scientific and Industrial Research, Pretoria, South Africa.

Wassmann, R., R.S. Lantin, H.U. Neue, L.V. Buendia, T.M. Corton, and Y. Lu. 2000. Characterization of methane emission from rice fields in Asia III. Mitigation options and future research needs. Nutr. Cycling Agroecosyst. 58:23–36. doi:10.1023/A:1009874014903

# Parameter-Induced Uncertainty Quantification of Regional $N_2O$ Emissions and $NO_3$ Leaching using the Biogeochemical Model LandscapeDNDC

**Steffen Klatt, David Kraus, Karl-Heinz Rahn, Christian Werner, Ralf Kiese, Klaus Butterbach-Bahl, and Edwin Haas***

Abstract

Numerical simulation models are increasingly used to estimate greenhouse gas emissions at site to regional and national scales and are outlined as the most advanced methodology (Tier 3) for national emission inventories in the framework of United Nations Framework Convention on Climate Change reporting. Process-based models incorporate the major processes of carbon and nitrogen cycles of terrestrial ecosystems like arable land and grasslands and are thought to be widely applicable at various spatial and temporal scales. The high complexity of ecosystem processes mirrored by such models requires a large number of model parameters. Many of those parameters are lumped parameters, describing simultaneously the effect of environmental drivers on, for example, microbial community activity and individual processes. Thus, the precise quantification of true parameter states is often difficult or even impossible. As a result model uncertainty is not solely originating from input uncertainty but also subject to parameter-induced uncertainty.

Abbreviations: BC, Bayesian calibration; EF, emission factor; GHG, greenhouse gas; HPC, high-performance computing; LHS, Latin hypercube sampling; MC, Monte Carlo; SOC, soil organic C.

Steffen Klatt (steffen.klatt@kit.edu), David Kraus (david.kraus@kit.edu), Karl-Heinz Rahn (rahn.karlheinz@yahoo.de), Christian Werner (christian.werner@senkenberg.de), Ralf Kiese (ralf.kiese@kit.edu), Klaus Butterbach-Bahl (klaus.butterbach-bahl@kit.edu), Edwin Haas, Karlsruhe Institute of Technology, Institute of Meteorology and Climate Research (IMK-IFU), Kreuzeckbahnstr. 19, 82467 Garmisch-Partenkirchen, Germany. *Corresponding author (Edwin.haas@kit.edu).

doi:10.2134/advagricsystmodel6.2013.0001


Synthesis and Modeling of Greenhouse Gas Emissions and Carbon Storage in Agricultural and Forest Systems to Guide Mitigation and Adaptation
S. Del Grosso, L. Ahuja, and W. Parton, Editors
Advances in Agricultural Systems Modeling, Volume 6. Lajpat R. Ahuja, Series Editor

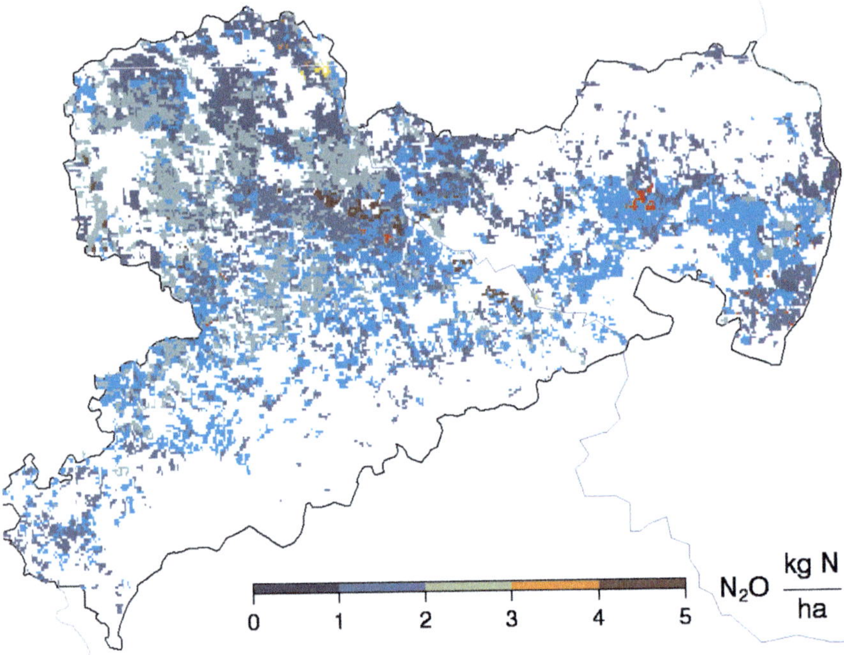

0
1
2
3
4
5
$N_2O$ $\frac{kg\ N}{ha}$


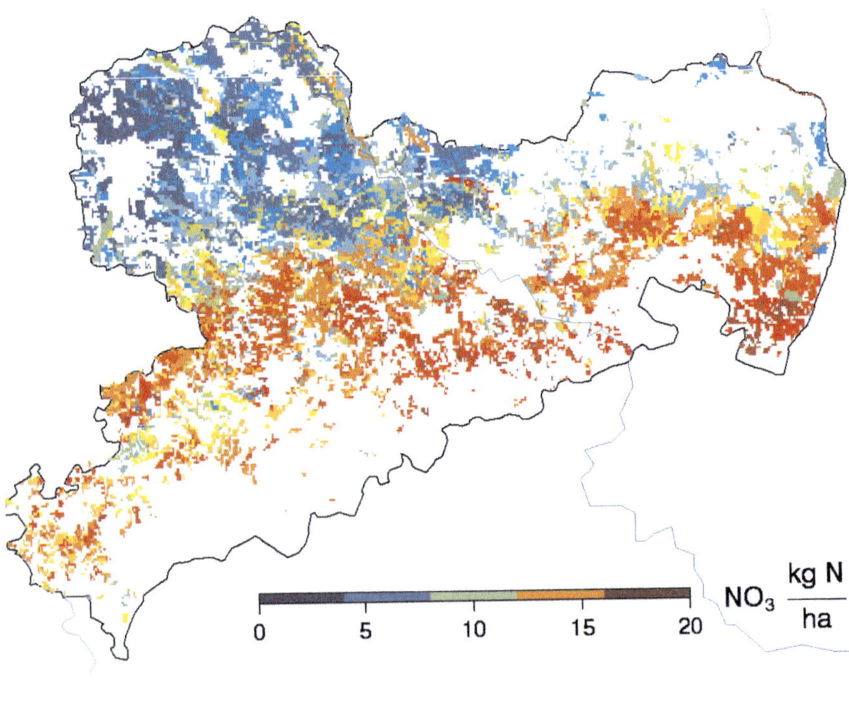

0
5
10
15
20
$NO_3$ $\frac{kg\ N}{ha}$

In this study we quantify regional parameter-induced model uncertainty on $N_2O$ emissions and nitrate ($NO_3$) leaching from arable soils of Saxony, Germany, using the biogeochemical model LandscapeDNDC. For this we calculated a regional inventory using a joint parameter distribution for key parameters describing microbial C and N turnover processes obtained by a Bayesian calibration (BC) study. We representatively sampled 400 different parameter vectors from the discrete joint parameter distribution comprising approximately 400,000 parameter combinations and used these to calculate 400 individual realizations of the regional inventory. The spatial domain (represented by 4042 polygons) is set up with spatially explicit soil and climate information and a region-typical 3-yr crop rotation consisting of winter wheat (*Triticum aestivum* L., canola [*Brassica juncea* (L.) Czern. subsp. *juncea*], and winter barley (*Hordeum vulgare* L.).

Average $N_2O$ emission from arable soils in the state of Saxony across all 400 realizations was 1.43 ± 1.25 kg N (ha yr)$^{-1}$ with a median value of 1.05 kg N (ha yr)$^{-1}$. The default IPCC emission factor (EF) approach (Tier 1) for direct emissions revealed a higher average $N_2O$ emission of 1.51 kg N (ha yr)$^{-1}$ due to fertilizer use. In the regional uncertainty quantification, the 20% likelihood range for $N_2O$ emissions was 0.79 to 1.37 kg N (ha yr)$^{-1}$ [50% likelihood: 0.46–2.05 kg N (ha yr)$^{-1}$; 90% likelihood: 0.11–4.03 kg N (ha yr)$^{-1}$]. Respective quantities were calculated for nitrate leaching where the IPCC EF methodology revealed higher average leaching rates [45.4 kg N (ha yr)$^{-1}$] than the ones simulated by LandscapeDNDC [29.32 ± 9.97 kg N (ha yr)$^{-1}$].

The method has proven its applicability to quantify parameter-induced uncertainty of simulated regional greenhouse gas emission and nitrate leaching inventories using process based biogeochemical models.

Quantifying source and sink strengths of soils by field measurements still remains a labor- and cost-intensive task. Due to the complicated measuring techniques, such field measurements can currently only be conducted at site scale and with less detail on larger scales (Groffman et al., 2006; Sutton et al., 2007). For that reason, sufficient measurement data with high temporal resolution and true representation of, e.g., a landscape, is very likely to be unavailable in the foreseeable future. This is most unfortunate taking into account the high spatial variability of greenhouse gas (GHG) emissions from soils of terrestrial ecosystems (Butterbach-Bahl et al., 2004). As a result, current practices for upscaling GHG emissions from site to regional or national scale, such as the IPCC emission factor approach

(IPCC, 2006), are prone to high uncertainty and as such fail to account for spatial variation of GHG emissions driven by variability of soil and climatic conditions.

Consequently, during the last decades a notable amount of effort was spent on developing biogeochemical simulation models capable of predicting GHG fluxes at different scales (Bell et al., 2012; Chirinda et al., 2010; Del Grosso et al., 2006; Gabrielle et al., 2006; Li, 2000). While first applications were conducted predominantly at site scale, biogeochemical models are increasingly used to simulate fluxes at landscape and regional scale (Del Grosso et al., 2006; Haas et al., 2013; Li et al., 2004; Werner et al., 2007; Zaehle et al., 2005). For such applications model input and driver data are generally obtained from GIS databases holding spatially explicit information. Model based regional and national inventories of soil derived GHG emissions, defined by IPCC as Tier 3, have recently been calculated for the United States, China, and Europe (Del Grosso et al., 2006; Leip et al., 2011; Qiu et al., 2009; Rochette et al., 2008; Schulze et al., 2010). However, for any simulation study to be admissible, the uncertainties of simulation results need to be included (IPCC, 2006).

Each computer model simulation will comprise different sources of uncertainty: input data, model parameters, and model structure (Refsgaard et al., 2007). Here, model inputs describe the simulation domain characteristics (e.g., soil and vegetation properties) and the model's driving forces (e.g., climate and agricultural management). Model parameters (e.g., potential decomposition rates) control the model processes and are assumed to be constant across domains. In contrast to inputs, parameters are often difficult to exactly quantify. When performing validation studies, or model calibration, measurement errors of the target entity have also to be taken into account. Input-induced uncertainty is the result of different factors, such as insufficient, inaccurate, or unrepresentative input data. Uncertainties induced by input data were previously investigated in GHG emission inventory calculations (Haas et al., 2013; Kiese et al., 2005; Li et al., 2004; Werner et al., 2007).

Parameter-induced uncertainty is caused by the imperfect knowledge, and thus uncertainty, in assigning values to parameters used in the description of biogeochemical processes, such as plant nutrient uptake, mineralization, nitrification, and denitrification (van Oijen et al., 2005, 2011). The imperfect knowledge results from measurement uncertainties and the usage of nonmeasureable empirical parameters. Using more complex models may not necessarily result in uncertainty reduction, especially taking into account the still obvious lack of understanding processes and controls of GHG emissions from soils (Butterbach-Bahl et al., 2013). More complex models usually include more parameters and often require more detailed input data, which is often beyond what has been recorded in most field and laboratory experiments.

Using the methodology proposed by van Oijen et al. (2005) and Kennedy and O'Hagan (2001) several researchers have quantified parameter-induced uncertainty for biogeochemical models at site scale (Klemedtsson et al., 2008; Knorr and Kattge 2005; Lehuger et al., 2009; Rahn et al., 2012; Svensson et al., 2008).

Furthermore, the level of detail in the description of biogeochemical processes involved in GHG production, consumption, and emission in models varies from rather simple (e.g., using empirical functions for moisture/temperature dependency of $N_2O$ emissions) to highly complex (e.g., considering explicitly soil microbial processes and microbial communities). As such, each model implementation is a different approximation of the reality, which is referred to as model structural uncertainty. The magnitude of such uncertainties can be assessed by applying different models using a single set of input followed by a comparison of simulation results for the individual models versus measurements (Cameron et al., 2013; Gibbons et al., 2008; Stange et al., 2000; van Oijen et al., 2011).

So far, attempts to quantify uncertainty of regional inventories calculated with biogeochemical simulation models were in fact sensitivity analyses performed on model inputs (Haas et al., 2013; Zaehle et al., 2005) and parameters (Haas et al., 2013; Zaehle et al., 2005). A consistent approach to assess parameter-induced uncertainty on the regional scale that builds on Bayesian inference techniques has only been done by Zhou et al. (2012) and in this study. While Zhou et al. (2012) investigated carbon residence times in forest ecosystems across the United States, we quantify the uncertainty of regional $N_2O$ emissions and nitrate leaching from arable systems in the state of Saxony, Germany, calculated using LandscapeDNDC (Haas et al., 2013). Del Grosso et al. (2010) demonstrated a mixed effect modeling analysis approach to estimate model structural and parameterization uncertainties combined with Monte Carlo simulations to additionally address input-induced uncertainties of $N_2O$ emissions on the national scale.

## Materials and Methods

### Model Description

LandscapeDNDC is a complex process-based biogeochemical model, which simulates plant growth, soil C and N cycling and the associated biosphere–atmosphere exchange of GHG ($N_2O$, $CO_2$, and $CH_4$), as well as soil hydrology, including nitrate leaching, on the basis of plant physiological, microbial and physicochemical processes. It combines and unifies functionality of the agricultural-DNDC model (Li et al., 2000) and the Forest-DNDC models (Kiese et al., 2011; Li et al., 2000; Stange et al., 2000). In contrast to the original DNDC, process parameter values in LandscapeDNDC are read as input supporting a straightforward way to employ different parameter sets.

LandscapeDNDC allows application at site and regional scales, the latter by driving single realizations of the model in each simulation unit of the discretized domain. It is designed to process large numbers of sites in parallel; that is, during a regional simulation, all sites are integrated synchronously forward in time. Regional simulations are perfectly suited for parallelization and can be deployed on various high-performance computing (HPC) systems. In the presented context, an efficient computation is essential since uncertainty quantification generally requires a substantial number of simulations to obtain meaningful results (Haas et al., 2013).

## Bayesian Model Uncertainty Quantification

Bayesian calibration is a well established method to quantify uncertainty induced by model parameters and has been used frequently in recent years in the realm of biogeochemical process-based modeling (Klemedtsson et al., 2008; Lehuger et al., 2009; van Oijen and Thomson 2010; Rahn et al., 2012; Svensson et al., 2008; van Oijen et al., 2005, 2011; Yeluripati et al., 2009).

By means of Bayes' theorem a relation between the likelihood, which quantifies the probability of the measurement data $D$ given a parameter $\theta$ and the prior distribution prob($\theta$) and posterior distributions prob($\theta | D$) of the parameter, can be derived:

$$\text{prob}(\theta | D) \propto \text{prob}(D | \theta)\ \text{prob}(\theta)$$

The prior distribution represents confirmed knowledge about the parameters, which is either obtained from previously conducted BC studies or estimated through expert knowledge. In the latter case typically a uniform distribution with parameter specific upper and lower bounds is chosen. Refraining from too limiting constraints on parameter priors by using such uninformative (e.g., uniform) priors helps explore a greater subspace of the parameter space, which positively affects the amount of information "retrieved" from the measurement data.

Instead of solely optimizing the $n$-dimensional parameter vector $\boldsymbol{\theta} = (\theta_1, \theta_2, \ldots, \theta_n)^T$ with respect to some metric, BC additionally yields the joint probability distribution, which contains information on the marginal distributions of each parameter $\theta_j$ ($j = 1,\ldots,n$) and their dependencies.

For analytically intractable problems the joint probability distribution can be obtained by Monte Carlo (MC) simulation techniques. Consequently, the number of simulation runs to obtain the posterior distribution is large (Van Oijen et al., 2005). An efficient approach to solving this problem is given by the iterative scheme of the Metropolis algorithm (van Oijen et al., 2005). Following this algorithm, the prospective joint parameter distribution is given by the stationary distribution (convergence) of a suitable Markov chain.

Rahn et al. (2012) conducted the outlined procedure with the LandscapeDNDC model at site scale. In this study the model was calibrated with soil fluxes of $N_2O$, NO, and $CO_2$. To assure the convergence of the Metropolis algorithm Rahn et al. (2012) generated four independent Markov chains, each starting at a different parameter value. The posterior distribution holds a total of approximately 400,000 parameter vectors (Rahn et al., 2012), which is used in this study to quantify parameter-induced uncertainty of simulation outputs of LandscapeDNDC at regional scale.

The most notable challenges in transferring the previously outlined approach of quantifying parameter-induced uncertainty from site to regional scale are the lack of regional measurement data on soil GHG emissions and the tremendous demand in computing resources if data were available. These circumstances render a BC attempt on the regional level not likely. Instead we can only apply the joint parameter distribution obtained from site scale to the regional scale, yielding an uncertainty estimation reflecting the current state of insight. This strategy has been already proposed by Lehuger et al. (2009) and was first used by Zhou et al. (2012).

Because the large number of about 400,000 parameter vectors representing the posterior distribution prevents reasonable application on the regional scale, we randomly selected, in steps of one thousand, a parameter vector from the posterior distribution, which resulted in 400 different parameter vectors. For each of these 400 parameter vectors we conducted a regional inventory simulation spanning four simulation years, with the first year being regarded as model spin-up and therefore discarded before analysis. Simulation outputs were assigned to different uncertainty levels defined by percentile bandwidths. For example the range between the 25th and the 75th percentile ( $P_{25}^{75}$ ) is associated with 50% likelihood.

To interpret the simulation results, we aggregated them with respect to time, parameter, and space: (i) simulation results for the target entities $N_2O$, $NO_3$, and biomass yields were averaged over the 3-yr simulation period (this is done for each parameter vector and polygon); (ii) for the 400 values per polygon we applied descriptive statistics revealing mean, median, and percentiles; and (iii) statistics on a regional scale were derived by averaging parameter related mean, median, and percentile values of all 4042 polygons.

Even though we deliberately chose a moderate number of parameter vectors to balance a sound representation of the joint parameter distribution and computing time, the total runtime still amounted to approximately 7200 CPU hours on a conventional computer. However the simulations were distributed among available computing resources and run in parallel according to the capabilities of the HPC system.

## Regional Application

The parameter-induced uncertainty of regional $N_2O$ emissions and $NO_3$ leaching was quantified within the framework of simulations with the LandscapeDNDC model for arable soils of the state of Saxony, Germany. Saxony covers an area of 18,416 km$^2$, and the elevation ranges from 80 to 400 m in the northern lowlands to 400 to 950 m in the southern mountainous area. Local climatic conditions in the northern part are warmer (mean annual temperature 8.5–10°C), while average temperatures in the south only reach 6 to 7.5°C. Average annual precipitations follow the same topographic pattern, ranging from 500 to 800 mm in the north and 900 to 1200 mm in the south. Spatial discretization into a total of 4042 polygons was prescribed by land use, variability in soil, and climate properties and was performed using GIS software (ESRI ArcGIS 10).

In LandscapeDNDC the vertical soil column is stratified into several soil layers of individual heights and associated with chemo-physical properties, such as soil texture, bulk density, carbon and nitrogen contents, pH, and parameters describing soil hydraulic conditions.

The regional soil input data were obtained from a spatially explicit soil database (BUEK200) supplied by the LfULG Saxony (Environmental, Agricultural, and Geological Service of the State of Saxony). The soil database has a resolution of 1:200,000 and includes land use, soil types, and soil attributes representing different vertical soil horizons on a polygonal basis.

Input data not explicitly supplied by the soil database (e.g., wilting point and maximum water holding capacity) were derived via pedotransfer functions from available information on soil texture. The soil organic carbon (SOC) content was derived from supplied, spatially explicit information of humus classifications (Finnern et al., 2005). Wetland and peat soils with high SOC (>8%), which cover only a minor part of the simulation domain, were excluded from the simulation.

Climate input data in daily resolution driving the regional simulations are obtained from a raster dataset with a spatial resolution of 1 km and includes minimum, maximum and average temperature as well as precipitation (RaKliDa, 2011, obtained from LfULG Saxony). The climate data were downscaled from simulations with the ECHAM model based on the IPCC SRES A1B scenario of present climate (1998–2001). For each polygon the climate data associated to the raster cell closest to the polygons center were used.

**Table 1. Agricultural activity data used for fertilizer application timing and amounts.**

| Year | Crop | Seeding | Harvest | N fertilization ($NH_4NO_3$) |
|---|---|---|---|---|
| 1998 | Summer barley | Apr. 1 | Aug. 15 | Apr. 14, 72 kg N<br>May 25, 72 kg N |
| 1999 | Rapeseed | Oct. 2 | July 30 | Apr. 25, 72 kg N<br>May 19, 72 kg N |
| 2000 | Winter wheat | Oct. 5 | Aug. 8 | Apr. 15, 90 kg N<br>May 25, 70 kg N |
| 2001 | Winter barley | Sept. 29 | July 30 | Mar. 25, 50 kg N<br>Apr. 15, 50 kg N<br>May 25, 50 kg N |

LandscapeDNDC is derived from the DNDC family of agricultural ecosystem models and can simulate the growth of all major commodity crops in Europe. The most typical crop rotation in Saxony, as affirmed by the LfULG Saxony, consists of three consecutively grown winter crops (canola, winter wheat, and winter barley). This crop rotation was used for all polygons. Simulation runs span 4 yr, with the first year cultivating summer barley to prevent unrealistically long bare soil conditions. Compared to other parts of Germany, in Saxony livestock breeding is rather extensive. Consequently, mineral N fertilization prevails and therefore the use of manure as fertilizers was neglected in our simulations. The timing and magnitude of fertilizer applications were based on expert knowledge, which was supplied by agricultural advisors of the LfULG (Table 1).

## Results

For the state of Saxony a mean $N_2O$ emission of 1.43 ± 1.25 SD kg N (ha yr)$^{-1}$ and a smaller median of 1.05 kg N (ha yr)$^{-1}$ is calculated. Applying the standard 2006 IPCC EF approach to our study predicts mean direct $N_2O$ emissions of 1.51 kg N (ha yr)$^{-1}$. This corresponds to total direct $N_2O$ emissions of 1.186 kt N yr$^{-1}$, while our modeling approach yields total emission of 1.166 kt N yr$^{-1}$. Furthermore, on regional scale our simulation study on parameter-induced uncertainty predicts likelihoods of 90% ($P_5^{95}$), 80% ($P_{10}^{90}$), and 50% ($P_{25}^{75}$) that simulated $N_2O$ emissions are in the ranges of 0.11 to 4.03, 0.18 to 3.29, 0.46 to 2.05 kg N (ha yr)$^{-1}$, respectively (Table 2). The spatial variation of median $N_2O$ emissions across the whole simulation domain—calculated from the 400 different results for each parameter vector per polygon—are illustrated in Fig. 1. Values range from 0.15 up to 5.3 kg N (ha yr)$^{-1}$ with highest values being associated with high SOC contents (data not shown). Spatial variation of $P_{25}^{75}$ of $N_2O$ emissions is presented in Fig. 2. The spatial pattern mainly follows the pattern of median $N_2O$ fluxes with higher median fluxes being associated with higher uncertainties. On average, values

**Table 2. Summary statistics and parameter-induced uncertainty of $N_2O$ emissions and nitrate leaching rates averaged (ha basis) over all 4042 polygons.**

| | Mean | SD | Median | $P_5^{95}$ | $P_{10}^{90}$ | $P_{25}^{75}$ |
|---|---|---|---|---|---|---|
| | kg N/(ha yr)$^{-1}$ | | | | | |
| $N_2O$ (IPCC) | 1.51 | – | – | – | – | – |
| $N_2O$ (sim) | 1.43 | 1.25 | 1.05 | 3.92<br>0.11–4.03 | 3.11<br>0.18–3.29 | 1.59<br>0.46–2.05 |
| $NO_3$ (IPCC) | 45.4 | – | – | – | – | – |
| $NO_3$ (sim) | 29.32 | 9.97 | 30.73 | 33.47<br>9.24–42.71 | 25.61<br>14.73–40.34 | 11.50<br>24.52–36.02 |

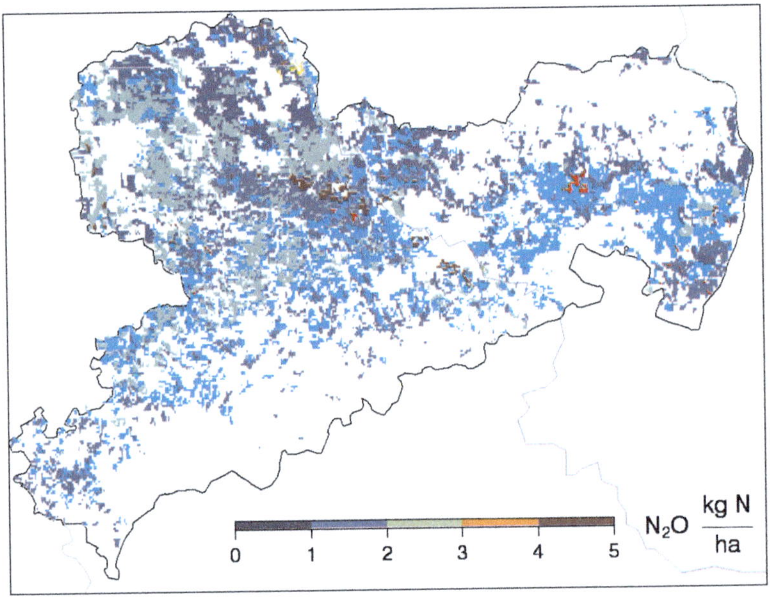


**Fig. 1. Spatial distribution of the median of the $N_2O$ emissions from agricultural soils in Saxony, Germany [kg $N_2O$-N (ha yr)$^{-1}$]. First, average emissions over the three simulation years were calculated for each of the 400 regional simulations. Then the mean was calculated across the 400 results for each polygon.**

of $P_{25}^{75}$ are 51% higher than median fluxes with maximum values up to 6 kg N (ha yr)$^{-1}$. The total area weighted mean of $N_2O$ emissions is 1.166 ± 1.018 kt N yr$^{-1}$, while the median is 0.858 kt N yr$^{-1}$ (Table 3). Taking into account parameter-induced uncertainty, likelihoods of 90% ($P_5^{95}$), 80% ($P_{10}^{90}$) and 50% ($P_{25}^{75}$) of total $N_2O$ emissions for the state of Saxony can be associated to variations of total $N_2O$ emissions in the ranges of 0.092 to 3.283, 0.147 to 2.685, and 0.379 to 1.686 kt N yr$^{-1}$, respectively (Table 3). The variability of total regional uncertainties for each of the three simulation years and the median are shown in Fig. 3, with the year

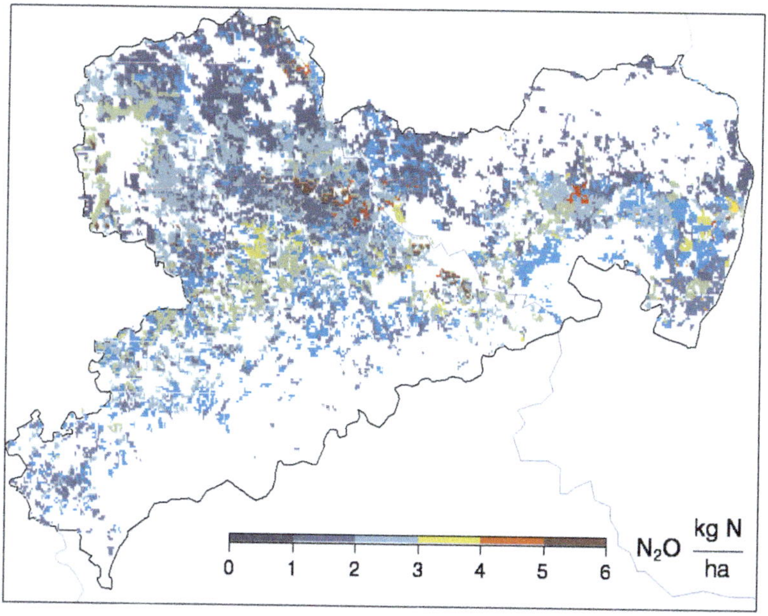


**Fig. 2. Spatial distribution of the uncertainties of the $N_2O$ emissions from agricultural soils in Saxony, Germany, expressed as the range between the 75th to 25th percentile. (Data processing done as in Fig. 1.)**

**Table 3. Summary statistics and parameter-induced uncertainty of regional $N_2O$ emissions and $NO_3$ leaching rates for the state of Saxony (total arable area: 7852.58 km²).**

| | Mean | SD | Median | $P_5^{95}$ | $P_{10}^{90}$ | $P_{25}^{75}$ |
|---|---|---|---|---|---|---|
| | kt N yr$^{-1}$ | | | | | |
| $N_2O$ (IPCC) | 1.186 | – | – | – | – | – |
| $N_2O$ (sim) | 1.166 | 1.018 | 0.858 | 3.191<br>0.092–3.283 | 2.538<br>0.147–2.685 | 1.307<br>0.379–1.686 |
| $NO_3$ (IPCC) | 35.651 | – | – | – | – | – |
| $NO_3$ (sim) | 22.845 | 7.760 | 24.000 | 26.026<br>7.119–33.145 | 19.978<br>11.403–31.381 | 8.886<br>19.174–28.060 |

2000 having a twofold higher uncertainty than the years 1999 and 2001. The differences across years mainly resulted from different crop growth characteristics, timing, and amount of N fertilization (Table 1), and residue application overlaid by differences in soil moisture and soil temperature. In addition, the classification of the total regional uncertainty shown in Fig. 4 illustrates a strongly positive skewed distribution.

The same statistical procedure is applied to quantify the uncertainty of simulated regional $NO_3$ leaching rates. The mean soil $NO_3$ leaching rate from arable

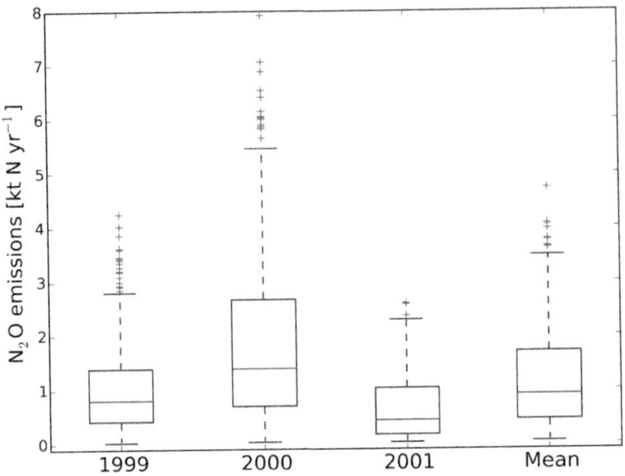


**Fig. 3. Yearly average $N_2O$ emission levels for agricultural soils in Saxony, Germany, and the associated uncertainties. The main body shows median, 25th, and 75th percentiles, and the whiskers are set to 1.5 IQR (interquartile range). Differences are due to different crop growth and nutrient availability (N fertilization and timing) and precipitation characteristics in individual years.**

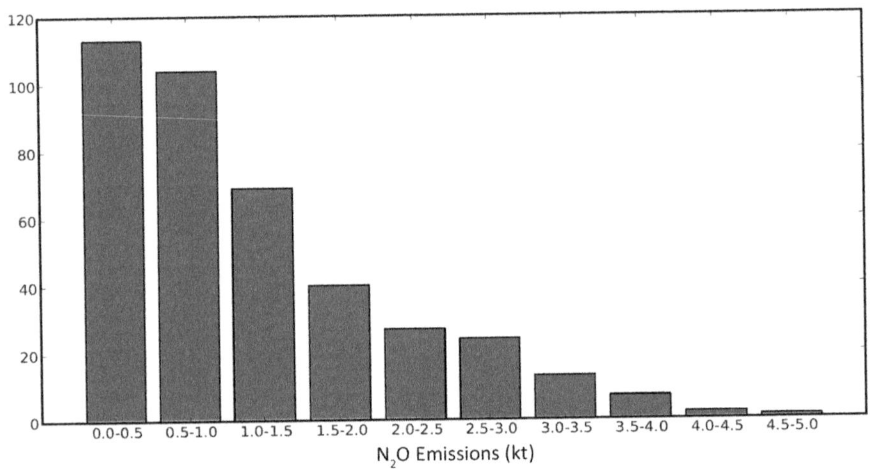


**Fig. 4. Histogram of total regional $N_2O$ emissions from agricultural soils in Saxony, Germany, based on 400 samples from the joint parameter distribution. Total regional $N_2O$ emissions are averages across the three simulation years for each regional simulation.**

systems in the state of Saxony is 29.3 ± 10.0 SD kg N (ha yr)$^{-1}$. Calculations of the median lead to a slightly larger value of 30.7 [kg N (ha yr)$^{-1}$] (Fig. 5). With the likelihoods of 90% ($P_{5}^{95}$), 80% ($P_{10}^{90}$), and 50% ($P_{25}^{75}$) the $NO_3$ leaching uncertainty bandwidths are 9.2 to 42.7, 14.7 to 40.3, and 24.5 to 36.0 kg N (ha yr)$^{-1}$ (Table 2). Spatial variation of median $NO_3$ leaching rates with values ranging from 9.1 to maximum leaching rates of more than 50 kg N (ha yr)$^{-1}$ are illustrated in Fig. 5.

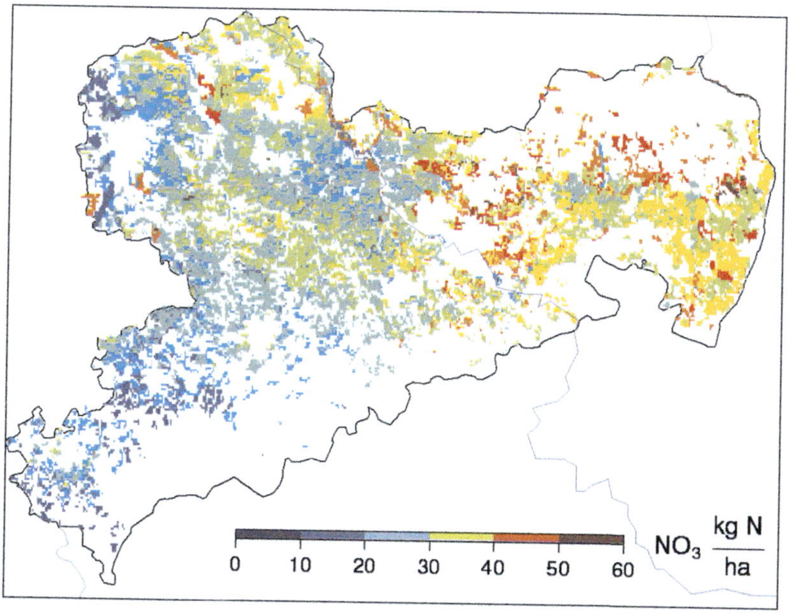


**Fig. 5. Spatial distribution of the median of the $NO_3$ leaching from agricultural soils in Saxony, Germany [kg $NO_3$–N (ha yr)$^{-1}$]. (Data processing done as in Fig. 1.)**

Highest values are found in regions with high annual rainfall and sandy soils (data not shown). In contrast to $N_2O$ emissions spatial patterns of $P_{25}^{75}$ for nitrate leaching (Fig. 6) do not follow the spatial pattern of median $NO_3$ leaching rates (correlation analysis does not reveal any significant relationship). Upscaling mean and median nitrate leaching rates to the whole simulation domain results in total losses of 22,845 ± 7759 SD and a median of 24,000 kt N yr$^{-1}$ (Table 3). These total losses correspond to approximately one fifth of the average fertilization of roughly 118,000 kt mineral N fertilizer per year. Taking into account parameter-induced uncertainty, likelihoods of 90% ( $P_{5}^{95}$ ), 80% ( $P_{10}^{90}$ ), and 50% ( $P_{25}^{75}$ ) can be associated to variations of total nitrate leaching rates in the ranges of 7119 to 33,145, 11,403 to 31,381, 19,174 to 28,060 kt N yr$^{-1}$, respectively (Table 3). Yearly and 3-yr mean total regional uncertainties are depicted in Fig. 7. Variations across the 3 yr can be attributed to different crop growth characteristics and the nutrient availability (N-fertilization and timing), as well as the precipitation regimes. The classification of the total nitrate leaching shown in Fig. 8 illustrates a slight negative skewed distribution.

Simulated yields are in good agreement with yield statistics from the LfULG. According to Trautmann and Fritsche (2006) average yields (kg DW) were 3300 (simulated: 3470.23 ± 220.29) for canola, 6200 (simulated: 5493.86 ±

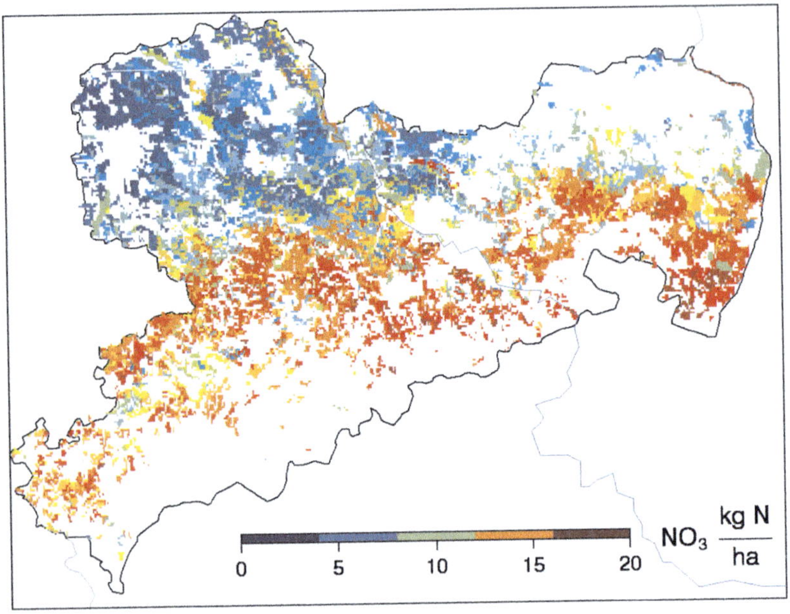


**Fig. 6. Spatial distribution of the uncertainties as the range between the 75th and 25th percentile corresponding to a 50% likelihood that the true rate of nitrate leaching from agricultural soils in Saxony, Germany, is within the simulated range. (Data processing done as in Fig. 1.)**

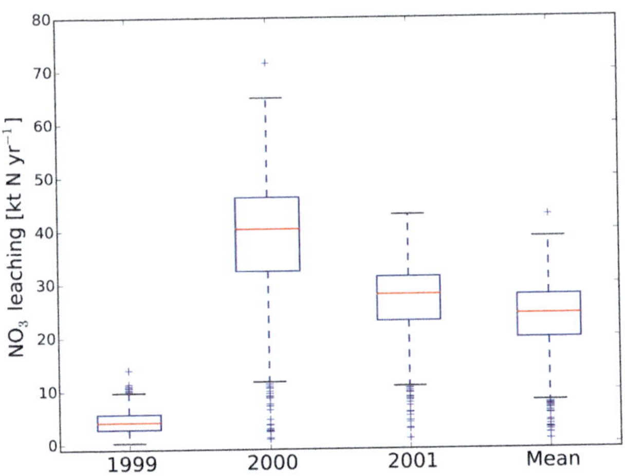


**Fig. 7. Yearly averaged rates of $NO_3$ leaching from agricultural soils in Saxony, Germany, and their uncertainties. The main body shows median, 25th, and 75th percentile, and the whiskers are set to 1.5 IQR (interquartile range). Differences are due to different crop growth and nutrient availability (N-fertilization and timing) and precipitation characteristics in individual years.**

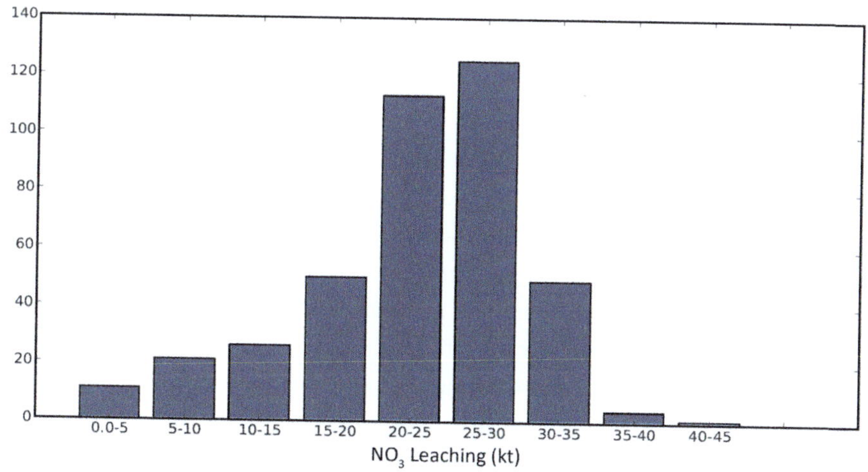


**Fig. 8. Histogram of total regional rates of $NO_3$ leaching from agricultural soils in Saxony, Germany, based on 400 samples from the joint parameter distribution.**

442.49) for winter wheat, and 5800 (simulated: 4693.84 ± 400.33) for winter barley. The slight underestimation (11.18%) of yields for winter grains is most likely related to the generalized use of only one single crop rotation for the full simulation domain, rather than more site specific crop rotations varying with soil and climatic conditions.

## Discussion

Upscaling of $N_2O$ emissions from site to regional or national scale is associated with uncertainty and highly depends on the scaling approach. Jungkunst et al. (2006) performed a review on $N_2O$ emission studies in Germany and found $N_2O$ emission uncertainties ranged from 0.18% to 15.54% of applied nitrogen as fertilizer, concluding there was a large uncertainty to the IPCC methodology. However, the application of more detailed studies, which account for more site-specific input, such as the IPCC Tier 2 approach for Germany (Dämmgen and Grünhage 2002; Dämmgen et al., 2006) would lead to a substantial reduction of uncertainty. Brown et al. (2001) performed a regional $N_2O$ emission analysis for the UK including an assessment of the involved regional uncertainty following IPCC 1997 methodology. The EF used was 1.25% with an uncertainty range from 0.25 to 2.25%. The study performed was based on a Latin hypercube sampling (LHS) of EF within given ranges and resulted in a 62% uncertainty of direct $N_2O$ emissions. In contrast to Brown et al. (2001), our study can quantify the probability of emission ranges for the regional inventory simulation based on a calibration versus observations.

Even though the notion of uncertainty quantification within the scope of regional-scale ecosystem modeling is frequently used, a clear definition, especially with regard to a precise delimitation to sensitivity, is still lacking. There are several studies using Latin hypercube sampling to assess parameter and input sensitivities related to GHG emissions and nitrate leaching (Brown et al., 2001; Hansen et al., 1999; Thorsen et al., 2001; van der Keur et al., 2008; Zaehle et al., 2005). Other sensitivity methodologies like the β-index method invented by Friend et al. (1993) was applied in several studies (Butterbach-Bahl et al., 2001, 2009; Haas et al., 2013; Kesik et al., 2005, 2005; Werner et al., 2007) whereas Del Grosso et al. (2006) performed a sensitivity analysis by defining different scenarios based on expert knowledge and compared the resulting regional $N_2O$ emission factors to IPCC methodology for assessing the sensitivity of each scenario.

In addition to the above-mentioned sensitivity-based analyses there are various studies using Bayesian inversion methods (Klemedtsson et al., 2008; Knorr and Kattge 2005; Lehuger et al., 2009; Nylinder et al., 2011; Rahn et al., 2012; Svensson et al., 2008; van Oijen et al., 2005; Zhou et al., 2012). In contrast to sensitivity studies, a common characteristic of these methodologies is to incorporate observations, to systematically investigate parameter interactions, and finally to derive confidence intervals for simulation outputs. For a regional uncertainty quantification of $N_2O$ emissions Lehuger et al. (2009) proposed to transfer the joint parameter distribution obtained by BC on the site to regional scale. This approach has been also applied by Zhou et al. (2012), aiming to quantify the carbon residence time in forest ecosystems across the United States.

For quantifying the parameter-induced uncertainty in our study we averaged simulation results over the 3-yr simulation time. However, depending on environmental conditions and agricultural management, uncertainties can vary across years (Fig. 3) and even days (Fig. 9). Because biogeochemical processes seem to be particularly sensitive to changes in parameter values at high substrate availability, highest uncertainties were, for example, observed after fertilization and/or harvest. For the latter, simulated soil $N_2O$ emissions responded to the decomposition of roots and harvest residues. For $N_2O$ Fig. 3 and especially Fig. 4 show the positive skewed distribution of regional $N_2O$ emissions stemming from the 400 different parameter vectors. Consequently, there are more parameter vectors revealing low emission levels resulting in a lower median of 1.05 kg N $(ha\ yr)^{-1}$ compared to the mean of 1.43 kg N $(ha\ yr)^{-1}$. This means that parameter vectors revealing higher emissions contribute disproportionally strong to total $N_2O$ emissions.

In this study total mean $NO_3$ leaching exceeds total mean $N_2O$ emissions approximately by a factor of 20 (factor of 28 for medians). In contrast to $N_2O$ emis-

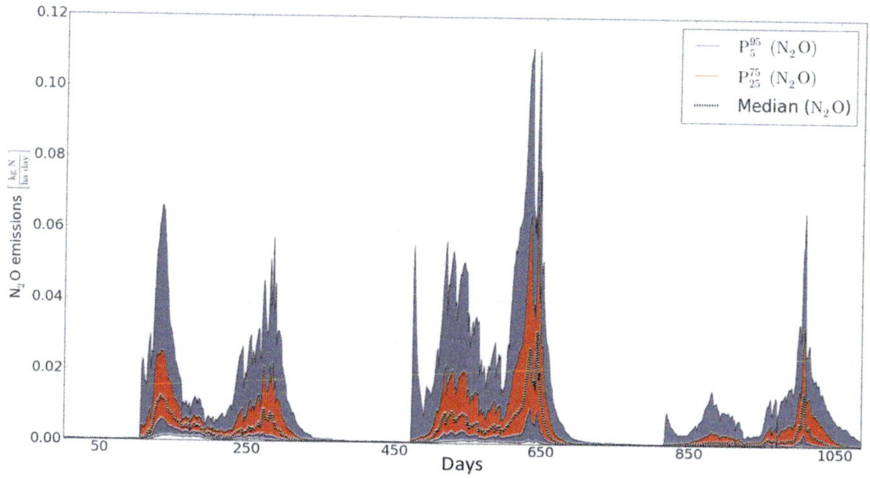


**Fig. 9. Daily average uncertainty bandwidths of $N_2O$ emission strengths [kg $N_2O$-N (ha d)$^{-1}$] for agricultural soils in Saxony, Germany, for the simulated 3-yr crop rotation (1999–2001). The arrows on the time axis indicate the timing of the fertilization events.**

sions relative uncertainties are considerably lower: $P^{50}$ ($NO_3$) > $P_{25}^{75}$ ($NO_3$); $P^{50}$ ($N_2O$) < $P_{25}^{75}$ ($N_2O$). Moreover, there are clear opposing trends compared to $N_2O$ with regard to temporal pattern and parameter dependencies. The parameter-induced regional uncertainty of yearly nitrate leaching were higher than for the $N_2O$ emissions (see Fig. 7) due to significant effects of subsoil properties such as bulk density or texture and crop type on simulated soil nitrate leaching. For the year 2000, both magnitude- and parameter-induced uncertainty of $N_2O$ emission and $NO_3$ leaching rates are highest. On the other hand year 1999 clearly exhibits the lowest values for $NO_3$, whereas values for $N_2O$ emissions are higher than in year 2001 (compare Fig. 6 and 7), which most likely is related to differences in climate affecting plant growth and nitrogen turnover and fluxes. The parameter dependencies of regional $NO_3$ leaching rates rather show a normal distribution (Fig. 8). It cannot be ruled out that the different behavior of $N_2O$ emissions and nitrate leaching in 1999 is caused by model initialization; however, due to general diverging parameter effects (normal distributed vs. positive skewed), there might be further influences that have not yet been considered. As already demonstrated by Kaiser and Ruser (2000) and Jungkunst et al. (2006) for Germany, other factors than fertilization rates, such as soil and climatic conditions influencing nitrogen turnover and plant N uptake, can be equally important for total N fluxes. Regarding such dependencies LandscapeDNDC, as a process-based model integrates many nonlinear relationships. One main actor might be given by plant growth, which strongly depends on temperature degree-days and therefore is positively connected to air temperature. Temperature as well as precipitation are both seen to contribute decisively to the variability of $N_2O$ emissions (Butterbach-Bahl et al., 2013) represented in LandscapeDNDC by detailed description of nitrogen metabolism (Li et al., 2000). Nitrate leaching rates mainly depend on precipitation and soil texture, which affect soil water and nutrient percolation.

Exploiting the Bayesian theory, the quality of the (regional) uncertainty quantification can be further improved by refining the joint parameter distribution. This can be accomplished through continuing the Bayesian inversion process with new information (i.e., observation data) on the site scale. In case this procedure yields drastically changed parameter distributions, it needs to be investigated if either inaccurate or insufficient input data or structural deficits in the model are causing such effects. In the latter case BC helps to identify and hopefully overcome model weaknesses. Because model parameters considered in our study form a subset of the complete set of model parameters used exclusively in a single component (soil biogeochemistry) common to all ecosystems, it is reasonable to incorporate high quality observations obtained at any ecosystem.

Other studies of LandscapeDNDC with focus on input uncertainty and sensitivity (Haas et al., 2013; Kiese et al., 2005, 2011; Werner et al., 2007) identified the most sensitive input quantities for $N_2O$ emissions and $NO_3$ leaching. Haas et al. (2013) highlighted regional input sensitivity associated to soil properties like SOC and pH, climatic input like precipitation, and management input like the amount of fertilizer application. The study clearly showed that there are strong nonlinear responses of simulation results related to changes in input data, but results are not comparable. Werner et al. (2007) additionally performed an uncertainty quantification of global $N_2O$ emissions on input data uncertainty by sampling subsets of grid cells of the global inventory. The study used the first quartile (Q1) and third quartile (Q3) of all calculated emissions to be the lower and upper bounds of input-induced model uncertainty revealing uncertainty ranges of −34.6 to 76.3% (see Fig. 10 of Werner et al., 2007) of the median. To assess the input uncertainty for global $N_2O$ emissions from tropical rainforest soils they upscaled the uncertainty boundaries onto the global inventory, assuming the set of samples is representative for the global inventory. Comparing input uncertainty of these studies with the regional scale parameter-induced uncertainty of this study is difficult due to large differences in the methodologies applied. However, both input uncertainty and parameter uncertainty for $N_2O$ emissions are mostly in the same range of 0.5 to 3 kg $N_2O$-N (ha yr)$^{-1}$.

Lindahl et al. (2008) addressed input-induced uncertainty quantification on landscape-scale pesticide leaching. Because the use of a full process-based model would have been too computationally demanding, they employed a meta-model.

Their applied method to quantify regional input-induced uncertainty follows a sequential Gaussian simulation approach, leading to a probability density function for each site.

Investigating input-induced uncertainty on the regional scale in particular with site-specific inputs in the sense of MC or LHS (Dalbey et al., 2008; Werner et al., 2007) one needs to consider the number of uncertain input quantities, their uncertainty ranges, and their heterogeneity across the domain, leading to an intractable number of simulation runs and data processing. To our understanding at present this is the only approach found in the literature on how to address the site-specific nature of the input quantities on the regional scale. No other study has reported use of the approach of a Bayesian calibration of regional site-specific input data. Some studies in hydrology deal with uncertainty of precipitation on catchment runoff (Kavetski et al., 2006a; Kavetski et al., 2006b; Vrugt et al., 2008), but these studies only considered non-site-specific input. Huard and Mailhot (2006) discussed the theory of BC for input-induced uncertainty and its impact on the model parameters. For reasons of feasibility this study made use of a simple model serving only academic purpose. It remains an open question how to transfer the applied method onto problems involving site-specific regional inputs.

## Conclusions

We provide a consistent approach to quantify parameter-induced uncertainties in the assessment of regional GHG emissions and $NO_3$ leaching. The approach is based on the application of results of Bayesian parameter-induced uncertainty quantification at the site to regional scale. Although simulation results are reasonable and comparable to other approaches, for example, the IPCC emission factor approach, it needs to be acknowledged that we only assessed the effect of parameter-induced uncertainties of the GHG emission component of LandscapeDNDC. However, biogeochemical models are composed of different components, such as microclimate, biogeochemistry, and physiology, which for a more realistic and exhaustive assessment need to be included at the same time in the Bayesian calibration procedure. Because this unavoidably increases the number of parameters, requirements on computational resources grow accordingly.

Furthermore, if more data become available, for example, via concerted national and international GHG monitoring networks such as ICOS (www.icos-infrastructure.eu, verified 10 Mar. 2015) or NEON (www.neoninc.org, verified 10 Mar. 2015), one should consider multi-site applications of BC, as well as the inclusion of further parameters, to get a more thorough and robust understanding of uncertainties when using biogeochemical models for calculation of regional and global GHG inventories.

## Acknowledgments

The authors would like to acknowledge the assistance and provision of data for the regionalization by the LfULG (Agricultural, Environmental, and Geological Service of the State of Saxony). Financial support for this study was obtained by the German Science Foundation (DFG) support contract BU1173/12-1, the EU IP ECLAIRE, and the FACCE-JPI MACSUR.

## References

Bell, M.J., E. Jones, J. Smith, P. Smith, J. Yeluripati, J. Augustin, R. Juszczak, J. Olejnik, and M. Sommer. 2012. Simulation of soil nitrogen, nitrous oxide emissions and mitigation scenarios at 3 European cropland sites using the ECOSSE model. Nutr. Cycling Agroecosyst. 92:161–181. doi:10.1007/s10705-011-9479-4

Brown, L., S. Armstrong Brown, S.C. Jarvis, B. Syed, K.W.T Goulding, V.R Phillips, R.W Sneath, and B.F Pain. 2001. An inventory of nitrous oxide emissions from agriculture in the UK using the IPCC methodology: Emission estimate, uncertainty and sensitivity analysis. Atmos. Environ. 35:1439–1449. doi:10.1016/S1352-2310(00)00361-7

Butterbach-Bahl, K., E.M. Baggs, M. Dannenmann, R. Kiese, and S. Zechmeister-Boltenstern. 2013. Nitrous oxide emissions from soils: How well do we understand the processes and their controls? Phil. Trans. R. Soc. B 368. doi:10.1098/rstb.2013.0122

Butterbach-Bahl, K., M. Kahl, L. Mykhayliv, C. Werner, R. Kiese, and C. Li. 2009. A European-wide inventory of soil NO emissions using the biogeochemical models DNDC/Forest-DNDC. Multiple values selected. Atmos. Environ. 43:1392–1402. doi:10.1016/j.atmosenv.2008.02.008

Butterbach-Bahl, K., M. Kock, G. Willibald, B. Hewett, S. Buhagiar, H. Papen, and R. Kiese. 2004. Temporal variations of fluxes of NO, $NO_2$, $N_2O$, $CO_2$, and $CH_4$ in a tropical rain forest ecosystem. Global Biogeochem. Cycl. doi:10.1029/2004GB002243

Butterbach-Bahl, K., F. Stange, H. Papen, and C. Li. 2001. Regional inventory of nitric oxide and nitrous oxide emissions for forest soils of southeast Germany using the biogeochemical model PnET-N-DNDC. J. Geophys. Res. 106:34155–34166. doi:10.1029/2000JD000173

Cameron, D.R., M. van Oijen, and C. Werner. 2013. Environmental change impacts on the C- and N-cycle of European forests: A model comparison study. Biogeosciences 10:1751–1773. doi:10.5194/bg-10-1751-2013

Chirinda, N., M.S. Carter, K.R. Albert, P. Ambus, J.E. Olesen, J.R. Porter, and S.O. Petersen. 2010. 2010. Emissions of nitrous oxide from arable organic and conventional cropping systems on two soil types. Agric. Ecosyst. Environ. 136:199–208. doi:10.1016/j.agee.2009.11.012

Dalbey, K., A.K. Patra, E.B. Pitman, M.I. Bursik, and M.F. Sheridan. 2008. Input uncertainty propagation methods and hazard mapping of geophysical mass flows. J. Geophys. Res. 113. doi:10.1029/2006JB004471.

Dämmgen, U., and L. Grünhage. 2002. Trace gas emissions from German agriculture as obtained from the application of simpler or default methodologies. Environ. Pollut. 117:23–34. doi:10.1016/S0269-7491(01)00162-2

Dämmgen U., M. Lüttich, H. Döhler, B. Eurich-Menden, and B. Osterburg. 2006. Calculations of Emissions from German Agriculture—National Emission Inventory Report (NIR) 2006 for 2004. Methods and Data. Landbauforschung Völkenrode Sonderheft 291:47–222.

Del Grosso, S.J., W.J. Parton, A.R. Mosier, M.K. Walsh, D.S. Ojima, and P.E. Thornton. 2006. DAYCENT national-scale simulations of nitrous oxide emissions from cropped soils in the United States. J. Environ. Qual. 35:1451–1460. doi:10.2134/jeq2005.0160

Del Grosso, S.J., S.M. Ogle, W.J. Parton, and F.J. Breidt. 2010. Estimating uncertainty in $N_2O$ emissions from US Cropland Soils. Global Biogeochem. Cycles 24. doi:10.1029/2009GB003544

Finnern, H., W. Grottenthaler, and D. Kühn. 2005. Bodenkundliche Kartieranleitung. 5th ed. Schweizerbart'Sche Verlagsbuchhandlung, Hannover, Germany.

Friend A, H. Schugart, and S.W. Running. 1993. A physiology-based gap model of forest dynamics. Ecology 74:792–797. doi:10.2307/1940806

Gabrielle, B., P. Laville, C. Henault, B. Nicoullaud, and J.C. Germon. 2006. Simulation of nitrous oxide emissions from wheat-cropped soils using CERES. Nutr. Cycling Agroecosyst. 74:133–146. doi:10.1007/s10705-005-5771-5

Gibbons, J.M., G.M. Cox, A.T.A. Wood, J. Craigon, S.J. Ramsden, D. Tarsitano, and N.M.J. Crout. 2008. Applying Bayesian model averaging to mechanistic models: An example and comparison of methods. Environ. Model. Softw. 23:973–985. doi:10.1016/j.envsoft.2007.11.008

Groffman, P.M., M.A. Altabet, J.K. Böhlke, K. Butterbach-Bahl, M.B. David, M.K. Firestone, A.E. Giblin, T.M. Kana, L.P. Nielsen, and M.A. Voytek. 2006. Methods for measuring denitrification: Diverse approaches to a difficult problem. Ecol. Appl. 16:2091–2122. doi:10.1890/1051-0761(2006)016[2091:MFMDDA]2.0.CO;2

Haas, E., S. Klatt, A. Fröhlich, P. Kraft, C. Werner, R. Kiese, R. Grote, L. Breuer, and K. Butterbach-Bahl. 2013. LandscapeDNDC: A process model for simulation of biosphere–atmosphere–hydrosphere exchange processes at site and regional scale. Landscape Ecol. 28:615–636. doi:10.1007/s10980-012-9772-x

Hansen, S., M. Thorsen, E.J. Pebesma, S. Kleeschulte, and H. Svendsen. 1999. Uncertainty in simulated nitrate leaching due to uncertainty in input data. A case study. Soil Use Manage. 15:167–175. doi:10.1111/j.1475-2743.1999.tb00083.x

Huard, D., and A. Mailhot. 2006. A Bayesian perspective on input uncertainty in model calibration: Application to hydrological model "abc.". Water Resour. Res. doi:10.1029/2005WR004661.

IPCC. 2006. 2006 IPCC Guidelines for National Greenhouse Gas Inventories: Vol. 4. Agriculture, Forestry and Other Land Use. Institute for Global Environmental Strategies (IGES), on behalf of the IPCC, Hayama, Japan

Jungkunst, H.F., A. Freibauer, H. Neufeldt, and G. Bareth. 2006. Nitrous oxide emissions from agricultural land use in Germany—A synthesis of available annual field data. J. Plant Nutr. Soil Sci. 169:341–351. doi:10.1002/jpln.200521954

Kaiser, E., and R. Ruser. 2000. Nitrous oxide emissions from arable soils in Germany—An evaluation of six long-term field experiments. J. Plant Nutr. Soil Sci. 163:249–259. doi:10.1002/1522-2624(200006)163:3<249::AID-JPLN249>3.0.CO;2-Z

Kavetski, D., G. Kuczera, and S.W. Franks. 2006a. Bayesian analysis of input uncertainty in hydrological modeling: 1. Theory. Water Resour. Res. 42. doi:10.1029/2005WR004368

Kavetski, D., G. Kuczera, and S.W. Franks. 2006b. Bayesian analysis of input uncertainty in hydrological modeling: 2. Application. Water Resour. Res. 42. doi:10.1029/2005WR004376

Kennedy, M.C., and A. O'Hagan. 2001. Bayesian calibration of computer models. J. R. Stat. Soc. B 63:425–464. doi:10.1111/1467-9868.00294

Kesik, M., P. Ambus, R. Baritz, N. Brüggemann, K. Butterbach-Bahl, M. Damm, J. Duyzer, L. Horváth, R. Kiese, B. Kitzler, A. Leip, C. Li, M. Pihlatie, K. Pilegaard, S. Seufert, D. Simpson, U. Skiba, G. Smiatek, T. Vesala, and S. Zechmeister-Boltenstern. 2005. Inventories of $N_2O$ and NO emissions from European forest soils. Biogeosciences 2:353–375. doi:10.5194/bg-2-353-2005

Kiese, R., H. Christoph, C. Werner, S. Wochele, R. Grote, K. Butterbach-Bahl. 2011. Quantification of nitrate leaching from German forest ecosystems by use of a process oriented biogeochemical model. Environ. Pollut. 159:3204–3214. doi:10.1016/j.envpol.2011.05.004

Kiese, R., C. Li, D.W. Hilbert, H. Papen, and K. Butterbach-Bahl. 2005. Regional application of PnET-N-DNDC for estimating the N2O source strength of tropical rainforests in the Wet Tropics of Australia. Glob. Change Biol. 11:128–144. doi:10.1111/j.1365-2486.2004.00873.x

Klemedtsson, L., P.-E. Jansson, D. Gustafsson, L. Karlberg, P. Weslien, K. von Arnold, M. Ernfors, O. Langvall, and A. Lindroth. 2008. Bayesian calibration method used to elucidate carbon turnover in forest on drained organic soil. Biogeochemistry 89:61–79. doi:10.1007/s10533-007-9169-0

Knorr, W., and J. Kattge. 2005. Inversion of terrestrial ecosystem model parameter values against eddy covariance measurements by Monte Carlo sampling. Glob. Change Biol. 11:1333–1351. doi:10.1111/j.1365-2486.2005.00977.x

Lehuger, S., B. Gabrielle, M. van Oijen, D. Makowski, J.-C. Germon, T. Morvan, and C. Hénault. 2009. Bayesian calibration of the nitrous oxide emission module of an agro-ecosystem model. Agric. Ecosyst. Environ. 133:208–222. doi:10.1016/j.agee.2009.04.022

Leip, A., M. Busto, and W. Winiwarter. 2011. Developing spatially stratified $N_2O$ emission factors for Europe. Environ. Pollut. 159:3223–3232. doi:10.1016/j.envpol.2010.11.024

Li, C. 2000. Modeling trace gas emissions from agricultural ecosystems. Nutr. Cycl. Agroecosyst. 58:259–276. doi:10.1023/A:1009859006242

Li, C., J. Aber, F. Stange, K. Butterbach-Bahl., and H. Papen. 2000. A process-oriented model of $N_2O$ and NO emissions from forest soils: 1. Model development. J. Geophys. Res. Atmos. 105:4369–4384. doi:10.1029/1999JD900949

Li, C., A. Mosier, R. Wassmann, Z. Cai, X. Zheng, Y. Huang, H. Tsurata, J. Boonjawat, and R. Lantin. 2004. Modeling greenhouse gas emissions from rice-based production systems: Sensitivity and upscaling. Global Biogeochem. Cycl. 18. doi:10.1029/2003GB002045

Lindahl, A.M.L., M. Söderström, and N. Jarvis. 2008. Influence of input uncertainty on prediction of within-field pesticide leaching risks. J. Contam. Hydrol. 98:106–114. doi:10.1016/j.jconhyd.2008.03.006

Nylinder, J., M. Stenberg, P.-E. Jansson, Å.K. Klemedtsson, P. Weslien, and L. Klemedtsson. 2011. Modelling uncertainty for nitrate leaching and nitrous oxide emissions based on a Swedish field experiment with organic crop rotation. Agric. Ecosyst. Environ. 141:167–183. doi:10.1016/j.agee.2011.02.027

Qiu, J., C. Li, L. Wang, H. Tang, H. Lit, and E. van Ranst. 2009. Modeling impacts of carbon sequestration on net greenhouse gas emissions from agricultural soils in China. Global Biogeochem. Cycl. 23. doi:10.1029/2008GB003180

RaKliDa, 2011. Rasterklimadaten (RaKliDa) zeigen mögliche Klimaentwicklungen bis zum Ende des Jahrhunderts auf. Environmental, Agricultural, and Geological Service of the State of Saxony. http://www.smul.sachsen.de/lfulg/18797.htm (accessed October 2011).

Rahn, K.-H., C. Werner, R. Kiese, E. Haas, and K. Butterbach-Bahl. 2012. Parameter-induced uncertainty quantification of soil $N_2O$, NO and $CO_2$ emission from Höglwald spruce forest (Germany) using the LandscapeDNDC model. Biogeosciences 9:3983–3998. doi: 10.5194/bg-9-3983-2012

Refsgaard, J.C., J.P. van der Sluijs, A.L. Hojberg, and P.A. Vanrolleghem. 2007. Uncertainty in the environmental modelling process—A framework and guidance. Environ. Model. Software 22:1543–1556. doi:10.1016/j.envsoft.2007.02.004

Rochette, P., D.E. Worth, E.C. Huffman, J.A. Brierley, B.G. McConkey, J. Yang et al. 2008. Estimation of $N_2O$ emissions from agricultural soils in Canada. II. 1990–2005 inventory. Can. J. Soil Sci. 88:655–669. doi:10.4141/CJSS07026

Schulze, E.D., P. Ciais, S. Luyssaert, M. Schrumpf, I.A. Janssens, B. Thiruchittampalam, J. Theloke, M. Saurat, S. Bringezu, J. Lelieveld, A. Lohila, C. Rebmann, M. Jung, D. Bastviken, G. Abril, G. Grassi, A. Leip, A. Greibauer, W. Kutsch, A. Don, J. Nieschulze, A. Börner, J. H. Gash, and A.J. Dolman. 2010. The European carbon balance. Part 4: Integration of carbon and other trace-gas fluxes. Glob. Change Biol. 16:1451–1469. doi:10.1111/j.1365-2486.2010.02215.x

Stange, F., K. Butterbach-Bahl, H. Papen, S. Zechmeister-Boltenstern, C. Li, and J. Aber. 2000. A process-oriented model of $N_2O$ and NO emissions from forest soils. 2. Sensitivity analysis and validation. J. Geophys. Res. Atmos. 105:4385–4398. doi:10.1029/1999JD900948

Sutton, M.A., E. Nemitz, J.W. Erisman, C. Beier, K. Butterbach Bahl, P. Cellier, W. de Vries, F. Cotrufo, U. Skiba, et al. 2007. Challenges in quantifying biosphere-atmosphere exchange of nitrogen species. Environ. Pollut. 150:125–139. doi:10.1016/j.envpol.2007.04.014

Svensson, M., P.-E. Jansson, D. Gustafsson, D. Berggren Kleja, O. Langvall, and A. Lindroth. 2008. Bayesian calibration of a model describing carbon, water and heat fluxes for a Swedish boreal forest stand. Ecol. Modell. 213:331–344. doi:10.1016/j.ecolmodel.2008.01.001

Thorsen, M., J.C. Refsgaard, S. Hansen, E. Pebesma, J.B. Jensen, and S. Kleeschulte. 2001. Assessment of uncertainty in simulation of nitrate leaching to aquifers at catchment scale. J. Hydrol. 242:210–227. doi:10.1016/S0022-1694(00)00396-6

Trautmann, U., and B. Fritsche. 2006. Ernteerträge in Sachsen. Statistic Service of Saxony, Kamenz, Germany.

van der Keur, P., J.R. Hansen, S. Hansen, and J.C. Refsgaard. 2008. Uncertainty in simulation of nitrate leaching at field and catchment scale within the Odense river basin. Vadose Zone J. 7. doi:10.2136/vzj2006.0186

van Oijen, M., D.R. Cameron, K. Butterbach-Bahl, N. Farahbakhshazad, P.-E. Jansson, R. Kiese, K.-H. Rahn, C. Werner, and J.B. Yeluripati. 2011. A Bayesian framework for model calibration, comparison and analysis: Application to four models for the biogeochemistry of a Norway spruce forest. Agric. For. Meteorol. 151:1609–1621. doi:10.1016/j.agrformet.2011.06.017

van Oijen, M., J. Rougier, and R. Smith. 2005. Bayesian calibration of process-based forest models: Bridging the gap between models and data. Tree Physiol. 25:915–927. doi:10.1093/treephys/25.7.915

van Oijen, M., and A. Thomson. 2010. Toward Bayesian uncertainty quantification for forestry models used in the United Kingdom Greenhouse Gas Inventory for land use, land use change, and forestry. Clim. Change 103:55–67. doi:10.1007/s10584-010-9917-3

Vrugt, J.A., C.J. Ter Braak, M.P. Clark, J.M. Hyman, and B.A. Robinson. 2008. Treatment of input uncertainty in hydrologic modeling: Doing hydrology backward with Markov chain Monte Carlo simulation. Water Resour. Res. doi:10.1029/2007WR006720

Werner, C., K. Butterbach-Bahl, E. Haas, T. Hickler, and R. Kiese. 2007. A global inventory of $N_2O$ emissions from tropical rainforest soils using a detailed biogeochemical model. Global Biogeochem. Cycl. doi:10.1029/2006GB002909

Yeluripati, J.B., M. Van Oijen, M. Wattenbach, A. Neftel, A. Ammann, W.J. Parton, and P. Smith. 2009. Bayesian calibration as a tool for initialising the carbon pools of dynamic soil models. Soil Biol. Biochem. 41:2579–2583. doi:10.1016/j.soilbio.2009.08.021

Zaehle, S., S. Stitch, B. Smith, and F. Hatterman. 2005. Effects of parameter uncertainties on the modeling of terrestrial biosphere dynamics. Global Biogeochem. Cycl. doi:10.1029/2004GB002395

Zhou, X., T. Zhou, and Y. Luo. 2012. Uncertainties in carbon residence time and NPP-driven carbon uptake in terrestrial ecosystems of the conterminous USA: A Bayesian approach. Tellus B 64. doi:10.3402/tellusb.v64i0.17223

# Modeling Greenhouse Gas Emissions from Enteric Fermentation

**Ermias Kebreab,* Luis Tedeschi, Jan Dijkstra, Jennifer L. Ellis, Andre Bannink, and James France**

## Abstract

Livestock directly contribute to greenhouse gas (GHG) emissions mainly through methane ($CH_4$) and nitrous oxide ($N_2O$) emissions. For cost and practicality reasons, quantification of GHG has been through development of various types of mathematical models. This chapter addresses the utility and limitations of mathematical models used to estimate enteric $CH_4$ emissions from livestock production. Models used in GHG quantification can be broadly classified into either empirical or mechanistic models. Empirical models might be easier to use because they require fewer input variables compared with mechanistic models. However, their applicability in assessing mitigation options such as dietary manipulation may be limited. The major driving variables identified for both types of models include feed intake, lipid and nonstructural carbohydrate content of the feed, and animal variables. Knowledge gaps identified in empirical modeling were that some of the assumptions might not be valid because of geographical location, health status of animals, genetic differences, or production type. In mechanistic modeling, errors related to estimating feed intake, stoichiometry of volatile fatty acid (VFA) production, and acidity of rumen contents are limitations that need further investigation. Model prediction uncertainty was also investigated, and, depending on the intensity and source of the prediction uncertainty, the mathematical model may inaccurately predict the observed values with more or less variability. In conclusion, although there are quantification tools available, global collaboration is required to come to a consensus on quantification protocols. This can be achieved through developing various types of models specific to region, animal, and production type using large global datasets developed through international collaboration.

Abbreviations: Ac, acetate; ADF, acid detergent fiber; ALU, Agricultural and Land Use National Greenhouse Gas Inventory; BW, body weight; DMI, dry matter intake; EE, ether extract; GE, gross energy; GEI, gross energy intake; GHG, greenhouse gas; IFSM, Integrated Farm Systems Model; IPCC, Intergovernmental Panel on Climate Change; NDF, neutral detergent fiber; Pr, propionate; RFI, residual feed intake; RSE, regression standard error; UQ, uncertainty quantification; VFA, volatile fatty acid.

Ermias Kebreab, Dep. of Animal Science, 2111 Meyer Hall, One Shields Ave., Univ. of California, Davis, CA 95616. *Corresponding author (ekebreab@ucdavis.edu). Luis Tedeschi, Dep. of Animal Science, Room 133 Kleberg, Texas A&M Univ. College Station, TX 77843-2471 (luis.tedeschi@tamu.edu). Jan Dijkstra (jan.dijkstra@wur.nl) and Jennifer L. Ellis (jen.ellis.stpierre@gmail.com), Animal Nutrition Group, Wageningen Univ., PO Box 338, 6700AH, Wageningen, The Netherlands. Jennifer L. Ellis and James France (jfrance@uoguelph.ca), Centre for Nutrition Modelling, Dep. of Animal and Poultry Science, Univ. of Guelph, Guelph, ON, N1G 2W1 Canada. Andre Bannink, Wageningen Univ. and Research Centre (UR) Livestock Research, 8219 PH Lelystad, The Netherlands (andre.bannink@wur.nl).

doi:10.2134/advagricsystmodel6.2013.0006


Synthesis and Modeling of Greenhouse Gas Emissions and Carbon Storage in Agricultural and Forest Systems to Guide Mitigation and Adaptation
S. Del Grosso, L. Ahuja, and W. Parton, Editors
Advances in Agricultural Systems Modeling, Volume 6. Lajpat R. Ahuja, Series Editor

518
804

Enteric fermentation, manure decomposition, and denitrification in soil are the main processes contributing directly to GHG emissions from livestock. These processes are the largest sources of GHG from any animal production system, with $CH_4$ and $N_2O$ being the main GHGs (Solomon et al., 2007). In most livestock-related GHG analyses, carbon dioxide ($CO_2$) emissions are assumed to have a zero net balance, as the C fixed by grazed or harvested plants in established grazing systems is considered to be respired by animals. However, in some livestock production systems, machinery use and associated $CO_2$ production could significantly contribute to the C balance. Livestock's contribution to the total GHG emissions varies from country to country, but at a global level it has been estimated to be as high as 18% of anthropogenic GHG emissions (Steinfeld et al., 2006). In 2012, ~8.1% of anthropogenic GHG emissions in the United States came from agriculture (USEPA, 2014). The US livestock sector was the second largest emitter of $CH_4$ (34.4% of the total anthropogenic $CO_2$–equivalent emission from agriculture) and animal manure was the third largest source of $N_2O$ (5% of the total $CO_2$–equivalent emission from agriculture) (USEPA, 2014). These two gases are highly potent GHGs (Solomon et al., 2007), with a global warming potential of 25 and 310 times, respectively, that of $CO_2$ over a 100-yr time horizon. Thus, enteric fermentation and manure decomposition, the processes responsible for these emissions, should be the primary target of GHG mitigation practices designed for livestock industries (Hristov et al., 2013). Enteric $CH_4$ emissions have been reported to contribute up to 70% of emissions in life-cycle analyses of ruminant production systems (Johnson et al., 2003; Vergé et al., 2008; Beauchemin et al., 2010). Consequently, most research has been focused on quantification and mitigation of enteric $CH_4$ emissions. However, enteric fermentation is a necessary process by which human-inedible fibers are converted into high-quality food products. Therefore, there are limits to reduction of enteric methane emissions without compromising productivity. Nitrous oxide emissions from livestock production systems are mostly from manure applied to soil, which is a subject of other chapters in this book. The objective of this chapter is to examine the utility and limitations of mathematical models used to estimate enteric $CH_4$ emissions from livestock production.

## Synopsis of the Current State of the Science

### Empirical versus Mechanistic Models

Because of the complex and often-expensive equipment required to measure GHG production from livestock, predictive equations are widely used to estimate GHG emissions. Models can be broadly classified into two groups: statistical or empirical models that relate nutrient intake to output (such as $CH_4$), directly, and dynamic mechanistic models that aim to simulate GHG emissions based on a mathematical description of underlying processes (e.g., rumen fermentation biochemistry for predicting enteric $CH_4$ emission).

### Empirical or Regression Models

Predictions of emissions are made using methods that range from simple fixed values estimated from literature (e.g., Intergovernmental Panel on Climate Change (IPCC), 2006) to statistical models using state-of-the-art statistical methodologies (e.g., Moraes et al., 2014). Table 1 summarizes empirical or regression equations from the literature (Eq. [1–24]). The advantages of empirical models include easier construction and less demand on input variables used for prediction. Therefore, depending on the objectives, statistical models can be more practical. The main disadvantage of an empirical model is that model parameters may not be biologically meaningful and may result in nutrient–emission relationships that are contrary to biological principles (e.g., Ellis et al., 2010). Empirical models are likely only applicable within the range of data on which the model was developed.

The IPCC, in its revised reference manual (IPCC, 2006), outlined two methodologies to estimate $CH_4$ emissions from livestock enteric fermentation, viz., Tier 1 and Tier 2 (Table 1). A third methodology, Tier 3, is also recommended for countries with detailed information on animal and diet characteristics. The IPCC Tier 1 method is a simplified approach that assigns default $CH_4$ emissions to distinct animal categories. Therefore, only readily available animal population data are needed to estimate emissions. For example, a North American dairy cow (*Bos Taurus*) is estimated to produce 128 kg $CH_4$ $yr^{-1}$ through enteric fermentation (Table 1) assuming an average milk production of 8400 kg $head^{-1}$ $yr^{-1}$. A limitation in this approach is that animals consume various amounts and types of feed, which influences $CH_4$ emissions significantly. Therefore, variability resulting from feed composition and amount would not be captured in Tier 1 methodology. In addition, the only mitigation that can be assessed using this methodology is reduction of animal numbers, which is not viable in most circumstances.

The IPCC Tier 2 method is appropriate for estimating $CH_4$ emissions when more detailed data about the livestock sector are available by using $CH_4$ emis-

**Table 1. Empirical methane prediction equations (sorted by date) (updated from Archibeque et al., 2012).**

| Eq. | Prediction equation for methane† | Reference |
|---|---|---|
| | $g\ d^{-1}$ | |
| [1] | $9 + 22.6 \times DMI$ | Kriss (1930) |
| [2] | $[-2.07 + 2.63 \times DMI - 0.105 \times DMI^2]/0.05565$ | Axelsson (1949) |
| [3] | $[1.3 + 0.112 \times (EDm) + MN \times (2.37 - 0.050 \times (EDm))]/100 \times GEI/0.05565$ | Blaxter and Clapperton (1965) |
| [4] | $[2.81 + 0.042 \times DOMI]/1.4$ | Murray et al. (1976) |
| [5] | $[3.41 + 0.511 \times NSC + 1.74 \times HC + 2.65 \times CEL]/0.05565$ | Moe and Tyrrell (1979) |
| [6] | Equation 1: $63 + 79 \times CF + 10 \times NFE + 26 \times CP - 212 \times EE$ | Kirchgeßner et al. (1995)‡ |
| [7] | Equation 2: $10 + 4.9 \times MY + 1.5 \times BW^{0.75}$ | |
| [8] | $DEI\ (0.094 + 0.028\ S_{ADFI}/T_{ADFI}) - 2.453\ (MN - 1) \times 18.1$ | Yan et al. (2000) |
| [9] | $[50 + 0.01 \times MY \times 365]/365 \times 1000$ | Corré (2002) |
| [10] | Linear 1: $[5.93 + 0.92 \times DMI]/0.05565$ | Mills et al. (2003) |
| [11] | Linear 2: $[8.25 + 0.07 \times MEI]/0.05565$ | |
| [12] | Linear 4: $[1.06 + 10.27 \times$ forage proportion in dietary DM $+ 0.87 \times DMI]/0.05565$ | |
| [13] | Nonlinear 1: $[56.27 - (56.27 + 0)\ e^{(-0.028 \times DMI)}]/0.05565$ | |
| [14] | Nonlinear 2: $[45.89 - (45.89 + 0)\ e^{(-0.003 \times MEI)}]/0.05565$ | |
| [15] | Nonlinear general $a - (a + b)\ e^{-[(0.0011 \times Starch(kg/d)/ADF] + 0.0045)]MEI}$ | |
| [16] | $[45.0 - 0.018 \times DMI\ (g\ kg^{-1}\ BW\ d^{-1})^2 - 1.84 \times C18{:}2\ (\%DM) - 84.2 \times C \geq 20\ (\%DM)] \times DMI\ (kg\ d^{-1}) \times 0.6802$ | Giger-Reverdin et al. (2003) |
| [17] | $20 \times$ Concentrate Intake (kg as fed $d^{-1}$) $+ 22 \times$ Corn Silage Intake (kg DM $d^{-1}$) $+ 27 \times$ Grass Intake (kg DM $d^{-1}$) | Schils et al. (2006) |
| [18] | Tier 1: 351 for North America, 321 for western Europe | IPCC (2006) |
| [19] | Tier 2: $[0.065 \times GEI]/0.05565$ | |
| [20] | Beef: $[2.94 + 0.059 \times MEI + 1.44 \times ADF - 4.16 \times$ lignin intake (kg $d^{-1}$)]/0.05565 | Ellis et al. (2007)§ |
| [21] | Dairy 1: $[8.56 + 0.14 \times$ forage (%DMI)]/0.05565 | |
| [22] | Dairy 2: $[3.23 + 0.81 \times DMI]/0.05565$ | |
| [23] | Combined beef and dairy: $[3.27 + 0.74 \times DMI]/0.05565$ | |
| [24] | $[(9.75 - 0.05 \times$ fecal DM digestibility %)/100 $\times GEI]/0.05565$ | FAO (2010) |

† $a$, theoretical maximum $CH_4$ output (g $d^{-1}$); $b$, minimum $CH_4$ output (g $d^{-1}$); ADF, acid detergent fiber (kg $d^{-1}$); BW, bodyweight (kg); C18:2, the quantity of linoleic acid (% of dry matter [DM]); CEL, cellulose (kg $d^{-1}$); CF, crude fiber (kg $d^{-1}$); CP, crude protein (kg $d^{-1}$); DEI, digestible energy intake (MJ $d^{-1}$); DMI, dry matter intake (kg $d^{-1}$); DOMI, digestible organic matter intake (g $d^{-1}$); EDm, energy digestibility at maintenance (% of gross energy); EE, ether extract (fat, kg $d^{-1}$); GEI, gross energy intake (MJ $d^{-1}$); HC, hemicellulose (kg $d^{-1}$); MEI, metabolizable energy intake (MJ $d^{-1}$); MN, multiple of maintenance; MY, milk yield (kg $d^{-1}$); NFE, nitrogen-free extract (kg $d^{-1}$); NSC, nonstructural carbohydrate (kg $d^{-1}$); $S_{ADFI}$, silage ADF intake (kg $d^{-1}$); $T_{ADFI}$, total ADF intake (kg $d^{-1}$).

‡ Intercept becomes 59 in Eq. [2] if the diet is based on corn silage.

§ Ellis et al. (2007) developed 14 equations for beef, 8 for dairy, and 10 combined; only the three best-fitting equations are presented here.

sions factors ($Y_m$; Table 1. Eq. [19]). The $Y_m$ represents the proportion of the animal's gross energy intake (GEI) that is lost as $CH_4$. The IPCC (2006) recommends a fixed $Y_m$ of 6.5 ± 1.0% for all categories of cattle, except feedlot cattle consuming diets with 90% or more concentrates, for which a $Y_m$ value of 3.0 ± 1.0% is recommended. Tier 2 is useful for quickly calculating expected emissions at a country level and is superior to Tier 1, as it takes feed energy intake into account. However, $Y_m$ does not have the capacity to fully describe changes in composition of the diet and has limited use when estimating the impact of varying nutritional strategies on enteric $CH_4$ emissions. For example, the Tier 2 model may not distinguish between an increase in energy intake caused by an increase in dry matter intake (DMI) and that caused by an increase in the fat content of the diet, both of which would have differing effects on the resulting enteric $CH_4$ emission (Ellis et al., 2010).

Bannink et al. (2011) developed a Tier 3 method for conditions in The Netherlands, with which it is possible to evaluate specific details of nutritional management. This is because the underlying mechanisms of enteric fermentation are represented dynamically. The Tier 3 approach used new VFA stoichiometry developed by Bannink et al. (2008) and integrated it with the COWPOLL model based on Dijkstra et al. (1992), which is discussed in the next section. The predictions from the Tier 3 approach of Bannink et al. (2011) were more in agreement with an older version of IPCC Tier 2 (IPCC, 1997; namely, $Y_m$ = 6.0%) than with the updated version (IPCC, 2006; $Y_m$ = 6.5%). Uncertainty of prediction by the Tier 3 model was estimated to be 15%, which is less than the estimated 20% for the IPCC Tier 2 approach. It is expected that most developed countries will move toward a Tier 3 approach using country-specific mechanistic models.

## Mechanistic Models

These models represent the underlying processes that control emissions and their interactions. A brief description of two mechanistic models that appear frequently in the literature, COWPOLL (Dijkstra et al., 1992; Mills et al., 2001; Bannink et al., 2006) and MOLLY (Baldwin et al., 1987a,b,c; Baldwin, 1995) is given. COWPOLL is a dynamic mechanistic model designed to simulate digestion, absorption, and outflow of nutrients in the rumen (Dijkstra et al., 1992). The model contains 19 state variables representing N, carbohydrate (neutral detergent fiber [NDF], starch, and sugar), lipid, and VFA pools. The model assumes continuous feeding using Michaelis–Menten or mass action kinetics. Enteric $CH_4$ production is estimated based on VFA stoichiometry developed by Bannink et al. (2006), which relates the VFA produced to the type of substrate fermented in the rumen. The assumption is that the hydrogen ($H_2$) produced in the rumen from fermentation of carbohydrate and protein is used (i) to support rumen microbial growth, (ii) for biohydrogenation of unsaturated fatty acids, and (iii) for production of glucogenic VFA (i.e., propionate and valerate). The remaining $H_2$ is used for reduction of $CO_2$ to $CH_4$ (Baldwin, 1995), and the prediction from rumen methanogenesis and hindgut fermentation is described by Mills et al. (2001). The model has been used to estimate enteric $CH_4$ emissions mostly from dairy cattle (Kebreab et al., 2008; Morvay et al., 2011; Alemu et al., 2011a). A version with an updated VFA stoichiometry that includes effect of rumen pH on stoichiometry of VFA formed on fermentation of soluble sugars and starch (Bannink et al., 2008) is used as a Tier 3 method in The Netherlands (Bannink et al., 2011). Recently, Ellis et al. (2011) have introduced modifications to the model to be able to better handle predictions for beef cattle.

MOLLY is a dynamic mechanistic model that simulates rumen digestion and whole-body metabolism in lactating dairy cows (Baldwin et al. 1987a,b,c; Baldwin 1995). The model was constructed assuming continuous feeding using Michaelis–Menten or mass action kinetics. The digestion element of the model is comprised of 15 state variables. Chemical composition of the diet is presented as starch, cellulose, hemicellulose, lignin, soluble carbohydrate, acetate, propionate, butyrate, crude protein (soluble and insoluble), nonprotein N, urea, ash (soluble and insoluble), lipid, organic acid, lactate, pectin, and fat. After microbial attachment and substrate hydrolysis, the rumen model uses stoichiometric coefficients to convert starch, soluble carbohydrates, and amino acids into VFA. The VFA stoichiometry is based on the equations developed by Murphy et al. (1982) and later updated by Argyle and Baldwin (1988), which relate the amount of VFA produced to the type of substrate fermented in the rumen. As in the COWPOLL model, excess $H_2$ produced is assumed to be used by methanogens for reduction of $CO_2$ to $CH_4$.

In addition to the stoichiometric differences described above, MOLLY and COWPOLL differ in the number of microbial pools: MOLLY uses one microbial pool whereas COWPOLL uses three pools (amylolytic, fibrolytic, and protozoa).

## Whole-Farm Models

Currently, there are a number of whole-farm models that simulate the impact of various management and nutritional strategies on GHG emissions. Most whole-farm models are a combination of empirical models, in light of their relative simplicity and fewer input requirements, although some have moved toward more detailed mechanistic representations (e.g., DairyNZ uses MOLLY). Thus, whole-farm models differ in their level of complexity and structure, and the predictive power of such models depends on the accuracy of the mathematical equations and data inputs used. Calculation software such as the Cool Farm

Tool (http://www.unilever.com/aboutus/supplier/sustainablesourcing/tools/) and Agricultural and Land Use National Greenhouse Gas Inventory (ALU) (http://www.nrel.colostate.edu/projects/ALUsoftware/index.html) are available online. These whole-farm models are based on IPCC (2006) equations, and they simulate emissions from various types of animals, manure storage, and soil using a factorial approach, that is, nutrient flows from one section (e.g., animals) does not necessarily follow to the next phase in the farm. The ALU has been developed with the help of USEPA for inventory calculations for countries using Tier 1 or 2 methodologies to facilitate capacity building (Ogle et al., 2016, this volume). More complex models, such as Holos (www.agr.gc.ca/holos-ghg) and the Integrated Farm Systems Model (IFSM, developed by USDA-ARS; Rotz et al., 2011), simulate whole-farm emissions of GHG, including $CH_4$, and evaluate the overall impact of management strategies used to reduce $CH_4$ emissions. The IFSM is a process-based whole-farm simulation model that incorporates soil processes, crop growth, tillage, planting and harvest operations, feed storage, feeding, herd production, manure storage, and economics. The IFSM uses the Mills et al. (2003) equation (Eq. [15] in Table 1) to predict $CH_4$ emissions from enteric fermentation. Evaluation of enteric $CH_4$ emissions, as determined by IFSM against independent data, is not yet available. Recently, Van Middelaar et al. (2013) evaluated the GHG reduction potential of increasing corn (*Zea mays* L.) silage at the expense of grass or grass silage in the diet of dairy cattle. They used a linear programming dairy farm model (maximizing labor income) with COWPOLL to quantify enteric $CH_4$ production and a life-cycle assessment. According to COWPOLL, increasing corn silage in the diet reduced enteric $CH_4$ emissions. However, plowing grassland for corn land resulted in significant one-off emissions of GHG and it took >40 yr before annual emission reduction has paid off these emissions related to land-use change.

## Model Comparisons

Several studies have evaluated the predictive potential of empirical and mechanistic models for enteric $CH_4$ production from cattle using independent data sources (Benchaar et al., 1998; Kebreab et al., 2006, 2008; Alemu et al., 2011a). Wilkerson et al. (1995) reviewed several empirical models and recommended adoption of the Moe and Tyrrell (1979) equation for dairy cows. Mills et al. (2003) also compared various empirical models from the literature as well as models developed by the authors using an extensive calorimetry dataset. Although the authors found that the Moe and Tyrrell (1979) equation provided reasonable predictions, their own nonlinear models not only improved predictions, but they were also less prone to misapplication. Linear models gave unrealistically high emission values as DMI increased, whereas nonlinear models yielded values approached theoretical maximum emissions and more biologically realistic predictions. Benchaar et al. (1998) compared the predictive capacity of two mechanistic and two linear models with a database constructed from literature data. Predictions from linear equations were poor; the models explained between 42 and 57% of the variation. The mechanistic models, on the other hand, explained more than 70% of the variation.

Ellis et al. (2010) evaluated empirical equations used within eight whole-farm models to predict $CH_4$ emissions from dairy cows. In general, predictions were poor with severe overprediction and underprediction of $CH_4$ production apparent in several models. This suggests that the evaluated equations cannot fully describe the underlying causes of variation. Simple, more generalized equations performed worse than those that attempted to represent some aspects of diet composition. Although some of the equations include basic aspects of the chemical composition of the diet and indirectly take account of the effect of feed intake level, other important descriptors that affect $CH_4$ production, such as fat fraction (review Beauchemin et al., 2009), were still lacking. Most of the equations also assume a constant $CH_4$ yield per unit of substrate (e.g., nonstructural carbohydrate), but variation in intake level may affect their rumen fermentability, pH, and, as a consequence, VFA profile and $CH_4$ release (Bannink et al., 2008). Although use of the IPCC Tier 2 in whole-farm models is adequate, it is based simply on GEI and does not fully describe changes in diet composition (Ellis et al., 2010). For example, GEI may increase as a result of greater feed intake, likely resulting in increased $CH_4$ emissions, but GEI may also increase if dietary fat content is increased, likely resulting in a decrease in $CH_4$ production. If a whole-farm model is attempting to describe the effect of a given mitigation strategy on $CH_4$ emissions and, subsequently, the effects on other aspects of the farm nutrient balance, empirical equations may not be robust enough, as they lack adequate sensitivity to change. The low prediction accuracy and poor prediction of variation in observed $CH_4$ production values may introduce substantial error into inventories of GHG emissions and lead to incorrect mitigation recommendations on a whole-farm level. However, mechanistic models require more detailed input data that may not be readily available, which hamper their inclusion in whole-farm models.

Kebreab et al. (2006) chose six models comprising two linear (Moe and Tyrrell, 1979; Mills et al., 2003), a nonlinear (Mills et al., 2003), IPCC Tier 1 and 2 (IPCC, 1997), and a mechanistic model (COWPOLL) for predicting $CH_4$ production and evaluated them using North American data. The linear models were recommended for use where nutrient information is limited and within the range in which the models were developed. The nonlinear model of Mills et al. (2003) could be used for extrapolation beyond the range of data used for developing the model, but the mechanistic model was recommended for assessment of mitigation options. Bannink et al. (2011) also demonstrated the less strong increase in $CH_4$ emissions predicted with the Dutch Tier 3 as compared with prediction with IPCC Tier 2 on changes in feed intake level and diet composition from 1990 to 2008. The Tier 1 model was found to be adequate for general inventory of $CH_4$ emissions but inadequate for assessing the mitigation options, as it relies on fixed default emissions factors. Although relatively less accurate in predicting the mean, the Tier 2 model is better suited than Tier 1 when information on feed intake is available (Kebreab et al., 2006).

Kebreab et al. (2008) evaluated two empirical models (IPCC, 2006; Moe and Tyrrell, 1979) and two mechanistic models (COWPOLL and MOLLY) for their predictive capacity using individual cattle measurements of $CH_4$ emissions. In dairy cattle, COWPOLL yielded the least predictive error. However, in feedlot cattle, MOLLY yielded the least predictive error. The predictions of the IPCC (2006) model agreed in large measure with observed values. The average observed $Y_m$ values were 5.63% of gross energy (GE) (a range of 3.78 to 7.43%) for the dairy cow database and 3.88% (a range of 3.36 to 4.56%) for the feedlot cattle database. The authors reported that using IPCC (2006) values might result in overestimation of emissions by 12.5% for dairy cattle and underestimation by 9.8% for feedlot cattle. In addition to providing more accurate estimates of emissions based on diet characteristics, mechanistic models, unlike empirical models, can help researchers assess mitigation options such as changing carbohydrate sources to reduce methane. More recently, using observations in lactating dairy cows, Alemu et al. (2011a) evaluated prediction accuracy of $CH_4$ production for various VFA stoichiometric approaches and for IPCC Tier 2. The Bannink et al. (2006) VFA stoichiometry resulted in the lowest prediction error, with $CH_4$ production underestimated by 4.8%. The IPCC Tier 2 approach had a considerably greater prediction error and $CH_4$ production was overestimated by 8.0%. Alemu et al. (2011b) compared empirical models and the VFA stoichiometry used in mechanistic models to estimate and assess trends in enteric $CH_4$ emissions from western Canadian beef cattle. The authors concluded that a more robust approach might be to use mechanistic models to estimate regional $Y_m$ values, which are then used as input for IPCC models for inventory purposes. The Cattle Enteric Fermentation Model (CEFM) used by the USEPA is an approximation to this approach but requires a mechanistic model to simulate $Y_m$ values to be adopted as a Tier 3 system similar to the Dutch system (Bannink et al., 2011). Use of a mechanistic model and VFA stoichiometry requires detailed nutrient composition input into the model, which may not be readily available.

# Major Driving Variables in Greenhouse Gas Quantification

## Feed Intake

Regardless of the type of model used, the major driving variables for predicting enteric $CH_4$ emissions include a measure of intake (dry matter [DM]), GE, metabolizable energy (e.g., Eq. [1, 2, 10, 11, 13, 14, 16, 19, 22, 23]), body weight (BW; Eq. [7, 16]), milk production (Eq. [7, 9]), proportion of forage (e.g., Eq. [12, 21]), and chemical composition of the diet such as NDF, acid detergent fiber (ADF), lipid, and nonstructural carbohydrate (e.g., Eq. [5, 15]). An analysis of 2574 observations on methane emissions produced the relationship depicted in Fig. 1 and given by Eq. [25] (SE in parentheses):

Enteric $CH_4$ (g $d^{-1}$) =
38.2 (1.64) + 16.2 (0.19) $\times$ DM intake, kg $d^{-1}$ ($R^2 = 0.85$; $P < 0.001$) [25]

The data were analyzed by treating the study effect as random (the LME4 package of R; http://www.r-project.org). Using values collected from the literature ($n = 377$), Hristov et al. (2013) established the following relationship:

Enteric $CH_4$ (g $d^{-1}$) =
2.54 (4.89) + 19.1 (0.43) $\times$ DM intake, kg $d^{-1}$ ($R^2 = 0.86$; $P < 0.001$) [26]

These two equations differ largely in their estimation of the intercept (feed intake = 0 kg $d^{-1}$), which has an influence on the slope. However, in both cases, the slope is positive and highly significant. Feed intake has a significant influence on prediction of $CH_4$ emissions, which is evident from the above equations and the number of empirical models that include intake as the basis for enteric $CH_4$ prediction (Table 1). Figure 1 shows that although there is a strong positive relationship between feed intake and enteric $CH_4$ emissions, for a given level of feed intake, there is a wide range of potential emissions (e.g., a cow consuming 15 kg $d^{-1}$ DM can produce 80 to 400 g $CH_4$ $d^{-1}$ or 5 to 27 g $CH_4$ $kg^{-1}$ DM), which indicates there are other variables that influence $CH_4$ emissions apart from DM intake.

Moraes et al. (2014) investigated some of the major driving variables cited in the literature for estimating enteric $CH_4$ emissions from dairy cattle. Model development was conducted using a sequential approach, with model complexity increasing at each level. The independent variables, which play a key role in predicting $CH_4$ emissions, were selected through a Bayesian model selection procedure. In this study, variables were selected under a linear hierarchical model. At the dietary level, diet characteristics (fiber fractions, crude protein, fat, lig-

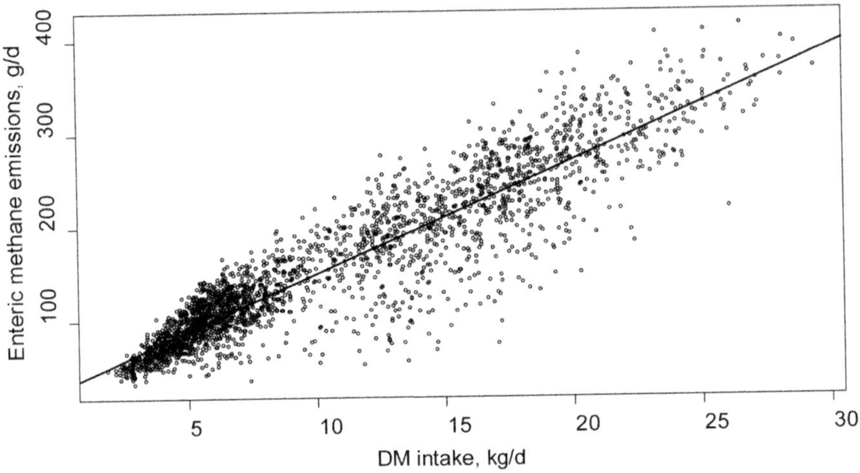


**Fig. 1. Relationship between dry matter (DM) intake (kg d$^{-1}$) and enteric methane emissions in dairy cattle (g d$^{-1}$).**

nin, and metabolizable energy) were allowed in the regression as well as GEI. At the animal level, milk composition (fat, protein, and nonfat soluble) and animal information (BW and breed) could also be selected, as well as variables from the dietary level. Not surprisingly, GEI was the key variable in predicting $CH_4$ emissions. Dietary proportions of NDF and ether extract (EE) or fat were present in the models selected at the dietary and animal levels of lactating cows. The probability of BW being included in the most probable model was 0.99. The high marginal probability of milk fat being included in the model suggests that milk fat percentage has an important role in predicting $CH_4$ emissions. Breed's marginal probability of differing from zero was low though ($P < 0.13$), so it was not used to build equations. Equations [27–29] were developed based on the procedure outlined above:

$$CH_4\ (\mathrm{MJ\ d^{-1}}) = 3.25\ (0.42) + 0.043\ (0.001) \times \mathrm{GEI\ (MJ\ d^{-1})} \quad [27]$$

$$CH_4\ (\mathrm{MJ\ d^{-1}}) = 0.2242\ (0.71) + 0.042\ (0.001) \times \mathrm{GEI\ (MJ\ d^{-1})} + 0.124\ (0.01) \times \mathrm{NDF\ (\%DM)} - 0.329\ (0.09) \times \mathrm{EE\ (\%DM)} \quad [28]$$

$$CH_4\ (\mathrm{MJ\ d^{-1}}) = -9.317\ (1.06) + 0.042\ (0.001) \times \mathrm{GEI\ (MJ\ d^{-1})} + 0.094\ (0.01) \times \mathrm{NDF\ (\%DM)} - 0.381\ (0.09) \times \mathrm{EE\ (\%DM)} + 0.008\ (0.001) \times \mathrm{BW\ (kg)} + 1.622\ (0.11) \times \mathrm{Milk\ Fat\ (\%)} \quad [29]$$

A *K*-fold cross-validation analysis was conducted and the root mean square prediction error (percentage of the observed mean) was 18.1, 17.9, and 15.6% for Eq. [27–29], respectively, while prediction based on the IPCC Tier 2 equation (IPCC, 2006) was 30.1%.

## Lipids

Another major driving variable for predicting enteric $CH_4$ emissions is fat (lipid) content of the diet. There is consensus that lipids (regardless of origin) suppress $CH_4$ production in the rumen. Some of the reductions in emissions could be due to reduced feed intake without compromising (or even increasing) productivity. For example, in a recent meta-analysis, Rabiee et al. (2012) reported a consistent decrease in DMI with all types of dietary fat examined (tallow, various calcium salts of fatty acids, oilseeds, prilled fat), and milk production was increased. Therefore, a combination of reduced intake and increased milk production resulted in net reduction of emissions per unit of product. Grainger and Beauchemin (2011) analyzed 27 studies and concluded that, within a practical feeding rate of less than 8% fat in the diet, a 10 g kg$^{-1}$ increase in dietary fat would decrease $CH_4$ yield by 1.0 g kg$^{-1}$ DMI in cattle and 2.6 g kg$^{-1}$ in sheep. The modes of action of lipids could be several, such as exchanging fermentable substrate for nonfermentable ones, direct microbial inhibition, and consuming $H_2$ in biohydrogenation. However, biohydrogenation of dietary fatty acids per se is a minor competitor for reducing equivalents compared with methanogenesis (Jenkins et al., 2008), but modeling of biohydrogenation and integrating results with methanogen activity might help predict dietary conditions when lipid compounds are effective without adding significant risk of milk fat depression or decreasing fibrolytic activity of rumen microbiota.

## Structural versus Nonstructural Carbohydrates

It is well established that structural carbohydrates (NDF, ADF) have greater methanogenic potential than nonstructural carbohydrates such as starch and sugars (Moe and Tyrrell, 1979). Wolin (1960) argued that the stoichiometry of rumen fermentation dictates that more $H_2$—and consequently relatively more $CH_4$—will be produced on fermentation of fiber than starch (in the latter case reducing equivalents can be used for propionate synthesis). Noziere et al. (2010) estimated that molar proportions of acetate (Ac), propionate (Pr), and butyrate would average, respectively, 66, 17, and 14 mol (100 mol)$^{-1}$ for NDF and 41, 44, and 12 mol (100 mol)$^{-1}$ for starch. Indeed, a 72 vs. 52% concentrate diet produced a 59% increase in rumen propionate concentration and a 44% drop in Ac/Pr ratio in lactating dairy cows, accompanied by milk fat depression from 4.20 to 3.20 (Agle et al., 2010).

Forages usually have greater NDF content than grains. Among forages, however, NDF content and digestibility vary significantly. Grasses, for example, have greater cell wall (i.e., NDF) content and NDF is generally more digestible than the NDF in legumes. One of the mitigation strategies to reduce $CH_4$ is to improve grass quality by reducing NDF content and increasing the sugar content. Ellis et al. (2012) simulated the effect of high-sugar grasses on $CH_4$ emissions. The authors reported that when water-soluble sugar content in grass increased at the expense of NDF, simulated milk yield increased and as a result $CH_4$ emissions (grams per kilogram of milk) tended to decrease. However, when water-soluble sugar content increased at the expense of protein content, simulated $CH_4$ emissions (grams per kilogram of milk) increased.

### Animals

To a lesser extent, there are also animal-related differences that influence $CH_4$ emissions. These include BW, milk yield and composition, breed, and adaptation to environmental stressors. From the animal-related factors, Moraes et al. (2014) showed BW and milk-fat composition to be the most significant ones (Eq. [29]). In ruminants, rumen pH plays an important role in the stoichiometry of VFAs. In a meta-analysis, Bannink et al. (2008) showed that the fermentation of sugars and starch would shift rumen fermentation toward production of propionate with decreasing rumen pH, comparable to earlier results by Argyle and Baldwin (1988).

## Knowledge Gaps

Although there is much research devoted to understanding GHG emissions and, to a lesser extent, prediction of emissions, there still are knowledge gaps and further research is needed to improve understanding and quantification of GHG emissions based on biological processes in livestock.

### Empirical Modeling

Most of the empirical models developed so far are based on high-producing animals in intensive agriculture. An assumption is that those animals are healthy and the emission prediction is not complicated by environmental or other factors. Most of the data used for prediction of emissions has been collected under highly controlled conditions in research stations. This ideal scenario may not reflect the reality in commercial livestock operations. For example, it is not known if GHG emissions will be impacted in an animal suffering from lameness or mastitis, which is common in commercial operations. There is also lack of knowledge on the effect of environmental stressors, such as heat or cold stress, on GHG emissions. Animals reduce their intake during heat stress, which will lead to fewer emissions, and increase their intake under cold stress, but if calculated per unit of DM intake, it is unclear whether animals have greater or similar $CH_4$ emissions.

Current empirical models may not be good predictors for pasture-based livestock systems and, in particular, animals raised in the tropics. Estimates of DMI are not reliable in such systems, and, therefore, prediction using current models may not be accurate. The uncertainty in estimates comes from calculating DMI rather than using prediction equations to estimate $CH_4$ emissions. There is a need to develop models that are based on data collected from such systems and for different types of tropical breeds. For example, Kennedy and Charmley (2012) measured $CH_4$ production by open-circuit gas exchange from 13 Brahman cattle (*Bos taurus indicus*) and, based on their results, recommended a significant downward revision of the $CH_4$ emissions attributable to the northern Australian beef herd grazing tropical pastures. Similar work should be conducted in other regions, such as Southeast Asia and Latin America, to develop more-appropriate empirical models that are targeted to the production system under investigation and enteric fermentation conditions in place.

Genetic selection has been investigated as one of the solutions to reduce enteric $CH_4$ emissions in ruminants. For example, some studies using beef cattle indicate that animals with low residual feed intake (RFI) (i.e., consume less feed than expected) can produce 28% less enteric $CH_4$ (Nkrumah et al., 2004; Hegarty et al., 2010). However, there is little evidence that efficient animals (low RFI) have a different $CH_4$ yield per unit of DM digested. In fact, Weber et al. (2013) reported that there was no statistically significant association between $CH_4$ and sire genomic breeding value for either RFI or progeny RFI phenotype. Genomic selection tools can further increase the reduction in $CH_4$ production, but effective application requires significant international effort and collaboration to bring together relevant data, which preferably include measurements of feed digestion.

## Mechanistic Modeling

Although current mechanistic models have greater prediction accuracy and a demonstrably more adequate representation of relationships than the empirical approach, the prediction accuracy of current models requires further improvement. Bannink et al. (2011) identified the most likely sources of uncertainty in current mechanistic models to be errors related to estimating feed intake, stoichiometry of VFA production, and acidity of rumen contents or other rumen fermentation conditions affecting rumen function. As shown earlier, feed intake is strongly and positively related to enteric $CH_4$ production. Therefore, errors in feed intake estimation will lead directly to errors in predicted $CH_4$. This applies particularly to empirical models used for pasture-based systems with high uncertainty surrounding pasture intake.

Current mechanistic models are based on the concept that end products of ruminal fermentation are mostly VFA, $H_2$, $CH_4$, ammonia, and $CO_2$. Prediction of enteric $CH_4$ is an outcome of simulated $H_2$ economy, which is represented by VFA stoichiometry and related $H_2$ production with feed fermentation. Thus, accuracy of predicted $CH_4$ production strongly depends on accuracy of quantifying VFA produced in the rumen. In most cases, VFA stoichiometry values are derived using meta-analysis of studies that quantified the molar proportions of VFA based on dietary inputs (e.g., Bannink et al., 2006). Alemu et al. (2011a) showed that representation of rumen fermentation using simple linear stoichiometric models to estimate the proportion of VFA showed poor correlation between measured and predicted VFA molar proportions ($R^2 < 0.23$), and there was variation among the stoichiometric models in predicting individual VFA. Similarly, Morvay et al. (2011) compared several VFA stoichiometric models and showed that there were large errors in estimating VFA molar proportions ($R^2 < 0.57$ for propionate, $R^2 < 0.45$ for acetate, and $R^2 < 0.32$ for butyrate). The authors demonstrated that most rumen VFA stoichiometric approaches explain a large part of the variation in VFA molar proportions among diets, particularly for acetate, but predictive power for propionate and butyrate differ largely among approaches. They recommend more research for an improved representation of rumen fermentation focused on improving our understanding of VFA molar proportions for a variety of dietary classes rather than gaining understanding of the mean only.

## Quantification of Model Uncertainty in Empirical Models

Uncertainty quantification (UQ) is an analytical procedure that provides insights to understand, characterize, and analyze uncertainties associated with model predictions. These uncertainties may originate from incomplete or faulty model structure (model inadequacy) or imprecise parameter estimates. For example, the model structure may not contain all necessary variables and the variability of the parameters of the independent variables may be too large. Another source of uncertainty can be caused by lack of statistical compliancy. In this case, the most common concerns are uncontrolled errors (residuals) have neither constant variance (heteroscedasticity) nor are they independent, high inflation factor among independent variables (independent variables are highly correlated), and the distribution of the error is not normally distributed either because of small sample size or biases in the database. There are different methodologies for UQ, and several tools have been developed to assist the UQ of mathematical models, mainly for engineering purposes (e.g., the Dakota project, http://dakota.sandia.gov/

index.html, and UCODE, http://igwmc.mines.edu/freeware/ucode/?CMSPAGE = igwmc/freeware/ucode/); UQ has had limited use in agriculture, specifically in animal science. It is difficult to separate and quantify the uncertainties caused by model inadequacy and parameter imprecision because the observed values can be predicted from many different combinations of parameters (independent variables). This constitutes the lack of identifiability, that is, whether the true value of a parameter can be determined based on the available data given the model structure (Boston et al., 2007; Tedeschi and Boston, 2010; Arendt et al., 2012).

Very often, published mathematical models (e.g., empirical equations) do not consider the uncertainty of the parameter estimates in making predictions, and few statistics, usually the same statistics, are used to assess the model adequacy. The most common statistics include mean bias, coefficient of determination ($R^2$), and mean square error of the prediction (Tedeschi, 2006). Alternative evaluation methods that take into account the distribution of the predictions of different models may be more adequate mainly when specific thresholds have been sought for policy-making decisions. In this case, the average prediction is of little interest; the variability of the model predictions is more important.

Regardless of the mathematical model complexity, the variables of a model can be collapsed and written in a generic equation of $n$ parameters and functions as shown in Eq. [30]. In Eq. [30], $Y$ represents the dependent variable (usually an observed variable of interest by the experimenter), $X$ represents the independent variables (usually model inputs), $\beta$ represents the parameter estimates and their standard errors (SE; estimate imprecision) for the $X$ variables, and $e$ represents the uncontrolled residual (model inadequacy), a random variable with mean zero and constant variance:

$$Y = \left(\beta_0 \pm SE_{\beta_0}\right) + \left[\left(\beta_1 \pm SE_{\beta_1}\right)X_1\right] + \left[\left(\beta_2 \pm SE_{\beta_2}\right)X_2\right] + \ldots + \left[\left(\beta_n \pm SE_{\beta_n}\right)X_n\right] + e \quad [30]$$

Equation [30] and the Monte Carlo simulation technique were used herein to illustrate the impact of parameter imprecision and model inadequacy on the predictions by a hypothetical model with one independent variable. In the first scenario (parameter imprecision), $SE_{\beta_0}$ and $SE_{\beta_1}$ were assigned to be either 10 or 20% of the average and fixed value for $\beta_0$ and $\beta_1$ (10 and 1.5, respectively). In the second scenario (model inadequacy), the SE of the residual ($e$) was assigned to either 5 or 10 units. The mean and SE of the $X_1$ variable was assumed to be 10 and 5, respectively; thus the SE of the residual was either the same or twice as variable as $X_1$. The simulation scenarios were (i) $SE_{\beta_0}$ and $SE_{\beta_1}$ of 10% and regression SE (RSE) of 5 units, (ii) $SE_{\beta_0}$ and $SE_{\beta_1}$ of 20% and RSE of 5 units, (iii) $SE_{\beta_0}$ and $SE_{\beta_1}$ of

10% and RSE of 10 units, and (iv) $SE_{\beta_0}$ and $SE_{\beta_1}$ of 20% and RSE of 10 units. Each simulation comprised 100 data points that were obtained assuming random distribution and no covariance between $\beta_0$ and $\beta_1$. One thousand simulations were used to obtain the mean and standard deviation (SD) of *Y*. Then, the mean and SD of *Y* were used to generate the density plots for each hypothetical scenario and the percentage of points within 10 and 40 were computed. Figure 2 depicts the results of this hypothetical comparative simulation.

The prediction uncertainty resulting from parameter imprecision or model inadequacy suggested that for our hypothetical simulations (Fig. 2), doubling the SE of the error from 5 to 10 units compared with the SE of the independent variable ($X_1$) of 5 units (simulations iii and iv) had a greater impact on the variability

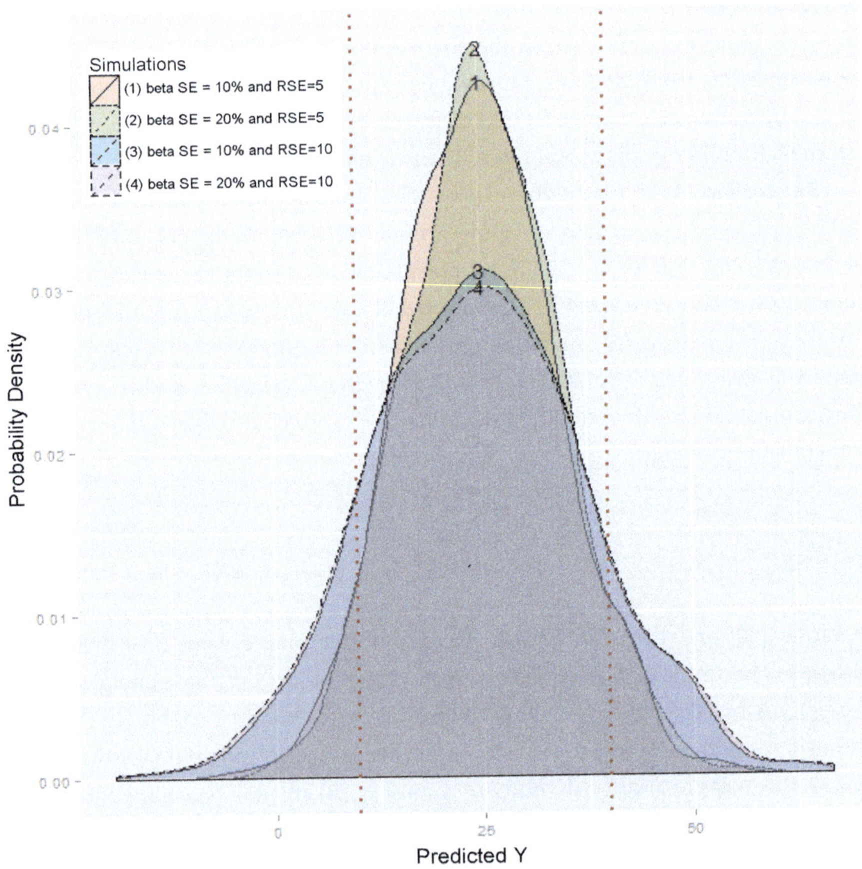


**Fig. 2. Probability density plots comparing the impacts of parameter imprecision and model inadequacy on the distribution of model-predicted values assuming standard error (SE) of the parameter coefficient of 10 and 20% of their coefficient estimate and regression SE (RSE) of 5 and 10 units. The vertical dotted lines represent the 10 and 40 values of the predicted variable.**

of the model predictions ($Y$) than doubling the SE of the parameter coefficients (simulations i and ii). The percentage of points within 10 and 40 (predicted $Y$ values) was 89, 89, 77, and 75%, respectively, for simulations i, ii, iii, and iv. In summary, depending on the intensity and source of the prediction uncertainty, the mathematical model may inaccurately predict the observed values with more or less variability.

## Future Work

In response to growing public concern to reduce anthropogenic impact on climate change, several research programs have been initiated with the aim of reducing GHG emissions. It is generally agreed that there is no "silver bullet" solution to the problem of climate change, and large emission reductions are unlikely to happen in the short term. However, there are incremental changes that can be applied to achieve reductions in GHG emissions from the livestock sector.

Data is being collected at an increasing pace from various production systems, and it would be advantageous to build a database of work that has been conducted or is still underway. Resources should be allocated at a global level to construct such a database so that various types of models can be constructed to help in understanding the system and in finding a solution to the global problem. Various researchers are collecting data, and, for small additional resource, they may be able to collect more variables in the system under investigation to use the data for modeling purposes. For example, Del Grosso et al. (2013) describe a publically available GRACEnet (Greenhouse Reduction through Agricultural Carbon Enhancement network) data management system. Recently, Hristov et al. (2013), under the auspices of the Food and Agriculture Organization (FAO) of the United Nations, brought together scientists from all over the world to review a large body of literature for technical options to mitigate GHG emissions in livestock production systems. Similarly, a group of scientists from different parts of the world, including developing countries, should put resources through local networks such as GRACEnet or global networks such as FAO or the Feed and Nutrition Network of the Global Research Alliance on Agricultural Greenhouse Gases to construct such a database. This database can then be used to develop models at different levels of complexity and assess mitigation options for GHG emission reduction at a local, regional, or global level.

# References

Agle, M., A.N. Hristov, S. Zaman, C. Schneider, P. Ndegwa, and V.K. Vaddella. 2010. Effect of dietary concentrate on rumen fermentation, digestibility, and nitrogen losses in dairy cows. J. Dairy Sci. 93:4211–4222. doi:10.3168/jds.2009-2977

Alemu, A.W., J. Dijkstra, A. Bannink, J. France, and E. Kebreab. 2011a. Rumen stoichiometric models and their contribution and challenges in predicting enteric methane production. Anim. Feed Sci. Technol. 166-167:761–778. doi:10.1016/j.anifeedsci.2011.04.054

Alemu, A.W., K. Ominski, and E. Kebreab. 2011b. Trends of enteric methane emissions from Manitoba beef cattle. Can. J. Anim. Sci. 91:305–321. doi:10.4141/cjas2010-009

Archibeque, S., K. Haugen-Kozyra, K. Johnson, E. Kebreab, and W. Powers-Schilling. 2012. Near-term options for reducing greenhouse gas emissions from livestock systems in the United States: Beef, dairy, and swine production systems. In: L.P. Olander, and A. Van de Bogert, editors, Technical Working Group on Agricultural Greenhouses Report 12-04. Nicholas Institute for Environmental Policy Solutions, Duke University, NC.

Arendt, P.D., D.W. Apley, and W. Chen. 2012. Quantification of model uncertainty: Calibration, model discrepancy, and identifiability. J. Mech. Des. 134:100908–100919. doi:10.1115/1.4007390

Argyle, J.L., and R.L. Baldwin. 1988. Modeling of rumen water kinetics and effects of rumen pH changes. J. Dairy Sci. 71:1178–1188. doi:10.3168/jds.S0022-0302(88)79672-1

Axelsson, J. 1949. The amount of produced methane energy in the European metabolic experiments with adult cattle. Ann. Rev. Agric. Colleges in Sweden 16:404–419.

Baldwin, R.L. 1995. Modeling ruminant digestion and metabolism. Chapman & Hall, London, UK.

Baldwin, R.L., J. France, D.E. Beever, M. Gill, and J.H.M. Thornley. 1987a. Metabolism of the lactating cow: 3. Properties of mechanistic models suitable for evaluation of energetic relationships and factors involved in the partition of nutrients. J. Dairy Res. 54:133–145. doi:10.1017/S0022029900025243

Baldwin, R.L., J. France, and M. Gill. 1987b. Metabolism of the lactating cow: 1. Animal elements of a mechanistic model. J. Dairy Res. 54:77–105. doi:10.1017/S002202990002522X

Baldwin, R.L., J.H.M. Thornley, and D.E. Beever. 1987c. Metabolism of the lactating cow. 2. Digestive elements of a mechanistic model. J. Dairy Res. 54:107–131. doi:10.1017/S0022029900025231

Bannink, A., J. France, S. Lopez, W.J.J. Gerrits, E. Kebreab, S. Tamminga, and J. Dijkstra. 2008. Modelling the implications of feeding strategy on rumen fermentation and functioning of the rumen wall. Anim. Feed Sci. Technol. 143:3–26. doi:10.1016/j.anifeedsci.2007.05.002

Bannink, A., J. Kogut, J. Dijkstra, J. France, E. Kebreab, A.M. Van Vuuren, and S. Tamminga. 2006. Estimation of the stoichiometry of volatile fatty acid production in the rumen of lactating dairy cows. J. Theor. Biol. 238:36–51. doi:10.1016/j.jtbi.2005.05.026

Bannink, A., M.W. van Schijndel, and J. Dijkstra. 2011. A model of enteric fermentation in dairy cows to estimate methane emission for the Dutch National Inventory Report using the IPCC Tier 3 approach. Anim. Feed Sci. Technol. 166-167:603–618. doi:10.1016/j.anifeedsci.2011.04.043

Beauchemin, K.A., H.H. Janzen, S.M. Little, T.A. McAllister, and S.M. McGinn. 2010. Life cycle assessment of greenhouse gas emissions from beef production in western Canada: A case study. Agric. Syst. 103:371–379. doi:10.1016/j.agsy.2010.03.008

Beauchemin, K.A., S.M. McGinn, C. Benchaar, and L. Holtshausen. 2009. Crushed sunflower, flax, or canola seeds in lactating dairy cow diets: Effects on methane production, rumen fermentation, and milk production Source. J. Dairy Sci. 92:2118–2127. doi:10.3168/jds.2008-1903

Benchaar, C., J. Rivest, C. Pomar, and J. Chiquette. 1998. Prediction of methane production from dairy cows using existing mechanistic models and regression equations. J. Anim. Sci. 76:617–627.

Blaxter, K.L., and J.L. Clapperton. 1965. Prediction of the amount of methane produced by ruminants. Br. J. Nutr. 19:511–522. doi:10.1079/BJN19650046

Boston, R.C., P.A. Wilkins, and L.O. Tedeschi. 2007. Identifiability and Accuracy: Two critical problems associated with the application of models in nutrition and the health sciences. In: M. Hanigan, editor, Mathematical modeling for nutrition and health sciences, Univ. Penn., Roanoke, VA. p. 161–193.

Corré, W.J. 2002. Agricultural land use and emissions of $CH_4$ and $NO_2$ in Europe. Report 40. Plant Res. Int., Wageningen, The Netherlands.

Del Grosso, S.J., J.W. White, G. Wilson, B. Vandenberg, D.L. Karlen, R.F. Follett, J.M.F. Johnson, A.J. Franzluebbers, D.W. Archer, H. Gollany, M. Liebig, J. Ascough, M. Reyes-Fox, L. Pellack, J. Starr, N. Barbour, M. Gutwein, and D. James. 2013. Introducing the GRACEnet/REAP data contribution, discovery and retrieval system. J. Environ. Qual. 42:1274–1280. doi:10.2134/jeq2013.03.0097

Dijkstra, J., H.D. Neal, D.E. Beever, and J. France. 1992. Simulation of nutrient digestion, absorption and outflow in the rumen: Model description. J. Nutr. 122:2239–2256.

Ellis, J.L., A. Bannink, J. France, E. Kebreab, and J. Dijkstra. 2010. Prediction of enteric methane production by dairy cows in whole farm models. Glob. Change Biol. 16:3246–3256. doi:10.1111/j.1365-2486.2010.02188.x

Ellis, J.L., J. Dijkstra, J. France, A.J. Parsons, G.R. Edwards, S. Rasmussen, E. Kebreab, and A. Bannink. 2012. Effect of high-sugar grasses on methane emissions simulated using a dynamic model. J. Dairy Sci. 95:272–285. doi:10.3168/jds.2011-4385

Ellis, J.L., J. Dijkstra, E. Kebreab, S. Archibeque, J. France, and A. Bannink. 2011. Prediction of methane production in beef cattle within a mechanistic digestion model. In: D. Sauvant, J. Van Milgan, P. Faverdin, and N. Friggens, editors, Modelling nutrient digestion and utilization in farm animals. Wageningen Academic Publishers, Wageningen, The Netherlands. p. 181–188.

Ellis, J.L., E. Kebreab, N.E. Odongo, B.W. McBride, E.K. Okine, and J. France. 2007. Prediction of methane production from dairy and beef cattle. J. Dairy Sci. 90:3456–3467. doi:10.3168/jds.2006-675

Food and Agriculture Organization of the United Nations (FAO). 2010. Greenhouse gas emissions from the dairy sector. FAO, Rome.

Giger-Reverdin, S., P. Morand-Fehr, and G. Tran. 2003. Literature survey of the influence of dietary fat composition on methane production in dairy cattle. Livest. Prod. Sci. 82:73–79. doi:10.1016/S0301-6226(03)00002-2

Grainger, C., and K.A. Beauchemin. 2011. Can enteric methane emissions from ruminants be lowered without lowering their production? Anim. Feed Sci. Technol. 166–167:308–320. doi:10.1016/j.anifeedsci.2011.04.021

Hegarty, R.S., D. Alcock, D.L. Robinson, J.P. Goopy, and P.E. Vercoe. 2010. Nutritional and flock management options to reduce methane output and methane per unit product from sheep enterprises. Anim. Prod. Sci. 50:1026–1033. doi:10.1071/AN10104

Hristov, A.N., J. Oh, C. Lee, R. Meinen, F. Montes, T. Ott, J. Firkins, A. Rotz, C. Dell, A. Adesogan, W.Z. Yang, J. Tricarico, E. Kebreab, G. Waghorn, J. Dijkstra, and S. Oosting. 2013. In: Gerber, P., B. Henderson, and H. Makkar, editors, Mitigation of greenhouse gas emissions in livestock production: A review of technical options for non-$CO_2$ emission. Food and Agric. Org. of the United Nations , Rome.

Intergovernmental Panel on Climate Change (IPCC). 1997. 1996 Revised IPCC guidelines for national greenhouse gas inventories: IGES, Hayama, Kanagawa, Japan.

Intergovernmental Panel on Climate Change (IPCC). 2006. Chapter 10. Emissions from livestock and manure management. guidelines for national greenhouse inventories. Vol. 4. Agriculture, forestry and other land use. IPCC, Geneva, Switzerland. p. 10.1–10.87.

Jenkins, T.C., R.J. Wallace, P.J. Moate, and E.E. Mosley. 2008. Board-invited review: Recent advances in biohydrogenation of unsaturated fatty acids within the rumen microbial ecosystem. J. Anim. Sci. 86:397–412. doi:10.2527/jas.2007-0588

Johnson, D.E., H.W. Phetteplace, A.F. Seidl, U.A. Schneider, and B.A. McCarl. 2003. Management variations for U.S. beef production systems: Effects on greenhouse gas emissions and profitability. In: Proceeding of the 3rd International Methane and Nitrous Oxide Mitigation Conference. Beijing, China, 17–21 Nov. 2003. Coal Institute, Beijing, China, p. 953–961.

Kebreab, E., J. France, B.W. McBride, N.E. Odongo, A. Bannink, J.A.N. Mills, and J. Dijkstra. 2006. Evaluation of models to predict methane emissions from enteric fermentation in North American dairy cattle. In: E. Kebreab, J. Dijkstra, W.J.J. Gerrits, A. Bannink, and J. France, editors, Nutrient utilization in farm animals: Modelling approach. CAB International, Wallingford, UK. p. 299–313.

Kebreab, E., K.A. Johnson, S.L. Archibeque, D. Pape, and T. Wirth. 2008. Model for estimating enteric methane emissions from US cattle. J. Anim. Sci. 86:2738–2748. doi:10.2527/jas.2008-0960

Kennedy, P.M., and E. Charmley. 2012. Methane yields from Brahman cattle fed tropical grasses and legumes. Anim. Prod. Sci. 52:225–239. doi:10.1071/AN11103

Kirchgeßner, M., W. Windisch, and H.L. Müller. 1995. Nutritional factors for the quantification of methane production In: W. Von Engelhardt, S. Leonhard-Marek, G. Breves, and D. Giesecke, editors, Ruminant physiology: Digestion, metabolism, growth and reproduction: Proceedings of the Eighth International Symposium on Ruminant Physiology. Ferdinand Enke Verlag, Stuttgart, Germany. p. 333– 348.

Kriss, M. 1930. Quantitative relations of the dry matter of the food consumed, the heat production, the gaseous outgo, and the insensible loss in body weight of cattle. J. Agric. Res. 40:283–295.

Mills, J.A.N., J. Dijkstra, A. Bannink, S.B. Cammell, E. Kebreab, and J. France. 2001. A mechanistic model of whole-tract digestion and methanogenesis in the lactating dairy cow: Model development, evaluation, and application. J. Anim. Sci. 79:1584–1597.

Mills, J.A.N., E. Kebreab, C.M. Yates, L.A. Crompton, S.B. Cammell, M.S. Dhanoa, R.E. Agnew, and J. France. 2003. Alternative approaches to predicting methane emissions from dairy cows. J. Anim. Sci. 81:3141–3150.

Moe, P.W., and H.F. Tyrrell. 1979. Methane production in dairy cows. J. Dairy Sci. 62:1583–1586. doi:10.3168/jds.S0022-0302(79)83465-7

Moraes, L.E., A. B. Strathe, J. G. Fadel, D. P. Casper, and E. Kebreab. 2014. Prediction of enteric methane emissions from cattle. Global Change Biol. 20:2140–2148.

Morvay, Y., A. Bannink, J. France, E. Kebreab, and J. Dijkstra. 2011. Evaluation of models to predict the stoichiometry of volatile fatty acid profiles in rumen fluid of dairy cattle. J. Dairy Sci. 94:3063–3080. doi:10.3168/jds.2010-3995

Murphy, M.R., R.L. Baldwin, and L.J. Koong. 1982. Estimation of stoichiometric parameters for rumen fermentation of roughage and concentrate diets. J. Anim. Sci. 55:411–421.

Murray, R.M., A.M. Bryant, and R.A. Leng. 1976. Rates of production of methane in the rumen and large intestine of sheep. Br. J. Nutr. 36:1–14. doi:10.1079/BJN19760053

Nkrumah, J.D., C. Li, J.A. Basarab, S. Guercio, Y. Meng, B. Murdoch, C. Hansen, and S.S. Moore. 2004. Association of a single nucleotide polymorphism in the bovine leptin gene with feed intake, feed efficiency, growth, feeding behavior and carcass quality and body composition. Can. J. Anim. Sci. 84:211–219. doi:10.4141/A03-033

Noziere, P., F. Glasser, and D. Sauvant. 2010. In vivo production and molar percentages of volatile fatty acids in the rumen: A quantitative review by an empirical approach. Anim. 5:403–414. doi:10.1017/S1751731110002016

Ogle, S.M., S. Spencer, M. Hartman, L. Buendia, L. Stevens, L. du Toit, and J.Witi. 2016. Developing national baseline GHG emissions and analyzing mitigation potentials for agriculture and forestry using an advanced national GHG inventory software system. In: S. Del Grosso, W. Parton, and L. Ahuja, editors, Synthesis and modeling of greenhouse gas emissions and carbon storage in agricultural and forest systems to guide mitigation and adaptation. Advances in Agricultural Systems Modeling 6. ASA, CSSA, and SSSA, Madison, WI. doi:10.2134/advagricsystmodel6.2013.0009

Rabiee, A.R., K. Breinhild, W. Scott, H.M. Golder, E. Block, and I.J. Lean. 2012. Effect of fat additions to diets of dairy cattle on milk production and components: A meta-analysis and meta-regression. J. Dairy Sci. 95:3225–3247. doi:10.3168/jds.2011-4895

Rotz, C.A., M.S. Corson, D.S. Chianese, F. Montes, S.D. Hafner, R. Jarvis, and C.U. Coiner. 2011. The integrated farm system model: Reference manual, version 3.4. USDA-ARS Pasture Systems and Watershed Management research unit, University Park, Pa. http://www.ars.usda.gov/SP2UserFiles/Place/80700000/ifsmreference.pdf (accessed 21 Feb. 2014).

Schils, R.L.M., A. Verhagen, H.F.M. Aarts, P.J. Kuikman, and L.B.J. Šebek. 2006. Effect of improved nitrogen management on greenhouse gas emissions from intensive dairy systems in The Netherlands. Glob. Change Biol. 12:382–391. doi:10.1111/j.1365-2486.2005.01090.x

Solomon, S., D. Qin, M. Manning, R.B. Alley, T. Berntsen, N.L. Bindoff, Z. Chen, J.M. Chidthaisong, G.C. Gregory, M. Hegerl, B. Heimann, B.J. Hewitson, F. Hoskins, A. Joos, J. Jouzel, V. Kattsov, U. Lohmann, T. Matsuno, M. Molina, N. Nicholls, G. Overpeck, V. Raga, J. Ramaswamy, M. Ren, R. Rusticucci, T.F. Somerville, P. Stocker, J. Whetton, R.A. Wood, and D. Wratt. 2007. In: Solomon, S., D. Qin, M. Manning, Z. Chen, M. Marquis, K.B. Averyt, M. Tignor, and H.L. Miller, editors, Technical summary. Climate change 2007: The physical science basis. Contribution of Working Group I to the Fourth Assessment Report of the Intergovernmental Panel on Climate Change. Cambridge Univ. Press, Cambridge, UK and New York.

Steinfeld, H., P. Gerber, T. Wassenaar, V. Castel, M. Rosales, and C. de Haan. 2006. Livestock's long shadow—Environmental issues and options. Food and Agric. Org. of the United Nations, Rome.

Tedeschi, L.O. 2006. Assessment of the adequacy of mathematical models. Agric. Syst. 89:225–247. doi:10.1016/j.agsy.2005.11.004

Tedeschi, L.O., and R. Boston. 2010. Identifiability and accuracy: A closer look at contemporary contributions and changes in these vital areas of mathematical modelling. In: D. Sauvant et al., editors, Modelling nutrient digestion and utilization in farm animals. Wageningen Academic Publishers, Wageningen, The Netherlands. p. 91–99 .

USEPA. 2014. Inventory of U.S. Greenhouse Gas Emissions and Sinks: 1990–2012. USEPA, Washington, DC. http://www.epa.gov/climatechange/Downloads/ghgemissions/US-GHG-Inventory-2014-Main-Text.pdf (accessed 21 July 2011).

Van Middelaar, C.E., P.B.M. Berentsen, J. Dijkstra, and I.J.M. De Boer. 2013. Evaluation of a feeding strategy to reduce greenhouse gas emissions from dairy farming: The level of analysis matters. Agric. Syst. 121:9–22. doi:10.1016/j.agsy.2013.05.009

Vergé, X.P.C., J.A. Dyer, R.L. Desjardins, and D. Worth. 2008. Greenhouse gas emissions from the Canadian beef industry. Agric. Syst. 98:126–134. doi:10.1016/j.agsy.2008.05.003

Weber, K.L., J.A. Rendon, C. Antwi, A.L. Van Eenennaam, and E. Kebreab. 2013. Methane production is not associated with sire groups or residual feed intake in feedlot Angus cattle. In: Proceedings of the Greenhouse Gases and Animal Agriculture Conference, Dublin, Ireland. 23–26 June 2013. Cambridge Univ. Press. Abstract 348.

Wilkerson, V.A., D.P. Casper, and D.R. Mertens. 1995. The prediction of methane production of Holstein cows by several equations. J. Dairy Sci. 78:2402–2414. doi:10.3168/jds.S0022-0302(95)76869-2

Wolin, M.J. 1960. A theoretical rumen fermentation balance. J. Dairy Sci. 43:1452–1459. doi:10.3168/jds.S0022-0302(60)90348-9

Yan, T., R.E. Agnew, F.J. Gordon, and M.G. Porter. 2000. Prediction of methane energy output in dairy and beef cattle offered grass silage-based diets. Livest. Prod. Sci. 64:253–263. doi:10.1016/S0301-6226(99)00145-1

# Mathematical Modeling of Greenhouse Gas Emissions from Agriculture for Different End Users

**Jon Hillier,* Mohammed Abdalla, Jessica Bellarby, Fabrizio Albanito, Arindam Datta, Marta Dondini, Nuala Fitton, Paul Hallett, Astley Hastings, Edward Jones, Matthias Kuhnert, Dali Nayak, Mark Pogson, Mark Richards, Jo Smith, Sylvia Vetter, Jagadeesh Yeluripati, and Pete Smith**

Abstract

Models are approximations of real systems. Although, in theory, there is no limit to the refinement and detail of a mathematical model, with greater refinement allowing more detailed representation of the physical system, there is a practical limit in terms of the model outputs, which is, in part, defined by the intended end use. This chapter describes a range of collaborative model studies across a range of model complexities from simple empirical models to complex simulation models. We illustrate with several applications of agricultural greenhouse gas (GHG) models and their value to different end users. We will thus demonstrate that a range of models and approaches is often required to develop our understanding as a stakeholder community. As such, we observe that the value of a model should not be judged by its precision alone but also its utility. The need to communicate understanding to multiple stakeholders has led to a number of exciting developments in recent times. For example, although a range of common techniques are often employed with regard to process-based models, recently, there have been several exciting initiatives to expand the utility of such methods for nonexperts, with an example being the ecosystem land-use model (ELUM) discussed in this chapter.

Abbreviations: BMP, best management practice; DNDC, denitrification–decomposition; DOC, dissolved organic C; DON, dissolved organic N; ECOSSE, Estimation of Carbon in Organic Soils—Sequestration and Emissions; EF, emission factor; ELUM, ecosystem land-use model; GHG, greenhouse gas; IPCC, Intergovernmental Panel on Climate Change; NCU, NitroEurope Calculation Unit; NEE, net ecosystem exchange; NPP, net primary productivity; PDF, probability distribution function; SOM, soil organic matter; SRC, short-rotation coppice; Windfarm-CAT, Carbon Assessment Tool for Wind Farms.

Jon Hillier, Mohammed Abdalla (mabdalla@abdn.ac.uk), Fabrizio Albanito (f.albanif@abdn.ac.uk), Marta Dondini (m.dondini@abdn.ac.uk), Nuala Fitton (n.fitton@abdn.ac.uk), Astley Hastings (astley.hastings@abdn.ac.uk), Matthias Kuhnert (matthias.kuhnert@abdn.ac.uk), Dali Nayak (d.nayak@abdn.ac.uk), Mark Richards (m.richards@abdn.ac.uk), Jo Smith (jo.smith@abdn.ac.uk), Sylvia Vetter (sylvia.vetter@abdn.ac.uk), and Pete Smith (pete.smith@abdn.ac.uk), Univ. of Aberdeen, Inst. of Biological and Environmental Sciences, 23 St Machar Drive, Aberdeen, AB24 3RY. Jessica Bellarby (jbellarby@gmail.com), Lancaster Environment Centre, Lancaster Univ., Bailrigg, Lancaster, LA1 4YQ, UK; Arindam Datta (a.datta@abdn.ac.uk), Centre for Environmental Studies, Earth Sciences and Climate Change Division, The Energy and Resources Inst., New Delhi, 110003, India. Paul Hallett (paul.hallett@abdn.ac.uk), Room 2.14, Cruickshank Building, Inst. of Biological and Environmental Sciences, Univ. of Aberdeen, Aberdeen, AB24 3UU, UK. Edward Jones (e.o.jones@abdn.ac.uk), 113 LSHTM, Keppel Street, London, WC1E 7HT. Mark Pogson (m.pogson@abdn.ac.uk), Engineering, Sports and Sciences Academic Group, The Univ. of Bolton, Deane Road, Bolton, BL3 5AB. Jagadeesh Yeluripati (jagadeesh.yeluripati@hutton.ac.uk), The James Hutton Inst., Craigiebuckler, Aberdeen AB15 8QH, UK. *Corresponding author (j.hillier@abdn.ac.uk).

doi:10.2134/advagricsystmodel6.2013.0038


Synthesis and Modeling of Greenhouse Gas Emissions and Carbon Storage in Agricultural and Forest Systems to Guide Mitigation and Adaptation
S. Del Grosso, L. Ahuja, and W. Parton, Editors
Advances in Agricultural Systems Modeling, Volume 6. Lajpat R. Ahuja, Series Editor

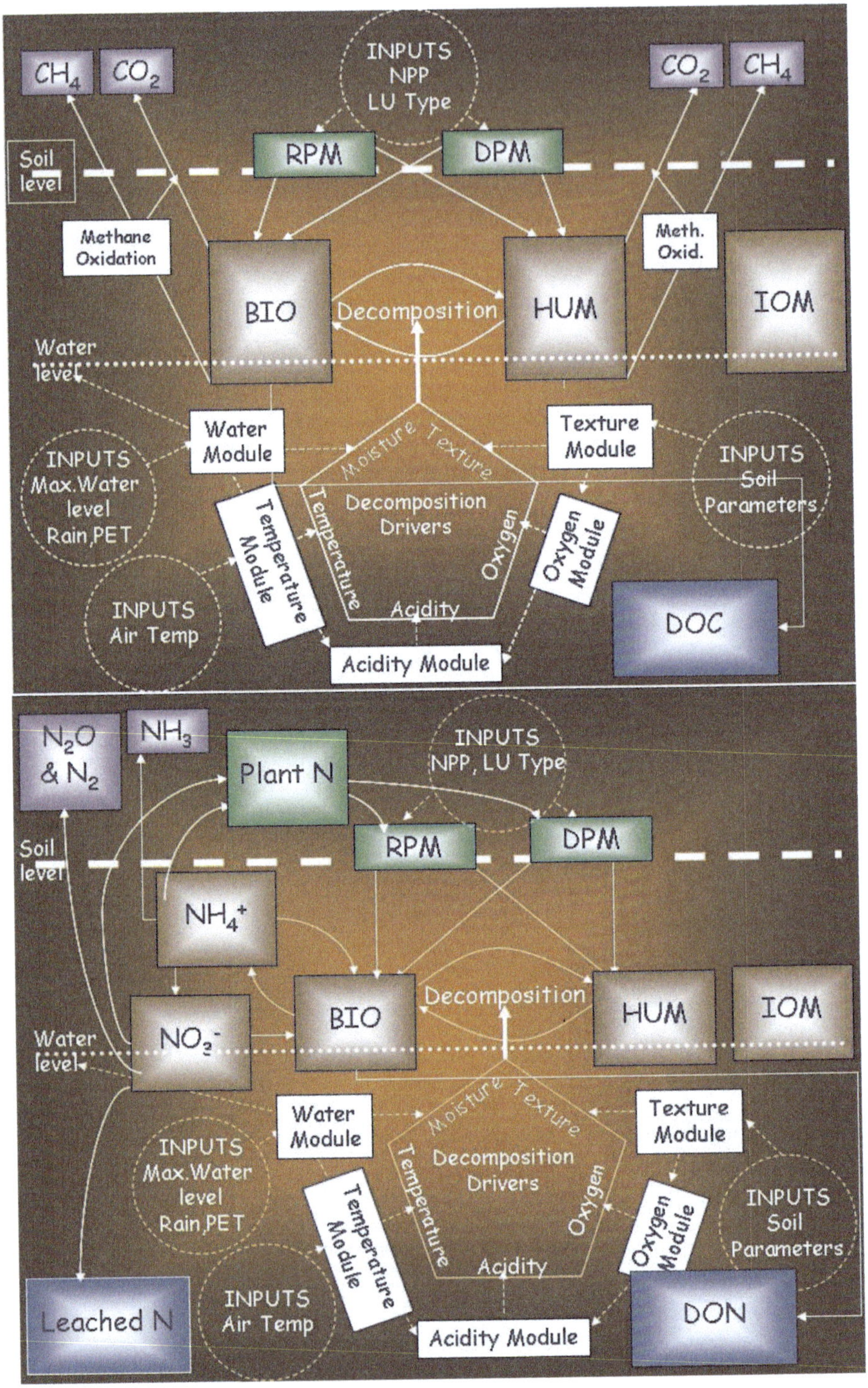

CH4
CO2
INPUTS
NPP
LU Type
CO2
CH4
Soil level
RPM
DPM
Methane Oxidation
Meth. Oxid.
BIO
Decomposition
HUM
IOM
Water level
Water Module
Texture Module
Moisture
Texture
INPUTS
Max.Water level
Rain,PET
Decomposition Drivers
INPUTS
Soil Parameters
Temperature Module
Temperature
Oxygen
Oxygen Module
INPUTS
Air Temp
Acidity
DOC
Acidity Module
N2O & N2
NH3
Plant N
INPUTS
NPP, LU Type
Soil level
RPM
DPM
NH4+
BIO
Decomposition
HUM
IOM
Water level
NO2-
Water Module
Moisture
Texture
Texture Module
INPUTS
Max.Water level
Rain,PET
Decomposition Drivers
Temperature Module
Temperature
Oxygen
Oxygen Module
INPUTS
Soil Parameters
Acidity
INPUTS
Air Temp
Leached N
DON
Acidity Module

Models are approximations of real systems and are, thus, imperfect representations of reality. Although, in theory, there is no limit to the refinement and detail of a mathematical model, with greater refinement allowing more detailed representation of the physical system, there is a practical limit in terms of the model outputs, which is partly defined by the intended end use. Modeling of greenhouse gas (GHG) emissions from agriculture, which are often highly dependent on the soil and its physical and biological properties, illustrates this point nicely. One may argue that very complex models are required for faithful representation of the real system as a result, in large part, of the dynamic nature of soils, which are mixed biological and physical systems and are, thus, heterogeneous even at microbial scale. However, representation of this complexity through dynamic linked-equation complexity can lead to propagation of errors through the model and result in a model that is impractical to apply and subject to inaccuracy. It is often assumed that the use of a more detailed method will provide a more precise prediction than a less detailed one. However, this is not justified; although process-based or mechanistic models have properties such as good temporal and spatial representation and use of mass balances to constrain model behavior to physical reality, they are also more susceptible than empirical models to numerical instabilities, error propagation, programmer error, and user error (Knorr, 2000; Zaehle et al., 2005). In addition, more complex models need more input data, which suggests a higher accuracy, but these additional data are not always available and may be based on assumptions, estimates, or interpolated data that import more uncertainty to the simulation results (Gottschalk et al., 2007). In light of this, the common assertion that mechanistic models are more likely to perform well outside their range of calibration appears unsafe. However, as will be illustrated later via several examples, the different characteristics of models of different complexity helps to define their appropriate end use. Empirical models are typically only constrained by the principle of parsimony (commonly known as Occam's Razor) and some globally inherent logic of the system (Crawley, 2007). That is, their purpose is to explain dependent variables as a function of independent ones or modeled variables as a function of input variables. The constraints on process-based or mixed-process empirical models are different, in that their complexity is, in theory, limited only by computational constraints. However, the functional forms employed therein impose the developers' understanding of the system or process being modeled.

In the context of predicting and managing GHG emissions from agriculture, it is important to understand the above. It is estimated that agriculture and associated land-use change contribute ~30% of global anthropogenic GHG emissions (Bellarby et al., 2008). A significant proportion of this is attributable to loss of soil carbon (C) or emissions of nitrous oxide ($N_2O$) from nitrification or denitrification processes associated with the use of mineral nitrogen (N) fertilizers or the release of organic N following mineralization of soil organic matter (SOM). These processes occur in a dynamic, heterogeneous medium, which may be modeled to different levels of detail; the level of detail selected affects the precision of model predictions. In modeling the impact of agriculture and land-use change on GHG emissions, it is important that such models can capture the turnover of C and N that occurs naturally in an environment undisturbed by anthropogenic interaction. This is important, as although GHG emissions from anthropogenic interventions such as agriculture and land-use change contribute around 30% of human-induced emissions, these are small compared with natural C processes, which are only ~1% of the naturally occurring annual C turnover of the terrestrial biosphere (IPCC, 2007).

The two extremes of mathematical modeling (as opposed to statistical modeling) presented in this chapter are referred to as empirical, or pure data-driven, and mechanistic, or process-based.

Empirical models seek to parameterize a hypothesized relationship between variables, typically known as the dependent and independent variables. The goal, in this case, is prediction of the value of the independent variable, not an explanation of the nature of the relationship between the variables. However, it is worth observing that the data that drives these models is generally collected because we have an expectation of the likely relationship based on a mechanistic understanding. So although models may not describe the complexity of the interactions, they are grounded in a mechanistic understanding of the process. The simplest kind of mathematical model is probably a linear one as used in the emission factor (EF) methods of the IPCC Guidelines for National Greenhouse Gas Inventories (IPCC, 2006). Mechanistic models, on the other hand, seek to model the causal effects; a system is modeled as component variables, with the causal links between certain variables parameterized based on observable biological or physical phenomena. This does not imply that mechanistic models do not contain simplifying assumptions. In the case of soils, even the most complex structural representations contain mathematical abstractions or simplified representations.

In the context of GHG emissions, the IPCC methodologies provide an accounting framework with three tiers (IPCC, 2006). Tier 1 and Tier 2 methods are usually linear models with coefficients defined as EFs. Tier 1 is the most simple default

**Table 1. Tiers of model complexity following the Intergovernmental Panel on Climate Change (IPCC) classification system.**

| Complexity | Models† | Description |
|---|---|---|
| Tier 1 | IPCC Tier 1 | Based on emission factors applied to activity data, with a specified form of equation, and include prescribed reference tables for additional driving data required to estimate emissions |
| Tier 2 | Tier 2 emission factors | Methods using the same form of equation as for Tier 1, but generic default emission factors are replaced by technology or region specific emission factors to better represent local conditions |
| Tier 3 | Process-based models, e.g. DNDC, DayCent, ECOSSE, RothC, etc. | Methods that depart from the specified form of equation in Tiers 1 and 2 and range from relatively simple relationships that are only slightly more complex than Tier 1 to more detailed methods including the spatial application of process based models with spatially disaggregated activity data |

† DNDC, denitrification–decomposition; ECOSSE, Estimation of Carbon in Organic Soils—Sequestration and Emissions.

method and is designed to be used in the absence of high-quality data, typically with EFs derived as global averages. Tier 2 allows the derivation of technology or region-specific EFs. Tier 3 is strictly defined as neither of the above, although, in practice, this usually (but not always) means process-based and mechanistic modeling methods and measurement. This is summarized in Table 1.

Many of the models we consider in the following text are specifically designed or applied to model a single source of emissions (albeit via a complicated set of interactions). However, understanding of single emissions sources at a given location, if used to determine for example whether—or how—GHG emissions should be reduced, is often required to put them in the context of other emissions from agricultural production. In life-cycle-analysis terms, this results in increasing the scope and breadth of the analysis. There are different nonexpert stakeholders who may have different requirements for the outputs of mathematical models. For example, while the goal of scientists may be to achieve a more accurate representation of the modeled system, farmers may wish to use outputs as proxies for good agricultural practice, supply-chain stakeholders may wish to reduce emissions from their products, and national governments may wish to determine optimal GHG emission reduction plans.

The authors all work in the field of GHG emissions from agriculture in the Soil Modeling group at the University of Aberdeen. In the following text, we illustrate, using examples from our own activities—some being applications of agricultural GHG models—and their value to different end users. In presenting a range of models and user groups, we will demonstrate that a range of models (in other words a range of mathematical representations of a system) is often required to develop our learning as a stakeholder community.

## Modeling Approaches and Their End Users

### Emission Factor Methods

Emission factor methods are generally derived from empirical observations or expert judgment and are intended to be used on crude or aggregated activity data for reporting GHG emissions to the United Nations Framework Convention on Climate Change based on a set of default EFs. The EFs are derived from a variety of methods ranging from systematic meta-analyses, such as that of Ogle et al. (2005), to the judgment of the expert Intergovernmental Panel on Climate Change (IPCC) technical expert panels. Their application is generally to enable country and global understanding of the major sources of emissions from agriculture including $N_2O$ emissions from N fertilizer use and methane from ruminant livestock digestion; this helps to target mitigation efforts and inform policy (e.g., Tubiello et al., 2013).

Emission factor methods have been applied in a few different contexts. Smith et al. (2008) used the method to identify mitigation opportunities and potentials from agriculture. These mitigation potentials are provided on a per-unit-area or per-head-livestock basis and do not explicitly account for site or technology specific variation in EFs. As such, they cannot be used with confidence for a particular region or agricultural product, but can be used to develop policy recommendations or identify mitigation practices that should be explored in greater depth or tried out at a regional level.

Emission factor methods may be incorporated into so-called C calculators to allow farmers to observe the impact of the whole set of farming practices on emissions related to a specific parcel of land or product. Recently, there have been a number of reviews of such calculators published, for example, Whittaker et al. (2013), Driver et al. (2010), and Sheane et al. (2010). We show one such example in Box 1.

As mentioned above, one application of EF methods is as a C-footprint tool for industrial activities. For this type of application, databases of EFs for a range of industrial activities are used to develop product C footprints to allow businesses to report on their environmental performance to consumers. However, as highlighted by the $N_2O$ case earlier in this section, the value of such models in providing geography-specific estimates of emissions is limited by their simplicity. For example, a model that is not responsive to the impact of soil texture or climatic conditions lacks credibility with farmers who are very aware of the importance of these characteristics inasmuch as they affect fuel, fertilizer, and pesticide use as well as yield. In summary, the farming community learns the importance of N as a contributor to GHGs as a factor in field-crop systems and the processing

**Box. 1. Cereals and Oilseeds Carbon Calculator.**

The Cereals and Oilseeds Carbon calculator (Fig. 1) was developed by the HGCA (Home Grown Cereals Authority, the cereal and oilseed division of the Agriculture and Horticulture Development Board). The objective in this example was primarily to empower farmers to explore the impact of their practices on GHG emissions from their farms and to identify methods to manage GHG emissions. The model makes a simple estimate of GHG emissions as a function of fertilizer rates, crop protection chemical use, and energy use up to the farm gate. The tool has been used by supply-chain stakeholders in the United Kingdom to benchmark performance in a rudimentary way and also used by individual farmers. For both user groups, the main benefit of the model appears to be educational, highlighting that for conventionally grown cereal crops the major contributor to farm gate emissions is N fertilizers. The tool was published with a growers' guide, which treated, in layperson language, some of the uncertainties associated with estimating in field emissions from N use. The tool also allows educated, but nonexpert, users to observe the link between practice and emissions and to screen for effective mitigation measures. However, the use of Tier 1 EF methods for $N_2O$ emissions, in particular, imposes limitations on the tool's applicability, as it cannot be used to indicate the impact of soil or climate properties on $N_2O$ emissions.

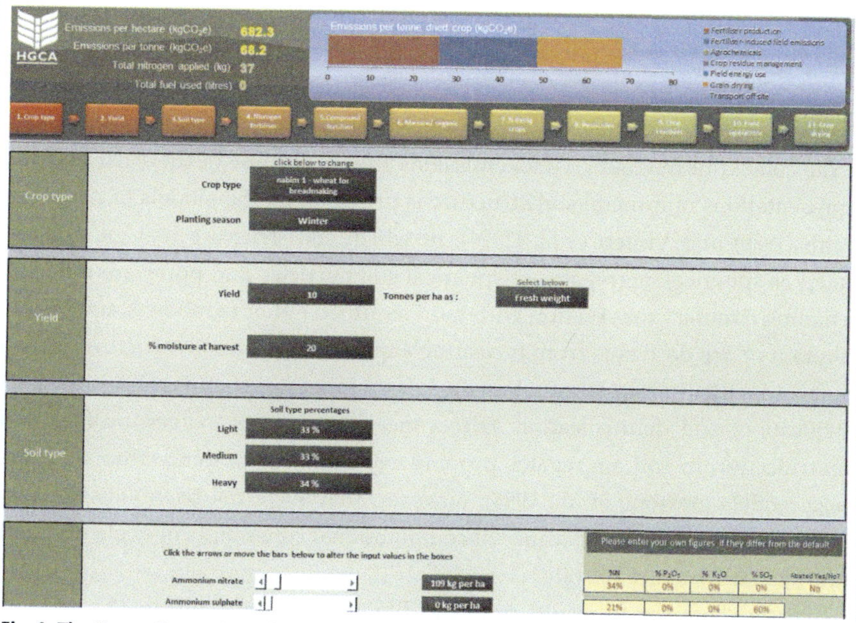

**Fig. 1. The Home Grown Cereals Authority (UK) carbon footprint decision support tool.**

community learns the need to employ models that are responsive to soil and climate conditions to develop credible estimates of emissions at the site level.

## Other Empirical Models

The EF methods are, by nature, linear models. However, there are many nonlinear empirical models, some of which indeed underlie the IPCC Tier 1 methods. For example, the default figure of 1% of applied N emitted as $N_2O$-N is based on a global, regionally disaggregated application of the Bouwman $N_2O$ emissions model (Bouwman et al., 2002). The Tier 1 methane EFs for enteric fermentation in ruminants are based on regional population of ruminants and the quality of the forage used with factors developed from measurements on animals with different feed intakes (e.g., Lassey et al., 1997). In the following two examples (Boxes 2 and 3) we show GHG calculators employing nonlinear empirical models.

One property of simple empirical accounting models is that they are deterministic; one set of driving data provides a unique set of outputs. This makes them attractive in the context of C markets where a transparent and accessible method to quantify C balance can be agreed upon and used to financially support C sequestration practices. However, as some evaluations have shown, their lack of sensitivity to key driving variables limits their ability to accurately model GHG emissions in certain cases.

## Process-Based Models

In the context of modeling GHG emissions from soils, one of the most detailed representations of mineral soil structure is provided by the models of Arah and Smith (1989) and Vinten et al. (1996), in which soil structure may be approximated as spherical aggregates of spherical soil particles and pores are modeled as having circular cross section with radii assumed either to follow standard distribution or are derivable from measurable quantities. The soil structure is then combined with differential equation models for processes such as solute transport, nitrification, and denitrification. Direct measurements of oxygen profiles and denitrification in soil aggregates provide experimental evidence that supports these models (Sextone et al., 1985); however, they have not been subsequently widely used, presumably because of computational constraints of explicitly modeling spatial structure of soils.

Further complexity results from the dynamic behavior of soils (Ahuja et al., 2006). Properties such as temperature and water content are major factors in GHG models that have been parameterized by a large body of evidence collected in the field and lab (Ball, 2013). Soil pore structure also changes with time, but as the processes have been largely ignored in soil science (Ahuja et al., 2006), the impacts to GHG emissions cannot be modeled. The biological processes that drive GHG production introduce further complications for modeling the impacts

**Box 2. The Cool Farm Tool**

The Cool Farm Tool (Hillier et al., 2011) has been developed as part of a collaboration between the University of Aberdeen, the Sustainable Food Lab, and several industry partners. It was designed to be both simple enough for general agricultural use but scientifically robust for calculating C emissions. It integrates emissions from multiple farm sources (fertilizer, agrochemicals, C stock change, livestock, energy use, transport, etc.) into a single package using models such as that of Bouwman et al. (2002) to derive $N_2O$ and NO emissions and Tier 1 EF methods.

Applications of this model have been numerous, highlighted by the following examples:

**Analysis of cotton emissions from India.** This work was supported by World Wildlife Fund (WWF)–India and Marks & Spencer (UK retail company) to identify the impact of best management practices (BMPs) on emissions from cotton production in the Warangal district of India (Agarwal and Jeffries, 2013). The BMPs include, for example, optimal and targeted use of fertilizers and pesticides, and it was demonstrated that BMP cotton grown under such management had total net GHG emissions approximately one-third of cotton grown under traditional cultivation methods. The tool is currently being employed by WWF–India and WWF–Pakistan using statistical regional agricultural census data to estimate GHG emissions from different cotton growing regions in India and Pakistan. An example shows the results for the Maharashtra State (Fig. 2) with different results depending on the major use of cotton type because of different management. Varieties used in the regions depend on suitability for the production region as well as on soil properties.

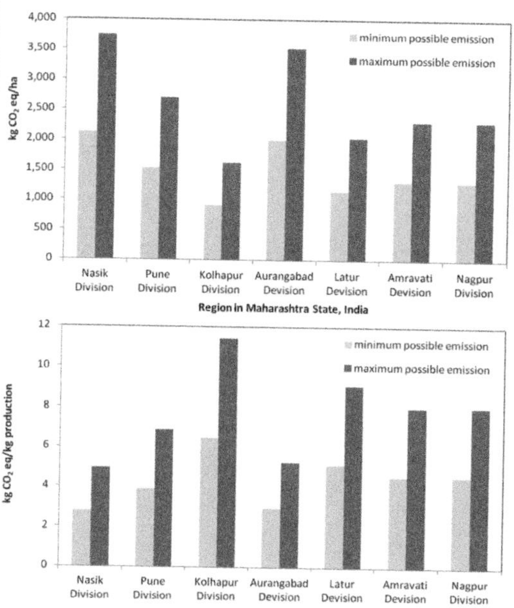


**Fig. 2. Range of greenhouse gas emissions of regions in Maharashtra State, India, depending on soil properties, cotton type, and management. (Top) Emissions per hectare; (bottom) emissions per kilogram production (cotton).**

**Analysis by PepsiCo under its 50-in-5 initiative.** PepsiCo (http://www.pepsico.co.uk/purpose/environmental-sustainability/agriculture) uses the Cool Farm Tool to account for the emissions from a number of raw materials. The tool, by informing users how practice relates to emissions, aids the company in working with its suppliers to identify and promote sustainable farming practices.

**Box 2. Continued.**

**Analysis with CIMMYT of socioeconomic farm survey data from smallholder maize producers in sub-Saharan Africa**. The data from farm surveys is not sufficient to populate a process-based model, but it is sufficient to provide region-sensitive prediction of emissions as a function of fertilizer use and yield. This enabled the identification of environmentally optimal doses of fertilizer application in sub-Saharan Africa regions combined with increased production and minimal emissions per unit production (Bellarby et al., 2014; Fig. 3).

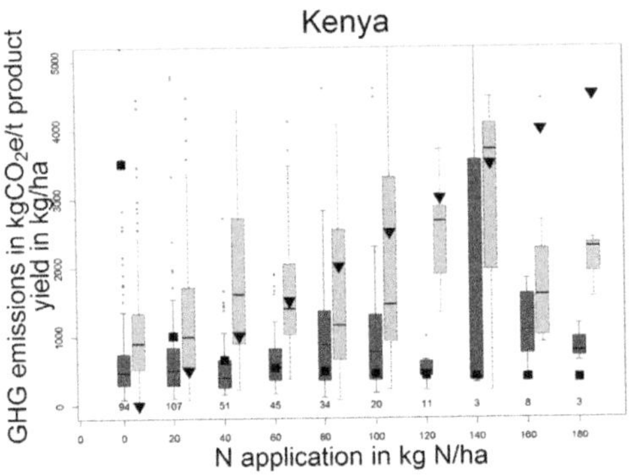


**Fig. 3. Boxplots of greenhouse gas (GHG) emissions per tonne maize (dark grey) and corresponding observed yield (light grey) for binned N application rates of single cropped maize by Kenyan and Ethiopian smallholders. Theoretic values for GHG emissions (black squares) and yield (black triangles) have been provided to enable judgment of adequacy of N supply vs. demand and its associated GHG emissions (Bellarby et al., 2014).**

**Analysis for screening optimal mitigation options using look-up tables and spatial screening of optimal mitigation options**. In this study (Hillier et al., 2012), global data on soil and climate were sufficient to analyze the expected impact of a range of mitigation options—for example, no-till, cover cropping, use of abated (reduced emission) fertilizers—and rank them as a function of broad soil and climate classification. The main finding of this study was that geographic location affects the relative ranking of the effectiveness of the mitigation options, thus reinforcing the point that one-size-fits-all policies are not optimal for agriculture.

of soil physical structure. However, soil environmental properties, such as water filled pore space and organic C, have been found to be more important to denitrification than the abundance or diversity of denitrifier organisms (Attard et al., 2011).

Most currently applied soil GHG flux models do not reflect this level of detail in modeling the spatial characteristics of soil but, rather, are conceptual pool models (described below). These model soil C in terms of its stability (half-life or some other parameter characteristic of turnover time). Rate modifiers alter the rates of decay, and plant matter enters the pools from plant exudates or is otherwise incorporated into the soil (e.g., through plowing, leaf litter, etc.). As an aside, one phenomenon these models typically exhibit is that the C content of mineral soils can often be assumed to reach a steady state over time if C inputs remain unchanged. This is not generally the case for peats and other organic soils. As soils

**Box 3. Windfarm-CAT.**

A second example of a simple empirical tool of this kind is the Scottish Government's Carbon Assessment Tool for Wind Farms (Windfarm-CAT) constructed on peats. Windfarm-CAT was constructed by Nayak et al. (2010), and further modified by Smith et al. (2011), to assess the C balance of wind farms on Scottish peatlands. The purpose of the calculator is to determine the C payback time of a wind farm sited on peatland by accounting for net soil organic C losses as a result of (i) infrastructure, that is, production, transportation, erection, operation, and dismantling of wind farms; (ii) backup power generation; (iii) loss of C-fixing potential of peatland; (iv) changes in C stored in peatland as a result of peat removal and changes in drainage; (v) carbon savings from improvement of habitat; and (vi) loss of C-fixing potential as a result of forestry clearance (Fig. 4). It is similar to the Cool Farm Tool in that it is a static model based on regression equations, and EFs chosen to ensure that input requirements are limited to the data available in a wind farm environmental impact assessment. It is currently being used by planning authorities (such as the Scottish Environmental Protection Agency), community interest groups, and the wind farm industry. Perceptions of this tool by the wind farm industry have been reviewed by the Scottish Government. Results suggest that the respondents believe the calculator provides a better understanding of what components of the wind farm construction increases GHG emissions, but some questioned whether the tool allowed them to improve wind farm design to reduce emissions. Around 50% of users report difficulties in providing some of the values needed to run the C calculator, with hydrological response, water table depth, and the effectiveness of restoration being most challenging. Thus, the tool provides insight into the lack of availability of driving variables influencing the important processes and leads to a greater understanding of the sensitivities by the stakeholder community.

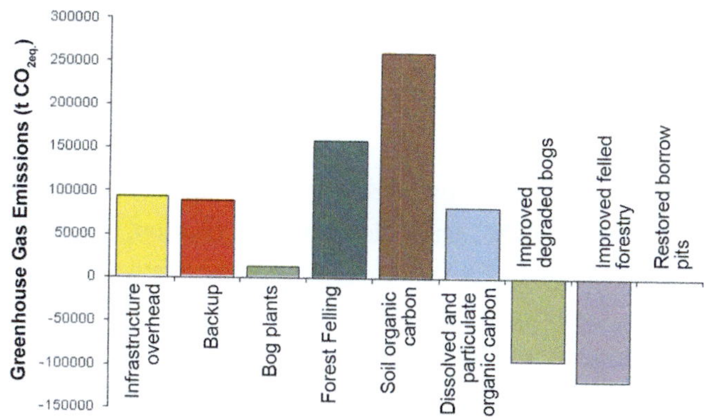


**Fig. 4. Outputs from the Carbon Assessment Tool for Wind Farms for an example calculation (after Nayak et al., 2010).**

that are frequently water saturated, they have a natural tendency to sequester C until limited by the natural physical geography of the region and simultaneously emit substantial amounts of methane (Belyea and Baird, 2006; Clymo et al., 1998).

When drained for agricultural use, the converse effect is that of substantial C and N losses through oxidation only partly compensated for by the reduction in methane emissions (Shannon and White, 1994; Van den Boss, 2003).

Bosatta and Ågren (1985, 1991) proposed a generic theory for the dynamics of C and N in the organic matter of the soil. This generic mathematical model considers the input of litter or organic material at various time intervals as separate cohorts of SOM with a quality distribution that degrades in a similar manner. The generic model of decomposition is based upon a set of three microbial functions: substrate utilization, the microbial utilization, and the microbial dispersion. The microbial dispersion describes dispersion of the C when it is returned to the substrate and is a measure of the lability, or quality, of the SOM. The microbial community is assumed to be C limited and the quality is associated with C availability. A similar model can be applied to N availability, and this introduces another set of functions to limit SOM degradation. This mathematical theory is sometimes called a continuum model or a cohort model (Smith et al., 1997; Smith, 2002). Process-based multicompartment and single-exponential models can be shown to be special cases of this generic model (Bosatta and Ågren 1991). The rate of decomposition and the partitioning of the SOM residues from each SOM pool at any time depends on the soil clay content, the soil water content, the pH, the Eh, the climatic conditions and whether SOM level is stable or is changing as a result of land use and vegetation change. These predictive variables are either implicitly or explicitly incorporated in all models.

In SOM models, the soil N cycle is generally modeled in connection with the C model (Li et al., 1992; Li 1995). Nitrogen released from decomposing SOM as ammonium ($NH_4^+$) or added to the soil may be nitrified to nitrate ($NO_3^-$). Carbon and N may be lost from the soil by leaching of $NO_3^-$, dissolved organic C (DOC) or dissolved organic N (DON), denitrification, volatilization, soil erosion, or crop offtake (Dawson and Smith, 2007). Conversely, C and N may be returned to the soil by plant inputs, inorganic fertilizers, atmospheric deposition, or organic amendments. The denitrification process is often modeled as a potential denitrification rate limited by soil pH, water filled pore space, dissolved organic N, and soil microbial activity. Both the denitrification–decomposition (DNDC) and CENTURY models incorporate N-cycle elements together with the C turnover, as does the Estimation of Carbon in Organic Soils—Sequestration and Emissions (ECOSSE) model, which is described in more detail in Box 4.

## Vegetation Growth and Vegetation Management Models

To understand GHG emissions from soils, one of the key components is the amount of organic matter input into the soil each year so that, in turn, the degradation of

**Box 4. ECOSSE.**

The ECOSSE model (Smith et al., 2010) was developed to simulate soils high in organic matter from concepts originally derived for mineral soils in the RothC (Jenkinson and Rayner, 1977; Jenkinson et al., 1987; Coleman and Jenkinson, 1996) and SUNDIAL (Bradbury et al., 1993; Smith et al., 1996) models. Following these established models, ECOSSE uses a pool-type approach, describing SOM as pools of inert organic matter, humus, biomass, resistant plant material, and decomposable plant material (Fig. 5). All of the major processes of C and N turnover in the soil are included in the model, but each of the processes is simulated using only simple equations driven by readily available input variables, allowing it to be developed from a field-based model to a national-scale tool without high loss of accuracy. To achieve this, measurements of soil C are used to interpolate the activity of the SOM and the plant inputs needed to achieve those measurements. Any data available describing soil water, plant inputs, nutrient applications, and timing of management operations are used to drive the model and better apportion the factors determining the interpolated activity of the SOM. However, if any of this information is missing, the model can still provide accurate simulations of SOM turnover because parameters are adjusted to attain the initial conditions of the soil using the input data provided. During the decomposition process, material is exchanged between the SOM pools according to first-order rate equations, characterized by a specific rate constant for each pool and modified according to rate modifiers dependent on the temperature, water content, crop cover, and pH of the soil. Under aerobic conditions, the decomposition process results in gaseous losses of carbon dioxide ($CO_2$), and under anaerobic conditions, losses as methane ($CH_4$) dominate. The N content of the soil follows the decomposition of the SOM using a stable C/N ratio defined for each pool at a given pH to determine if N must be mineralized or immobilized to maintain that ratio (Fig. 6). Nitrogen is released from decomposing SOM as $NH_4^+$ and may then be nitrified to $NO_3^-$ depending on soil conditions. Carbon and N may be lost from the soil as $NO_3^-$, DOC, or DON by leaching, which is described by simple piston flow. Denitrification, volatilization, and crop offtake are also described using simple empirical equations. The model allows C and N to be returned to the soil as plant inputs, inorganic fertilizers, atmospheric deposition, or organic amendments.

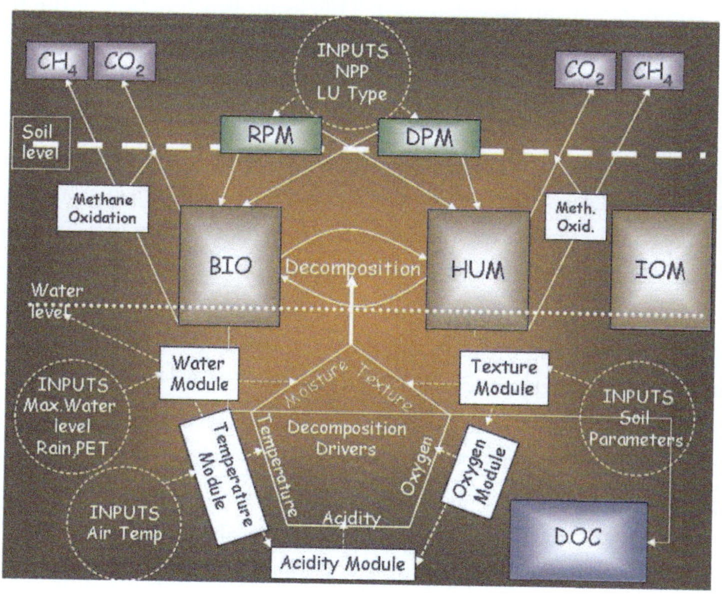


**Fig. 5. Structure of the carbon components of Estimation of Carbon in Organic Soils—Sequestration and Emissions (ECOSSE) (after Smith et al., 2010).**

**Box continued.**

**Box 4. Continued.**

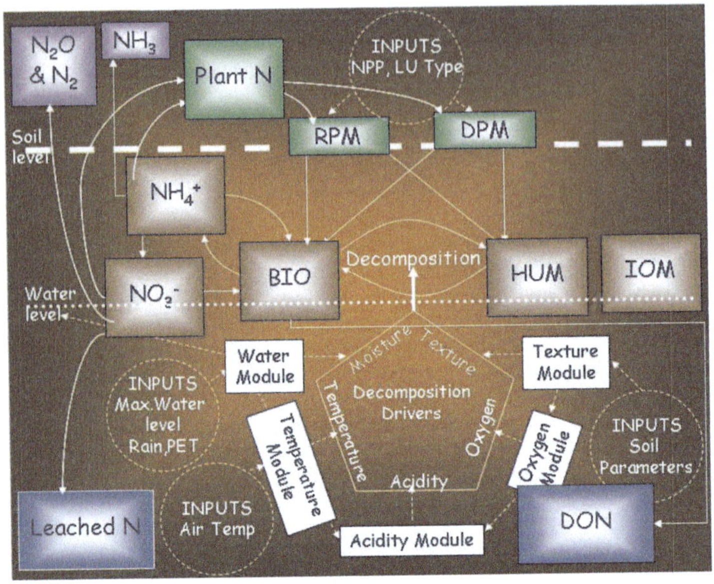


**Fig. 6. Structure of the (a) carbon components and (b) nitrogen components of Estimation of Carbon in Organic Soils—Sequestration and Emissions (ECOSSE) (after Smith et al., 2010).**

C and N components can be calculated. This input is a function of the biomass productivity, whether it is cropland, woodland, grassland, or a natural habitat. This productivity can be modeled as an annual net primary productivity (NPP) as, for example, in the MIAMI model (http://www.juergen-grieser.de/downloads/NetPrimaryProduction/npp.htm), which is a function of annual mean temperature and rainfall. It can also be modeled as an empirical growth curve based upon a defined maximum yield with allocation to grain root and shoot as in DNDC (Li et al., 1994). Alternatively, it could be a process-based model that includes phenomic traits; reacts to nutrient, soil, and climatic conditions; and uses a Monteith and Moss (1977) photosynthesis model that requires knowledge of incident radiation, leaf area, and the efficiency of photosynthesis.

Models are particularly useful for extrapolating current knowledge to explore the impact of crop expansion using treatments that currently have limited spatial coverage. For example, there are limited field trials of bioenergy crops. For bioenergy crops (particularly dedicated ones such as short-rotation coppice [SRC] willow (*Salix* spp.), poplar (*Populus* spp.), and *Miscanthus* spp., process-based models provide a low-cost method of extrapolating our current knowledge to explore scenarios for future expansion. In such cases, it is likely that the processes are

**Box 5. MiscanFor Model.**

MiscanFor (Hastings et al., 2009) is a crop-growth model that calculates yield mass based on meteorological and soil data inputs. It is based on a model by Clifton-Brown et al. (2000), which follows the radiation use efficiency approach of Monteith and Moss (1977). The model was initially parameterized for *Miscanthus ×giganteus* J. M. Greef & Deuter ex Hodk. & Renvoize , which is of particular interest as an energy crop because of its relatively low input requirements and high yields under a range of conditions (Ercoli at al., 1999), but has subsequently been parameterized for other perennial crops. The radiation intercepted by the canopy is dependent on several parameters including temperature and available soil water, which are modeled using characterizations of the processes of evapotranspiration, soil moisture content, photosynthetically active radiation, plant physiological time clock, water and temperature stress, and kill events (Hastings et al., 2009). Data required to drive the model include meteorological data (mean temperature, temperature range, cloud cover, and precipitation) and soil data for field capacity and wilt point. The model is widely used in Miscanthus breeding programs to understand yields at germplasm collection sites and to choose breeding parents for suitability to different climates and environments. It is also used commercially to determine energy yields and GHG emissions from its use for bioenergy.

modeled, by and large, as for food crops; however, the calibration is performed using different quantity and quality of data (e.g., well characterized experimental sites but with less geographic spread). Examples of such models are MiscanFor (Box 5) and Forest Growth–SRC. Below we provide examples of models used to spatially extrapolate management impacts on plant growth.

In the discussion below, we explore various uses of process-based models.

## Evaluation Against Observed Data

The most straightforward and intuitive use of process-based models is as a test of scientific understanding. Once the component processes have been characterized, they can be applied to independent data to test whether the model can mimic observation. A model evaluation can be used to evaluate understanding of the systems to which they are applied. In other words, if a model can consistently mimic experimental data over a range of conditions, then we have confidence that the key processes are modeled and well characterized. Poor model performance or a need to systematically calibrate a model for a particular environment or system illustrates a potential knowledge gap or a defect in the characterization of the embedded processes.

Although generic in concept, most process-based models suffer from the drawback that both rate modifiers and pools are difficult to measure, particularly at scale. In addition, the model pools of organic C or N are based upon organic matter stability, whereas experimental measurements are based upon particle size, bulk density, and chemical composition. The relationship between these two is complex and difficult to quantify. As such, they often require some site-level calibration in addition to a spin-up or initialization cycle. The dilemma, with regard to calibration, is whether to perform the calibration in order to provide the best match to the observed flux data or, for example, estimates of the conceptual pool sizes. The former raises questions about the validity of the model, and the latter is technically demanding and subject to considerable uncertainty regarding input variables. A thorough evaluation would evaluate against both flux data and pool sizes.

## Sensitivity Analyses and Projections

Sensitivity analysis identifies which model components exert the most influence on the model results (Boxes 6 and 7). When analyzing numerical models, one can test how the model outputs vary with changes in the model inputs or internal model parameters. It explores how outputs vary as a function of inputs or model parameters (Smith et al., 2012; Hastings et al., 2010). Once the sensitivity of model outcomes to variation in inputs has been established, a model can be applied across a range of environmental conditions to explore the impact of, for example, management decisions or global change predictions. For example, although it is possible to run simulations to approximate the expected impacts of climate change (e.g., the elevated $CO_2$ and temperature studies conducted under the free-air $CO_2$ enrichment experiments [Triggs et al., 2004; Weigel et al., 2005; Weigel and Manderscheid, 2012]), these are limited and costly. Simulation using process-based models represents a low-cost alternative. The ability to characterize the underlying process allows quantification of the expected effect as a function of the inputs. As such, the model can be used as a form of evidence to inform policy in advance of implementation. In this way it informs potential land management interventions on GHG emissions and helps guide BMP decisions.

The GHG Europe study illustrates how process-based models can be used to project the impacts of future climate change. They can similarly be used to explore counterfactuals to assess the impact of current activities. For example, the project, supported by the UK Department of Energy and Climate Change, explores counterfactual cases where the cultivation of biomass for bioenergy replaces land abandonment. Current global C stocks (above- and below-ground) under dominant land uses are estimated from Tier 1 global biomass carbon and harmonized world soil database maps and aboveground C stocks under land abandonment have been analyzed using CENTURY.MC1 dynamic

**Box 6. Model validation example, GHG Europe.**

The EU-funded GHG Europe project provides an example of model validation against measurements. The ECOSSE, DNDC, and DayCent models have been evaluated in GHG Europe. The DNDC and DayCent models were applied and compared in the modeling of GHG emissions in different ecosystems on mineral soils in Ireland. These studies were funded by the Ireland Environmental Protection Agency, Department of Agriculture, and the Ireland National Council for Forest Research and Development. In a direct test of model performance (comparing modeled prediction with observation), it was found that (i) the DNDC model works better to estimate $CO_2$ and $N_2O$ emissions from Ireland's arable soils than from grassland soils because it appears to overestimate soil water and the impacts of soil organic C under grasslands; and (ii) the DayCent model, though requiring some calibration for Ireland conditions, simulates $N_2O$ fluxes more consistently on grasslands (Abdalla et al., 2010).

The models were then employed in the above studies to project the impact of climate change downscaled data from global climate models for different emission scenarios (HadGEM2-ES [Environment Canada, 2011] with A1B and A2 scenarios [Jones et al., 2011], and GCM3.1 [Environment Canada, 2011] with RCP4.5 and RCP8.5 scenarios [IPCC, 2014]). In this example the model results indicate that elevated temperature and precipitation resulting from climate change will increase N mineralization and total denitrification rates leading to greater fluxes of $N_2O$ (Abdalla et al., 2009, 2011, 2012). On the other hand, aboveground biomass is expected to simultaneously increase by up to 48% because of the overall effects of precipitation, increased temperature, and $CO_2$ concentration. This will consequently increase the N demand by the vigorously growing grass. Thus $N_2O$ flux from grassland does not significantly increase unless greater quantities of N are used in managing the grasslands. The field-observed and simulated net ecosystem exchange (NEE) values (2003–2007) from three widespread ecosystems in southeast Ireland (forest, arable, and grassland) were investigated using the DNDC model. The field and forest DNDC models successfully estimated changes in seasonal and annual NEE from the three ecosystems. However, differences in NEE were found to be primarily land-cover specific (Fig. 7).

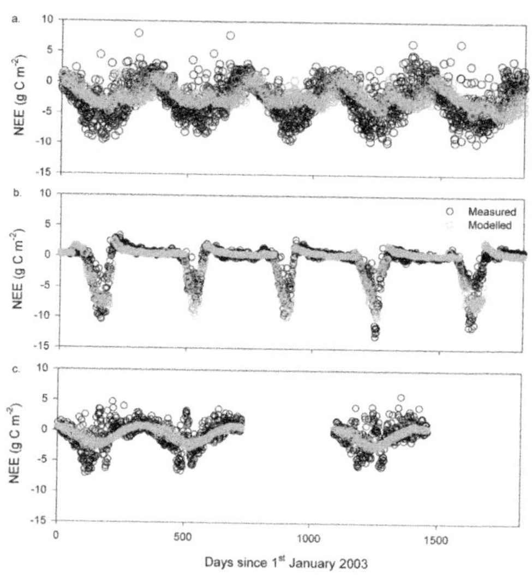


**Fig. 7. Measured and simulated net ecosystem exchange (NEE) for the (a) forest plantations, (b) arable system, and (c) grassland, during the experimental period (2003–2007) (Adapted from Abdalla et al., 2013).**

**Box continued.**

**Box 6. Continued.**

The GHG Europe project also studied ecosystem respiration, $CH_4$, and $N_2O$ from a number of European peatland sites. These sites are spread over four northern European countries: Auchencorth Moss (Scotland, UK), Horstermeer (the Netherlands), Fäjemyr (Sweden), Degerö Stormyr (Sweden), Kaamanen (Finland), and Lompolojänkkä (Finland). Full site descriptions can be found in Drewer et al. (2010), Hendriks et al. (2007), Lund et al. (2007), Sagerfors et al. (2008), Maanavilja et al. (2010), and Aurela et al. (2007), respectively. The sites have different climatic conditions, vegetation, and management. Average annual temperatures and precipitation ranged from −1.4 to 10°C and from 441 to 1155 mm, respectively. For this, work simulations using the ECOSSE model were compared with eddy covariance and chamber measurements (Fig. 8). The results suggest that the ECOSSE model adequately simulated changes in seasonal and annual ecosystem respiration, demonstrating that the goal of ECOSSE in simulating peatland soils was well implemented. A sensitivity analysis of the ECOSSE model highlighted that increasing air temperature, increasing pH, decreasing precipitation, and lowering water table depth could significantly increase ecosystem respiration.

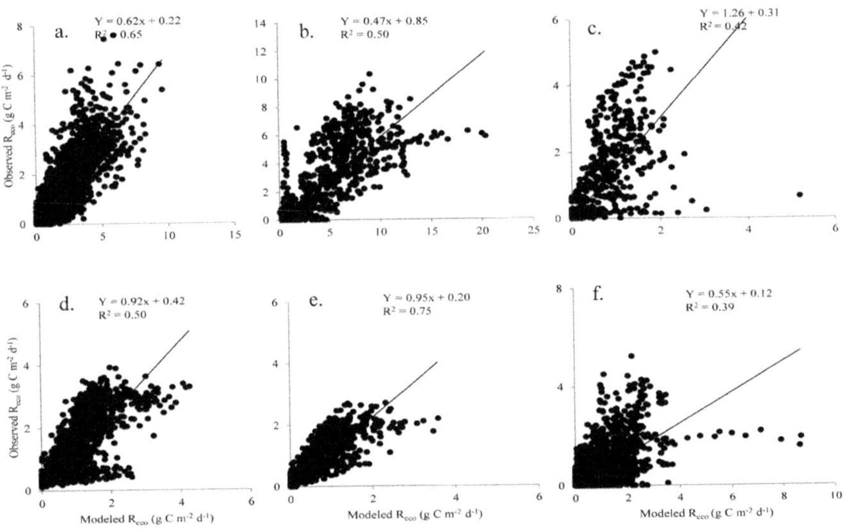


**Fig. 8. Regression relationships between the observed and modeled ecosystem respiration ($R_{eco}$) form (a) Auchencorth moss, (b) Horstermeer, (c) Lompolojänkkä, (d) Fäjemyr, (e) Kaamanen, and (f) Degerö.**

global vegetation model scenarios, while RothC (Jenkinson and Rayner, 1977; Jenkinson et al., 1987; Coleman and Jenkinson, 1996) is used for belowground C. For bioenergy scenarios the LPJmL model simulations from Beringer et al. (2011) are used to obtain expected Miscanthus and short-rotation forestry yields, respectively.

**Box 7. Model sensitivity analysis example, InveN$_2$Ory.**

An example of model sensitivity analysis was conducted under the project InveN$_2$Ory (Fitton et al., 2014), funded by UK Department for Environment Food and Rural Affairs. DailyDayCent was employed to predict $N_2O$ emissions in the United Kingdom. The performance of the model was first evaluated for four different winter wheat (*Triticum aestivum* L.) cropland sites in the United Kingdom (Boxworth, Terrington, Betley, and Middleton). It was found that the DailyDayCent model was able to provide a reasonable approximation of both annual emissions of $N_2O$ and crop yields across the six cropland experimental sites (Fig. 9). However, in some cases, modeled annual emissions fell outside the range of their corresponding measured values. There could be numerous explanations for this discrepancy apart from a defect of the model (structural error), for example, insufficient information on the key climate or soil inputs and also insufficient land management information especially where N fertilizers or organic manure was applied in the period immediately preceding the experimental measurements.

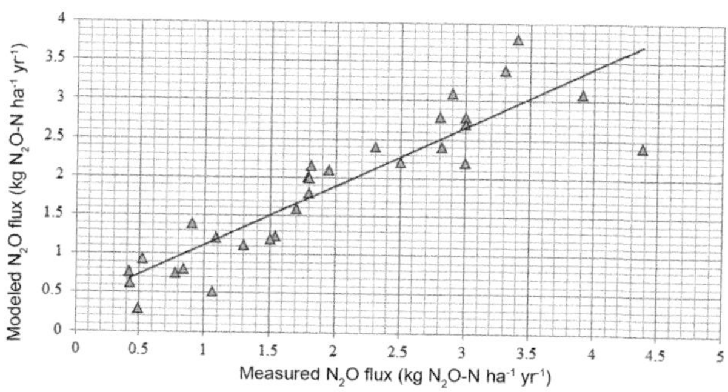


**Fig. 9. A one-to-one plot of measured vs. modeled cumulative fluxes for all experimental plots used in the InveN$_2$Ory project**

The study also included a sensitivity analysis of the estimation of the uncertainty of soil and meteorological (temperature and precipitation) input parameters on simulated $N_2O$ emissions at different sites conducted in the following way (Fig. 10):

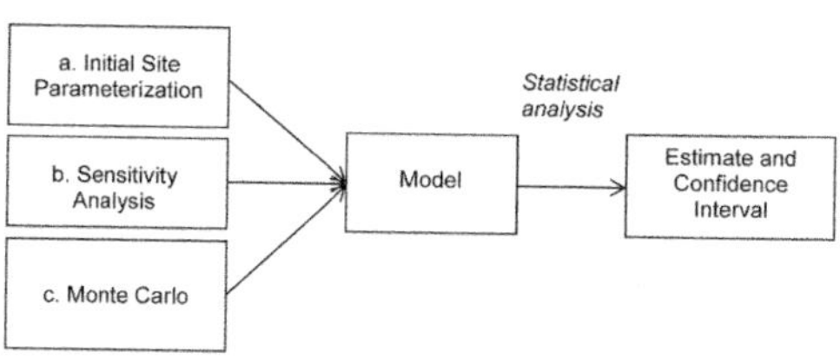


**Fig. 10. Illustration of the uncertainty analysis method.**

1. The model was evaluated against available measured $N_2O$ fluxes and other crop parameters as above.
2. The uncertainty in the key model inputs—soil pH, clay content, bulk density, temperature, and precipitation—was expressed as a probability distribution function (PDF), where the PDF of uncertainty was defined using the site-level input value as the midpoint. For each input variable, a normal distribution of their uncertainties was assumed. The range of the PDFs for each input variable can be user defined; however, for this experimental design the variability was reflected on a site level. Therefore, the range of values used was based on both expert survey and published estimates by Hastings et al. (2009). Following this, a 10-step increase and decrease of equal intervals ($n$ = 10) from the midpoint was applied, with the minimum and maximum values set, based on the percentage differences, kept the same for each site.

**Box continued.**

**Box 7. Continued.**

3. A point-sensitivity analysis was performed to define the behavior of the modeled $N_2O$ emissions to a variation of each single input over the above range, while all remaining inputs were fixed at the site level value
4. A full Monte Carlo analysis using the PDFs of each input parameter was then undertaken. Input parameters were randomly sampled 10,000 times, and from this, an estimate of confidence intervals was obtained of the range of values of $N_2O$ fluxes for each experimental site.

## Uncertainty Analysis

Uncertainty analysis determines how the variability in the input is propagated through the model and quantifies how this is translated into variability (uncertainty) in the model output. Process-based models define multiple links between soil processes on different C and N pools. An advantage of this is that soil process feedbacks are simulated. For example, an increase in solar radiation or rainfall may increase plant growth leading to increased soil C inputs, and, subsequently, these increased organic C inputs increase microbial activity and stimulate N cycling. On the other hand, since the models are integrated and iterative, errors in the setting of parameters and driving variables propagate through the iteration of the model and lead to increased uncertainty in outputs. If the output variability is high with a changing input, the input is termed important. A sensitivity analysis can be used to identify the inputs with the highest correlation between input and output. These methods do not necessarily identify the same inputs. A model is always sensitive to the important inputs that contribute most to model output uncertainty, since the variability will not appear in the model output unless the model is also sensitive to the input. However, an input to which the model is sensitive is not necessarily important; it does not necessarily contribute to output uncertainty since its value may be known precisely (Gottschalk et al., 2007). Another important component of uncertainty in simulation is model structural uncertainty; even if inputs are precisely known, outputs are still uncertain because models are not perfect representations of reality as a result of omitted or simplified processes represented within the model.

A key goal of the InveN$_2$Ory project was to test model accuracy for application across a range of UK sites. The use of sites with varying conditions and the testing of model sensitivity to parameters achieved this validation. By testing the model accuracy and uncertainty, and by validating model simulation of $N_2O$ fluxes at each site, the model is demonstrated to be appropriate for assessing spatial estimates of the range of $N_2O$ fluxes within the United Kingdom. As a result of the variability in the spatial and temporal fluxes of $N_2O$, biogeochemical models have become a useful tool for policy makers. More specifically, ecosystem models, used in predictive form as described above, can allow for national inventories to report emissions using Tier 3 methodology. Such methods exploit some underlying qualities of process-based models, subject to the constraint that the range of inputs does not exceed the range over which we have confidence in the characterization of the embedded processes.

## Bayesian Calibration

Another approach to model calibration is Bayesian calibration (Box 8). The Bayesian approach (Gelman et al., 2004), in essence, admits uncertainty around the parameterization of models as well as in the driving data and the evaluation data. Characterization of the prior distribution of both variables and model parameters allows adjustment of these in the context of the experimental data. As such, adequate performance of the model at site level can often be achieved. This is not, however, guaranteed and may, to some extent, compensate for structural error in the models. The fact that such calibration improves the modeling at site level may indeed be an indication of the structural flaws in the models and thus informs our scientific understanding of the process involved.

The above examples show applications where process-based models have been applied and interpreted by scientists and developers. The complexity of process-based models often makes this a lengthy exercise. Once a level of validation has been achieved, analyses may then be conducted before the outputs are summarized for other stakeholders (e.g., policy makers). A drawback of this process is that the stakeholder community does not have direct access to the model or its immediate interpretation. An alternative approach to overcome this drawback is to develop a package or platform with the end-user stakeholder in mind that embeds a process-based model. In the final example we demonstrate just such an exercise.

The ELUM project (Box 9) represents a simplified metamodeling approach, where simple relationships between input and outputs variables are derived from model simulations. Assuming that a process-based model can be applied with confidence to the system in question (for example, that it has been calibrated and validated) then it can be employed to generate credible data mimicking that system. This data, though synthetic, can have a higher resolution than experimental data since it is less costly to obtain. Such data can also be hypothetical (e.g., climate projections). Having been extracted from dynamic simulation models, these simple relationships can be embedded in more practical software solutions for other user groups, although perhaps with the loss of some information regarding

**Box 8. Bayesian calibration example.**

Yeluripati et al. (2009) demonstrated a method that uses Bayesian calibration by means of the accept–reject algorithm (Gelman et al., 2004) and used this method to calibrate the initial distribution of soil organic C pools against observed soil respiration measurements. The Bayesian calibration method quantified and reduced the uncertainties in initial C distribution. The method combines two sources of information: (i) prior information about the model parameters (as defined above) and (ii) measured data on model output variables to obtain a revised probability distribution (or posterior probability distribution) for the parameters. The calibration is an application of Bayes' theorem, according to which the posterior probability of any parameter vector is proportional to the product of its prior probability and its corresponding data likelihood (Eq. [1]).

$$\log(L) = \sum_{d=1}^{n} \left[ -0.5 \left( \frac{\mathrm{OD} - \mathrm{MD}}{\mathrm{UD}} \right)^2 - \log(\mathrm{UD}) \right] \quad [1]$$

where OD is mean observed soil winter $CO_2$ emissions on each sampling day, MD is model-simulated soil $CO_2$ emission, UD is error or uncertainty in the measurements, and $n$ is the number of data points.

In this way, initial size of the C pools embeds both our initial expectation of the pool sizes combined with a requirement for adequate model performance at site level.

After successful calibration at different sites across Europe (Fig. 11), the DailyDayCent model was used for the comprehensive assessment of future changes in C balance of European croplands (EU27) from 1970 to 2028 considering the effects of IPCC Special Report on Emissions Scenarios A1B emission scenario. The total land area in the EU27 was classified into around 41,000 homogeneous units (NitroEurope Calculation Units [NCUs]). The model was run in each NCU from 1971 to 2000 (baseline) and 2001 to 2028 using climate under the A1B emission scenarios. The model was used to produce geographical patterns of NPP, soil heterotrophic respiration, and NEE under contemporary and future climate (Fig. 12).

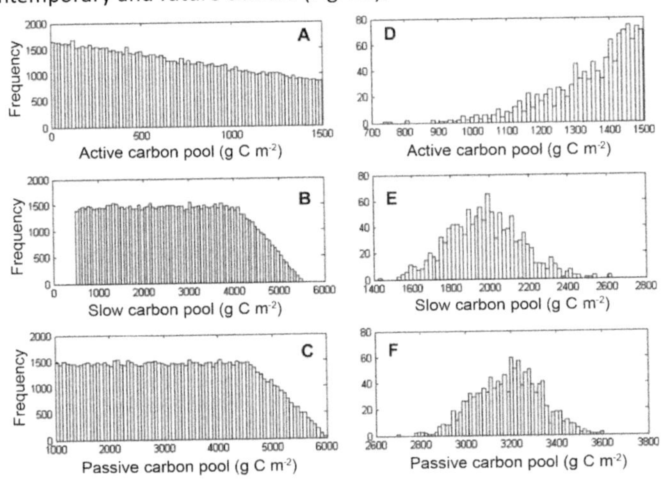


**Fig. 11. Samples from the prior ($n$ = 105) and posterior ($n$ = 1179) parameter probability distributions for the C1, C2, and C3 pools of the DayCent model. The figure illustrates prior marginal parameter distribution of (A) active C pool, (B) slow C pool, (C) passive C pool, and posterior parameter distribution of (D) active C pool, (E) slow C pool, and (F) passive C pools.**

Box continued.

**Box 8. Continued.**

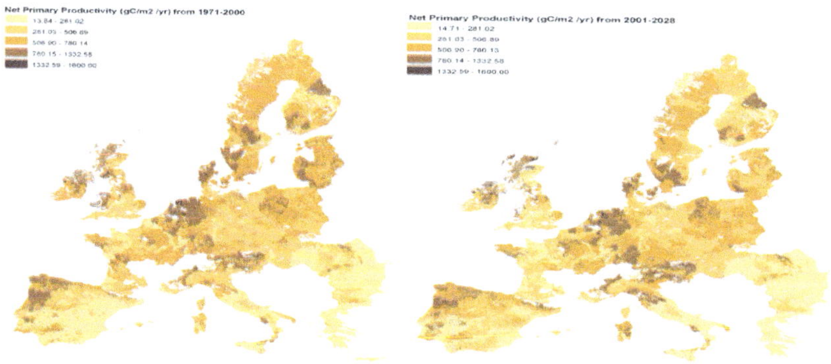


**Fig. 12. Net primary productivity (NPP) simulated by DailyDayCent model. (Left) Average annual NPP from 1971 to 2000. (Right) Average annual NPP from 2001 to 2028.**

The uncertainties in the present and future C balance associated with the choice of crop rotation were then assessed by running DailyDayCent with five different crop rotations with the same climate variables. The results suggest that European croplands are near to C neutral during 1970 to 2000; soil C losses resulting from future climate change reduce or even offset C sequestration resulting from growth enhancement induced by climate warming and $CO_2$ fertilization in Mediterranean region. At European country level, arable systems are a net weak C source; many individual countries are weak C sink, while on the European scale under the A1B scenario, European agricultural soils will lose C and act as weak sources. Although being close to C neutral today, it emerges that European agriculture C balance is sensitive to management practice and climate change. The conclusion is that more focused guidelines at individual country level can be derived from results presented in this study and would help in reducing these effects.

variability. Such applications are still relatively new but show considerable promise if used appropriately.

# Summary

Mathematical modeling is still a young and fast-evolving discipline. Our conventional understanding and terminology has developed along with classifications such as strategic, tactical, individual-based, bottom-up, top-down, systems, etc., which defined the functional aspects and characteristics of models. Related to this classification, mathematical models of physical and biological phenomena also, typically, vary significantly in complexity from the very concise classical population biology models (e.g., predator–prey models) to complex and dynamic models reaching tens of thousands of lines of code (e.g., ecosystem models). Understanding of methods to evaluate and interpret mathematical models is improving.

**Box. 9. A simplified metamodeling example, the ELUM project.**

The ELUM project was commissioned and funded by the Energy Technologies Institute to produce a framework for predicting the sustainability of bioenergy deployment across the United Kingdom. The project closely combines field studies with computational modeling in order to map the UK bioenergy GHG balance from soil respiration up to 2050. The example illustrates embedding process-based models in a method that is accessible to nonexperts.

The ECOSSE model was first used to estimate GHG changes resulting from land-use change for various UK bioenergy crops using biomass production data generated for Miscanthus (MiscanFor), short-rotation forestry (ESC-CARBINE; Pyatt et al., 2001; Thompson and Matthews, 1989), SRC (ForestGrowth–SRC; Tallis et al., 2013), and first-generation crops (MIAMI; Lieth, 1975). Model simulations were verified at site level against experimental data obtained from the ELUM project.

The product of this project includes output from multiple simulations of ECOSSE using projected climate data across the United Kingdom. These have then been stored in a look-up table, which can be accessed via an interface intended for use by nonexpert users to obtain location-specific estimates of GHG fluxes caused by land-use transitions to grow bioenergy crops. A look-up table was used, as it provides faster and more accurate results than the alternative option of a simplified soil organic matter model.

The tool presents analysis of spatial and temporal results, compares the effects of different transitions, and produces outputs in a format that can be opened in a GIS so that users are able to postprocess results according to their interests. Likely, users of the tool include the following:

- Energy, environmental, and agricultural policy makers
- Soil and climate scientists
- Energy producers and suppliers
- Investors and traders in energy or C credits
- Land owners and managers

In practice, the user data is keyed to look up the nearest-neighbor prerun simulations in the look-up table. The process and a mock-up of the interface are shown in Fig. 13 and 14.

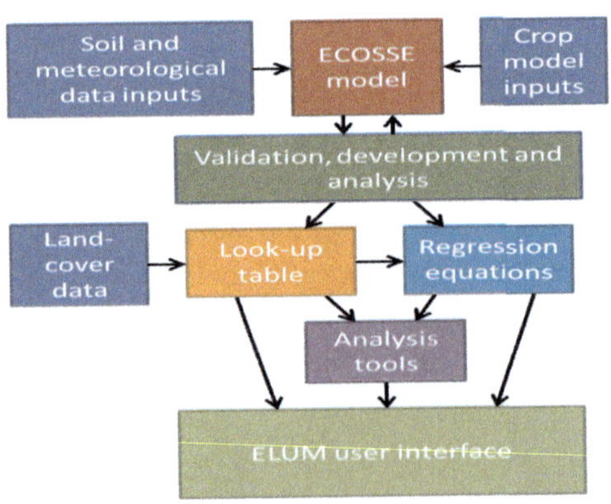


**Fig. 13. Schematic for the process of developing the ELUM user interface.**

**Box 9. Continued.**

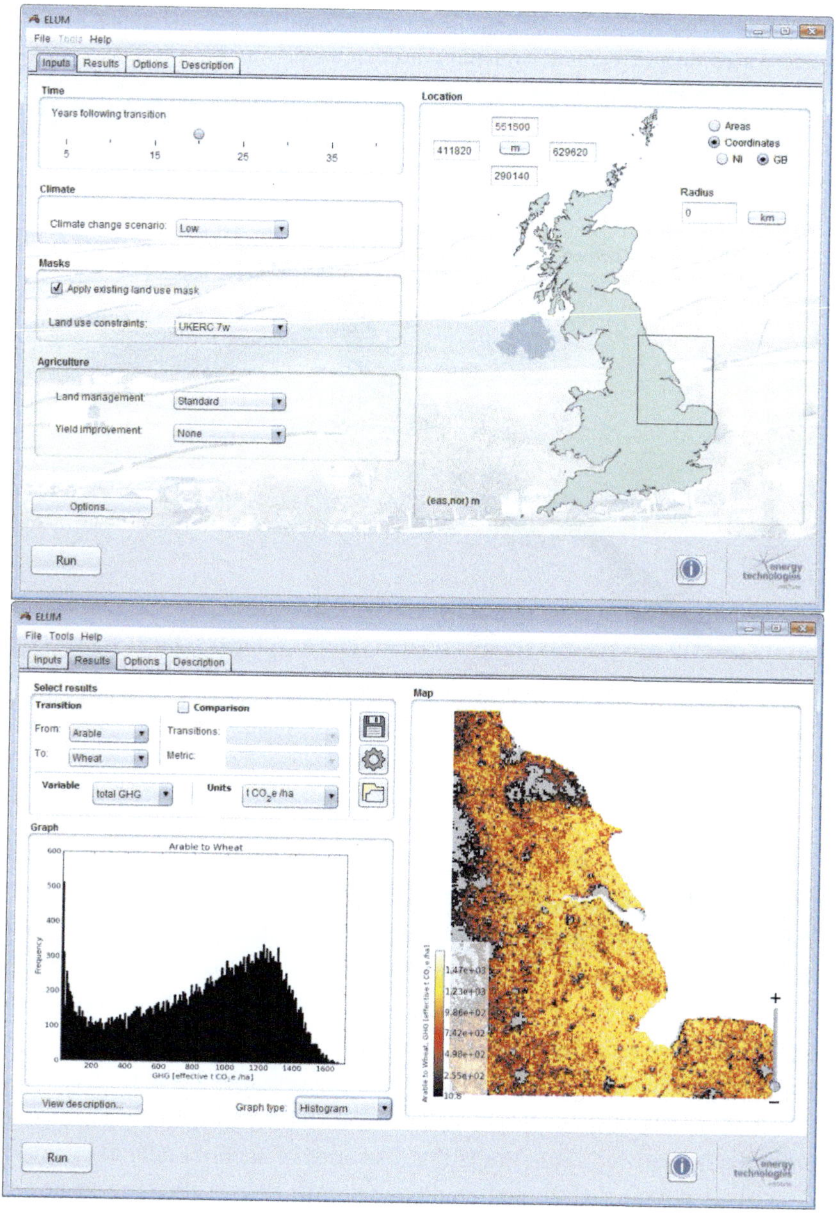


**Fig. 14. Early prototype of the ELUM user interface.**

**Box 9. Continued.**

Depending on the options selected by users, results from the look-up table may be rescaled according to regression relationships that are derived from ECOSSE simulations for the response of certain model outputs to key input data; this is performed in order to reduce the storage space required for the look-up table for unusual user selections. Figure 15 shows an example of the $N_2O$ flux as a function of rainfall. This output is used to inform the user on the likely impact of practice or environmental changes.

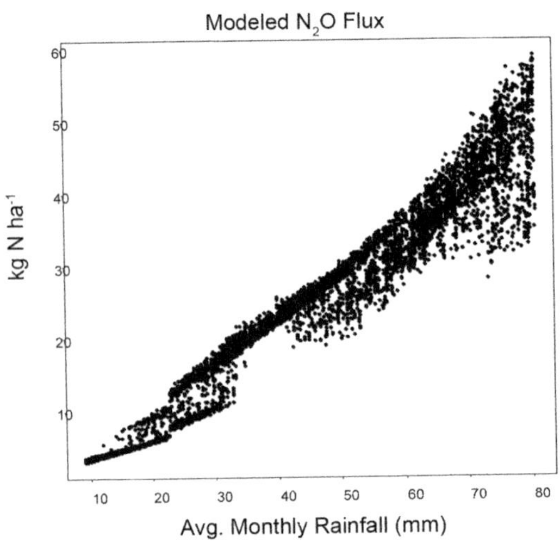


**Fig. 15. Curvilinear relationship of modeled $N_2O$ emissions as a function of monthly rainfall (mm).**

In the above, we explored a range of modeling approaches and their applications. It might be expected that with greater model complexity comes greater modeling precision but at the cost of usability. Detailed, process-based models perform well for situations where input data can be specified with certainty, but, because of their complexity, expert users must interpret them before being translated into a usable form for policy makers or laypersons, for example. However, neither of the above is necessarily true. Often, performance at site level cannot be guaranteed even with quite detailed models, and a lengthy calibration and evaluation phase is required. This limits the relevance of such models to businesses of C markets without additional credible deterministic methods to ensure reliability. As such, there is a current role for empirical models that have the benefit of being purely data driven, and, because of their relatively simple structure, are transparent in terms of their algorithms. Simple EF methods currently have, perhaps, wider applicability, since they can be embedded in a protocol such as the national inventory reporting methods of the IPCC and the value of each EF has a natural meaning as an indicator of importance of emissions sources.

In recent times, there have been several exciting initiatives to overcome some of the conventional limitations of process-based models. For example, the implementation of the CENTURY model in the COMET-VR platform (http://www.comet2.colostate.edu/), meta-analyses such as that of Leip et al. (2011) on calibrated applications of DNDC in Europe, or the ELUM model interface discussed above using a prerun look-up table based on ECOSSE simulations together with point sensitivity analyses.

In this chapter several models and applications are described, and their value to various user groups is discussed. The following quote is well known to statisticians: "All models are false—some are useful" (Box and Draper, 1987). We should remember that models are only an attempt to describe some aspects of real systems, and that the value of a model should not be judged by its precision alone.

## References

Abdalla, M., M. Jones, J. Yeluripati, P. Smith, J. Burke, and M. Williams. 2010. Testing DayCent and DNDC model simulations of $N_2O$ fluxes and assessing the impacts of climate change on the gas flux and biomass production from a humid pasture. Atmos. Environ. 44:2961–2970. doi:10.1016/j.atmosenv.2010.05.018

Abdalla, M., S. Kumar, M. Jones, J. Burke, and M. Williams. 2011. Testing DNDC model for simulating soil respiration and assessing the effects of climate change on the $CO_2$ gas flux from Irish agriculture. Global Planet. Change 78:106–115. doi:10.1016/j.gloplacha.2011.05.011

Abdalla, M., K. Rueangritsarakul, M. Jones, B. Osborne, M. Helmy, B. Roth, J. Burke, P. Nolan, P. Smith, and M. Williams. 2012. How effective is reduced tillage-cover crop management in reducing $N_2O$ fluxes from arable crop soils? Water Air Soil Pollut. 223:5155–5174. doi:10.1007/s11270-012-1268-4

Abdalla, M., M. Saunders, A. Hastings, M. Williams, P. Smith, B. Osborne, G. Lanigan, and M. Jones. 2013. Simulating the impacts of land use in Northwest Europe on Net Ecosystem Exchange (NEE): The role of arable ecosystems, grasslands and forest plantations. Sci. Total Environ. 1:325–336. doi:10.1016/j.scitotenv.2012.12.030

Abdalla, M., M. Wattenbach, P. Smith, P. Ambus, M. Jones, and M. Williams. 2009. Application of the DNDC model to predict emissions of $N_2O$ from Irish agriculture. Geoderma 151:327–337. doi:10.1016/j.geoderma.2009.04.021

Agarwal, B., and B. Jeffries, editors. 2013. Cutting cotton carbon emissions: Findings from Warangal, India. World Wildlife Fund, India.

Ahuja, L.R., L.W. Ma, and D.J. Timlin. 2006. Trans-disciplinary soil physics research critical to synthesis and modeling of agricultural systems. Soil Sci. Soc. Am. J. 70:311–326. doi:10.2136/sssaj2005.0207

Arah, J.R.M., and K.A. Smith. 1989. Steady-state denitrification in aggregated soils: a mathematical model. J. Soil Sci. 40:139–149. doi:10.1111/j.1365-2389.1989.tb01262.x

Attard, E., S. Recous, A. Chabbi, C. De Berranger, N. Guillaumaud, J. Labreuche, L. Philippot, B. Schmid, and X. Le Roux. 2011. Soil environmental conditions rather than denitrifier abundance and diversity drive potential denitrification after changes in land uses. Glob. Change Biol. 17:1975–1989. doi:10.1111/j.1365-2486.2010.02340.x

Aurela, M., T. Riutta, T. Laurila, J.P. Tuovinen, T. Vesala, E.S. Tuittila, J. Rinne, S. Haapanalen, and J. Laine. 2007. $CO_2$ exchange of a sedge fen in southern Finland-the impact of a drought period. Tellus B Chem. Phys. Meterol. 59:826–837. doi:10.1111/j.1600-0889.2007.00309.x

Ball, B.C. 2013. Soil structure and greenhouse gas emissions: A synthesis of 20 years of experimentation. Eur. J. Soil Sci. 64:357–373. doi:10.1111/ejss.12013

Bellarby J, B. Foereid, A. Hastings, and P. Smith. 2008. Cool farming: Climate impacts of agriculture and mitigation potential. Greenpeace International, Amsterdam.

Bellarby, J., C. Stirling, S.H. Vetter, M. Kassie, F. Kanampiu, K. Sonder, P. Smith, and J. Hillier. 2014. Identifying secure and low carbon food production practices: A case study in Kenya and Ethiopia. Agric., Ecosyst. Environ. 197:137–146.

Belyea, L.R., and A.J. Baird. 2006. Beyond "The limits to peat bog growth": Cross-scale feedback in peatland development. Ecol. Monogr. 76:299–322. doi:10.1890/0012-9615(2006)076[0299:BTLTPB]2.0.CO;2

Beringer, T., W. Lucht, and S. Schaphoff. 2011. Bioenergy production potential of global biomass plantations under environment and agricultural constraints. GCB Bioenergy 3:299–312.

Bouwman A.F., L.J.M. Boumans, and N.H. Batjes. 2002. Modeling global annual $N_2O$ and NO emissions from fertilized fields. Global Biogeochem. Cycles 16:28-1–28-9

Bosatta, E., and G.I. Ågren. 1985. Theoretical analysis of decomposition of heterogeneous substrates. Soil Biol. Biochem. 17:601–610. doi:10.1016/0038-0717(85)90035-5

Bosatta, E., and G.I. Ågren. 1991. Dynamics and carbon and nitrogen in the organic matter of the soil: A generic theory. Am. Nat. 138:227–245. doi:10.1086/285213

Box, G.E.P., and N.R. Draper. 1987. Empirical model building and response surfaces. John Wiley & Sons, New York.

Bradbury, N.J., A.P. Whitmore, P.B.S. Hart, and D.S. Jenkinson. 1993. Modelling fate of nitrogen in crop and soil in the years following application of $^{15}N$-labelled fertilizer to winter wheat. J. Agric. Sci. 121:363–379. doi:10.1017/S0021859600085567

Clifton-Brown, J.C., B.M. Neilson, I. Lewandowski, and M.B. Jones. 2000. The modelled productivity of *Miscanthus ×giganteus* (GREEF et DEU) in Ireland. Ind. Crops Prod. 12:97–109. doi:10.1016/S0926-6690(00)00042-X

Clymo, R.S., J. Turunen, and K. Tolonen. 1998. Carbon accumulation in peatland. Oikos 81:368–388. doi:10.2307/3547057

Coleman, K., and D.S. Jenkinson. 1996. RothC-26.3 - A model the turnover of carbon in soil. In: D.S. Powlson, P. Smith, and J.U. Smith, editors, Evaluation of soil organic matter models using existing long-term datasets. NATO ASI Series I. Vol. 38. Springer, Berlin. p. 237–246.

Crawley, M.J. 2007. The R book. John Wiley & Sons Ltd. Chichester, UK.

Dawson, J.J.C., and P. Smith. 2007. Carbon losses from soil and its consequences for land-use management. Sci. Total Environ. 382:165–190. doi:10.1016/j.scitotenv.2007.03.023

Drewer, J., A. Lohila, M. Aurela, T. Laurila, K. Minkkinen, T. Penttila, K.J. Dinsmore, R.M. McKenzie, C. Helfter, C. Flechard, M.A. Sutton, and U.M. Skiba. 2010. Comparison of greenhouse gas fluxes and nitrogen budgets from an ombotrophic bog in Scotland and a minerotrophic sedge fen in Finland. Eur. J. Soil Sci. 61:640–650. doi:10.1111/j.1365-2389.2010.01267.x

Driver, K., K. Haugen-Kozyra, and R. Janzen. 2010. Agriculture sector greenhouse gas practices and quantification review: Phase I report. Appendix A: Quantification tools and reference materials. Leading Carbon Ltd, ClimateCHECK Corporation, and KHK Consulting, LC Ltd., Alberta, Canada. p. 46–50.

Environment Canada. 2011. Climate modeling and analysis. Environment Canada, Victoria, Canada. http://www.ec.gc.ca/ccmac-cccma (accessed 27 Oct. 2015)

Ercoli, L., M. Mariotti, A. Masoni, and E. Bonari. 1999. Effect of irrigation and nitrogen fertilization on biomass yield and efficiency of energy use in crop production of Miscanthus. Field Crops Res. 63:3–11. doi:10.1016/S0378-4290(99)00022-2

Fitton, N., A. Datta, K. Smith, J.R. Williams, A. Hastings, M. Kuhnert, C.F.E. Topp, and P. Smith. 2014. The challenge of modelling nitrogen management at the field scale: simulation and sensitivity analysis of $N_2O$ fluxes across nine experimental sites using DailyDayCent. Environ. Res. Lett. 9:095003. doi:10.1088/1748-9326/9/9/095003

Gelman, A., J.B. Carlin, H.S. Stern, and D.B. Rubin. 2004. Bayesian data analysis. 2nd ed. Chapman & Hall–CRC Press, Boca Raton, FL.

Gottschalk, P., M. Wattenbach, A. Neftel, J. Fuhrer, M. Jones, G. Lanigan, P. Davis, C. Campbell, J.F. Soussana, and P. Smith. 2007. The role of measurement uncertainties for the simulation of grassland net ecosystem exchange (NEE) in Europe. Agric. Ecosyst. Environ. 121:175–185. doi:10.1016/j.agee.2006.12.026

Hastings, A., J. Clifton-Brown, M. Wattenbach, C.P. Mitchell, and P. Smith. 2009. The development of MISCANFOR, a new Miscanthus crop growth model: Towards more robust yield predictions under different climatic and soil conditions. GCB Bioenergy 1:154–170. doi:10.1111/j.1757-1707.2009.01007.x

Hastings, A., M. Wattenbach, W. Eugster, C. Li, N. Buchmann, and P. Smith. 2010. Uncertainty propagation in soil greenhouse gas emission models: An experiment using the DNDC model and at the Oensingen cropland site. Agric., Ecosyst. Environ. 136:97–110.

Hendriks, D.M.D., J. van Huissteden, A.J. Dolman, and M.K. van Der Molen. 2007. The full greenhouse gas balance of an abandoned peat meadow. Biogeosciences 4:411–424. doi:10.5194/bg-4-411-2007

Hillier, J., F. Brentrup, M. Wattenbach, C. Walter, T. Garcia-Suarez, L. Mila-i-Canals, and P. Smith. 2012. Which cropland greenhouse gas mitigation options give the greatest benefits in different world regions? Climate and soil-specific predictions from integrated empirical models. Glob. Change Biol. 18:1880–1894. doi:10.1111/j.1365-2486.2012.02671.x

Hillier, J., C. Walter, D. Malin, T. Garcia-Suarez, L. Mila-i-Canals, and P. Smith. 2011. A farm-focused calculator for emissions from crop and livestock production. Environ. Model. Software 26:1070–1078. doi:10.1016/j.envsoft.2011.03.014

IPCC. 2006. 2006 IPCC guidelines for national greenhouse gas inventories. Prepared by the National Greenhouse Gas Inventories Programme. IPCC, Geneva.

IPCC. 2007. Climate change 2007: Chapter 2: The physical science basis. Contribution of Working Groups I to the Fourth Assessment Report of the Intergovernmental Panel on Climate Change. IPCC, Geneva.

IPCC. 2014. Climate change 2014: Synthesis report. R.K. Pachauri and L.A. Meyer, editors, core writing team. Contribution of Working Groups I, II, and III to the Fifth Assessment Report of the Intergovernmental Panel on Climate Change. IPCC, Geneva, Switzerland.

Jenkinson, D.S., P.B.S. Hart, J.H. Rayner, and L.C. Parry. 1987. Modelling the turnover of organic matter in long-term experiments at Rothamsted. INTECOL Bull. 15:1–8.

Jenkinson, D.S., and J.H. Rayner. 1977. The turnover of organic matter in some of the Rothamsted classical experiments. Soil Sci. 123:298–305. doi:10.1097/00010694-197705000-00005

Jones, C.D., J.K. Hughes, N. Bellouin, S.C. Hardiman, G.S. Jones, and J. Knight. 2011. The HadGEM2-ES implementation of CMIP5 centennial simulations. Geosci. Model Dev. 4:543–570.

Knorr, W. 2000. Annual and interannual $CO_2$ exchanges of the terrestrial biosphere: Process-based simulations and uncertainties. Glob. Ecol. Biogeogr. 9:225–252. doi:10.1046/j.1365-2699.2000.00159.x

Lassey, K.R., M.J. Ulyatt, R.J. Martin, C.F. Walker, and I.D. Shelton. 1997. Methane emissions measured directly from grazing livestock in New Zealand. Atmos. Environ. 31:2905–2914. doi:10.1016/S1352-2310(97)00123-4

Leip, A., M. Busto, and W. Winiwarter. 2011. Developing spatially stratified $N_2O$ emission factors for Europe. Environ. Pollut. 159:3223–3232. doi:10.1016/j.envpol.2010.11.024

Li, C. 1995. Modeling impact of agricultural practices on soil C and $N_2O$ emissions. In: R. Lal, J. Kimble, E. Levine, and B.A. Stewart, editors, Soil management and greenhouse effect. CRC Lewis Publishers, Boca Raton, FL. p. 101–112.

Li, C., S. Frolking, and T.A. Frolking. 1992. A model of nitrous oxide evolution from soil driven by rainfall events: I. Model structure and sensitivity. J. Geophys. Res. 97:9759–9776. doi:10.1029/92JD00509

Li, C., S. Frolking, and R. Harriss. 1994. Modeling carbon biogeochemistry in agricultural soils. Global Biogeochem. Cycles 8:237–254. doi:10.1029/94GB00767

Lieth, H. 1975. Modeling the primary productivity of the world In: H. Lieth, and R.H. Whittaker, editors, Primary productivity of the biosphere. Springer, Berlin. p. 237–263.

Lund, M., A. Lindroth, T.R. Christensen, and L. Strom. 2007. Annual $CO_2$ balance of a temperate bog. Tellus 59B:804–811. doi:10.1111/j.1600-0889.2007.00303.x

Maanavilja, L., T. Riutta, M. Aurela, M. Pulkkinen, T. Laurila, and E.S. Tuittila. 2010. Spatial variation in $CO_2$ exchange at a northern aapa mire. Biogeochemistry 104:325–345. doi:10.1007/s10533-010-9505-7

Monteith, J.L., and C.J. Moss. 1977. Climate and the efficiency of crop production in Britain. Philos. Trans. R. Soc. Lond. B Biol. Sci. 281:277–294. doi:10.1098/rstb.1977.0140

Nayak, D.R., D. Miller, A. Nolan, P. Smith, and J.U. Smith. 2010. Calculating carbon budgets of wind farms on Scottish peatlands. Mires and Peat 4:1–23

Ogle, S.M., F.J. Breidt, and K. Paustian. 2005. Agricultural management impacts on soil organic carbon storage under moist and dry climatic conditions of temperate and tropical regions. Biogeochemistry 72:87–121. doi:10.1007/s10533-004-0360-2

Pyatt, G., D. Ray, and J. Fletcher. 2001. An ecological site classification for forestry in Great Britain. Bulletin 124. Research Division, Forestry Commission, Edinburgh.

Sagerfors, J., A. Lindroth, A. Grelle, L. Klemedtsson, P. Weslien, and M. Nilsson. 2008. Annual $CO_2$ exchange between a nutrientpoor, minerotrophic, boreal mire and the atmosphere. J. Geophys. Res. 113:G01001–G01015.

Sextone, A.J., N.P. Revsbech, T.B. Parkin, and J.M. Tiedje. 1985. Direct measurement of oxygen profiles and denitrification rates in soil aggregates. Soil Sci. Soc. Am. J. 49:645–651. doi:10.2136/sssaj1985.03615995004900030024x

Shannon, R.D., and J.R. White. 1994. A three year study of controls on methane emissions from two Michigan peatlands. Biogeochemistry 27:35–60. doi:10.1007/BF00002570

Sheane, R., P. Sheane, P. Holmes-Ling, and K. Stewart. 2010. A Scottish farm-based greenhouse gas accounting tool Scotland. Scottish Government, Edinburgh.

Smith, P. 2002. Organic matter modeling. In: R. Lai, editor, The encyclopedia of soil science. 1st ed. Marcel Dekker Inc., New York. p. 917–924.

Smith, P., F. Albanito, M. Bell, J. Bellarby, S. Blagodatskiy, A. Datta, M. Dondini, N. Fitton, H. Flynn, A. Hastings, J. Hillier, E. Jones, M. Kuhnert, D. Nayak, M. Pogson, M. Richards, G. Sozanska-Stanton, S. Wang, J. Yeluripati, E. Bottoms, C. Brown, J. Farmer, D. Feliciano, C. Hao, A. Robertson, S. Vetter, H.M. Wong, and J. Smith. 2012. Systems approaches in global change and biogeochemistry research. Phil. Trans. R. Soc. B 367:311–321. doi:10.1098/rstb.2011.0173

Smith, P., D. Martino, Z. Cai, D. Gwary, H. Janzen, P. Kumar, B. McCarl, S. Ogle, F. O'Mara, C. Rice, B. Scholes, O. Sirotenko, M. Howden, T. McAllister, G. Pan, V. Romanenkov, U. Schneider, S. Towprayoon, M. Wattenbach, and J. Smith. 2008. Greenhouse gas mitigation in agriculture. Philos. Trans. R. Soc. Lond. B Biol. Sci. 363:789–813. doi:10.1098/rstb.2007.2184

Smith P, J.U. Smith, D.S. Powlson, W.B. McGill, J.R.M. Arah, O.G. Chertov, K. Coleman, U. Franko, S. Frolking, D.S. Jenkinson, L.S. Jensen, R.H. Kelly, H. Klein-Gunnewiek, A.S. Kamarov, C. Li, J.A.E. Molina, T. Mueller, J.W. Parton, J.H.M. Thornley, A.P. Whitmore. 1997. A comparison of the performance of nine soil organic matter models using datasets from seven long term experiments. Geoderma 80:163–225.

Smith, J.U., N.J. Bradbury, T.M. Addiscott. 1996. SUNDIAL: A user-friendly, PC-based system for simulating nitrogen dynamics in arable land. Agron. J. 88:38–43.

Smith, J.U., P. Gottschalk, J. Bellarby, S. Chapman, A. Lilly, W. Towers, J. Bell, K. Coleman, D.R. Nayak, M.I. Richards, J. Hillier, H.C. Flynn, M. Wattenbach, M. Aitkenhead, J.B. Yeluripurti, J. Farmer, R. Milne, A. Thomson, C. Evans, A.P. Whitmore, P. Falloon, and P. Smith. 2010. Estimating changes in national soil carbon stocks using ECOSSE: A new model that includes upland organic soils. Part I. Model description and uncertainty in national scale simulations of Scotland. Clim. Res. 45:179–192. doi:10.3354/cr00899

Smith, J.U., P. Graves, D.R. Nayak, P. Smith, M. Perks, B. Gardiner, D. Miller, A. Nolan, J. Morrice, S. Xenakis, S. Waldron, and S. Drew. 2011. Carbon implications of windfarms located on peatlands: Update of the Scottish Government carbon calculator tool. http://www.scotland.gov.uk/Topics/Business-Industry/Energy/Energy-sources/19185/17852-1/CSavings/V2UpdReport (accessed 27 Oct. 2015)

Tallis, M.J., E. Casella, P.A. Henshall, M.J. Aylott, T.J. Randle, J.I.L. Morison, and G. Taylor. 2013. Development and evaluation of ForestGrowth-SRC a process-based model for short rotation coppice yield and spatial supply reveals poplar uses water more efficiently than willow. GCB Bioenergy 5:53–66. doi:10.1111/j.1757-1707.2012.01191.x

Thompson, D.A., and R.W. Matthews. 1989. The storage of carbon in trees and timber. Forestry Commission Research Information Note 160. Forestry Commission, Edinburgh, UK

Triggs J.M., B.A. Kimball, P.J. Pinter, G.W. Wall, M.M. Conley, T.J. Brooks, R.L. LaMorte, N.R. Adam, M.J. Ottman, A.D. Matthias, S.W. Leavitt, R.S. Cerveny. 2004. Free-air $CO_2$ enrichment effects on the energy balance and evapotranspiration of sorghum.

Tubiello, F., M. Salvatore, S. Rossi, A. Ferrara, N. Fitton, and P. Smith. 2013. The FAOSTAT database of greenhouse gas emissions from agriculture. Environ. Res. Lett. 8:015009.

Van den Boss, R. 2003. Restoration of former wetlands in the Netherlands: Effect on the balance between $CO_2$ sink and $CH_4$ source. Neth. J. Geosci. 82:325–332.

Vinten, A.J.A., K. Castle, and J.R.M. Arah. 1996. Field evaluation of models of denitrification linked to nitrate leaching for aggregated soil. Eur. J. Soil Sci. 47:305–317. doi:10.1111/j.1365-2389.1996.tb01404.x

Weigel, H.J., and R. Manderscheid. 2012. Crop growth responses to free air $CO_2$ enrichment and nitrogen fertilization: Rotating barley, ryegrass, sugar beet and wheat. Eur. J. Agron. 43:97–107. doi:10.1016/j.eja.2012.05.011

Weigel, H.J., A. Pacholski, S. Burkart, M. Helal, O. Heinemeyer, B. Kleikamp, R. Manderscheid, C. Fruehauf, G.R. Hendrey, K. Lewin, and J. Nagy. 2005. Carbon turnover in a crop rotation under free air CO2 enrichment (FACE). Pedosphere 15:728–738.

Whittaker, C., M.C. McManus, and P. Smith. 2013. A comparison of carbon accounting tools for arable crops in the United Kingdom. Environ. Model. Softw. 46:228–239. doi:10.1016/j.envsoft.2013.03.015

Yeluripati, J.B., M. van Oijen, M. Wattenbach, A. Neftel, A. Ammann, W.J. Parton, and P. Smith. 2009. Bayesian calibration as a tool for initialising the carbon pools of dynamic soil models. Soil Biol. Biochem. 41:2579–2583. doi:10.1016/j.soilbio.2009.08.021

Zaehle, S., S. Sitch, B. Smith, and F. Hatterman. 2005. Effects of parameter uncertainties on the modeling of terrestrial biosphere dynamics. Global Biogeochem. Cycles 19: doi:10.1029/2004GB002395

# Quantifying Greenhouse Gas Emissions from Agricultural and Forest Landscapes for Policy Development and Verification

**Christina Tonitto,* Noel P. Gurwick, and Peter B. Woodbury**

## Abstract

The emergence of policies to reduce greenhouse gas (GHG) emissions motivated a rapid shift in ecosystem modeling, especially concerning agricultural and forest management effects on nitrous oxide ($N_2O$) emissions and soil carbon (C). First, we review models for estimating GHG fluxes in agricultural and forest landscapes. Second, we investigate the interplay of models with institutions that develop and implement climate change policy. Limited GHG flux observations constrain our understanding of process controls, especially with respect to $N_2O$ flux. Even when simulation models explicitly include $N_2O$ flux, this capability is infrequently applied or validated. In the policy arena, tools used to estimate aboveground forest biomass C rely on robust inventory data. Such data are scarce outside of the industrialized world. Compared with biomass C, soil C storage in both forest and agricultural systems is difficult to quantify. Widely used estimates depend on critical assumptions such as the extent to which erosion contributes to C loss from fields and the mechanisms by which C is stabilized in soil, both of which are the subject of serious debate. As evidenced by at least one instance we explore, institutions dedicated to policy development and implementation can rapidly achieve a nuanced understanding of these limitations, given sufficient engagement of the research community. Our review illuminates challenges that affect land-based GHG mitigation initiatives as a result of current knowledge gaps, institutional capacity, and decision-making frameworks. We suggest key points that should be communicated to model users and raise questions that could be fruitfully addressed by model developers and consumers of modeled estimates.

Abbreviations: AB32, Assembly Bill 32; ACR, American Carbon Registry; APEX, Agricultural Policy/Environmental eXtender; C-AGG, Coalition for Agriculture and Greenhouse Gases; CAR, Climate Action Reserve; CDM, Clean Development Mechanism; CMPP, cropland management project protocol; CO2e, CO2 equivalents; CSA, Climate Smart Agriculture; DAYCENT, Daily CENTURY; DNDC, Denitrification–Decomposition; EF, emission factor; EPIC, Environmental Policy Integrated Climate; EUETS, EU Emissions Trading System; FIA, Forest Inventory and Analysis; FVS, Forest Vegetation Simulator; GHG, greenhouse gas; GPP, gross primary productivity; LCA, life-cycle analysis; LTE, long-term experiment; NI, nitrification inhibitor; NPP, net primary productivity; OM, organic matter; RGGI, Regional Greenhouse Gas Initiative; RZWQM2, Root Zone Water Quality Model 2; SL, slow-release; SOC, soil organic carbon; SOM, soil organic matter; T-AGG, Technical Working Group on Agriculture and Greenhouse Gases; UNFCCC, United Nations Framework Convention on Climate Change; WFPS, water-filled pore spaces.

Christina Tonitto, Cornell International Institute for Food, Agriculture and Development, Cornell University, B75 Mann Library, Ithaca, NY 14853. *Corresponding author (ctonitto@cornell.edu). Noel P. Gurwick, Smithsonian Environmental Research Center, 647 Contees Wharf Rd., Edgewater, MD 21037 (noel.gurwick@gmail.com). Peter B. Woodbury, Section of Soil and Crop Sciences, School of Integrative Plant Science, 1017 Bradfield Hall, 306 Tower Rd., Cornell University, Ithaca, NY 14853 (peterwoodbury@cornell.edu).

doi:10.2134/advagricsystmodel6.2013.0007


Synthesis and Modeling of Greenhouse Gas Emissions and Carbon Storage in Agricultural and Forest Systems to Guide Mitigation and Adaptation
S. Del Grosso, L. Ahuja, and W. Parton, Editors
Advances in Agricultural Systems Modeling, Volume 6. Lajpat R. Ahuja, Series Editor

In this chapter, we draw together research and thinking centered on land use management and climate change mitigation. Three topics have their roots in ecosystem science. First, a long history of experimentation and modeling has examined changes in soil organic carbon (SOC) over time and the response of SOC to different management practices. Second, more recent efforts have measured and modeled nitrous oxide ($N_2O$) emissions from agricultural soils, and some of these studies have targeted the influence of management practices on GHG emissions. Third, a rich history of measuring tree growth across the United States has provided the foundation for estimating accumulation of carbon in aboveground biomass in forests. The empirical measurements and associated models that developed in each of these areas have recently been called into service to estimate GHG balances in agricultural fields and forests and to project changes in GHG budgets that would result from specific land management changes.

Our investigation differs from previous analyses in three respects. In our review of models and model projections, we consider both complex process-based simulation models and simpler empirical models. We discuss model strengths and weaknesses given the data available to parameterize them. Furthermore, we discuss how limitations in both data availability and conceptual understanding affect how different models represent ecosystem processes. For example, while models of soil carbon cycling have taught us a great deal and sometimes effectively project the trajectory of SOC over time, they remain limited by fundamental gaps in our understanding of processes that stabilize or destabilize SOC (Kramer et al., 2012, references therein; Kleber and Johnson, 2010; Guggenberger and Kaiser, 2003). While we are not the first to illuminate limitations to understanding these processes (e.g., Baker et al., 2007, for soil carbon responses to tillage practices in agriculture), our analysis is more synthetic, including both $N_2O$ and SOC and both agricultural and forest management. In addition, we place our model comparisons in historical contexts, showing how disciplinary roots of different models have influenced the complexity with which they represent different ecosystem processes (e.g., hydrology versus soil biogeochemistry).

In addition to this technical overview, we discuss how particular institutions that develop market-based GHG trading systems have used available science, illustrate how groups of research scientists and policy developers have influenced each other, and discuss the institutional commitments that have structured those interactions. This kind of analysis should help us make the best use of ecosystem science to support land use management for climate change mitigation.

Finally, by placing our analysis of data and model limitations directly adjacent to a summary of selected policy initiatives, we show how policies can benefit from model projections, but we also show the limitations of current models and data sets and the need for further development of both models and data.

Our intent is to ground policy application of cutting-edge research tools in an understanding of their limitations, to emphasize that the natural evolution of models in response to new scientific knowledge conflicts with the need for policy methodologies to be standardized and implemented over multiple years or decades, and to reinforce the need for continued model development apart from the immediate policy applications. This level of analysis supports the research community's responsibility to provide best-available research for near-term management and policy decisions while simultaneously helping to retain a focus on studies that build deeper understanding over longer time frames.

The remainder of this chapter is divided into four sections. We begin by describing policy initiatives that have created demand for modeling tools that quantify net GHG emissions from agricultural and forest land management. We then review models for agricultural landscapes, including quantification of carbon dioxide ($CO_2$), methane ($CH_4$), and $N_2O$ fluxes, with a particular focus on $N_2O$ because of agriculture's dominant role as an $N_2O$ source as well as the persistent challenges in modeling $N_2O$ emissions. Our review complements recent work by Shepherd et al. (2011), which outlined the broad process-level capabilities of 30 currently available simulation models in agricultural systems, but did not focus on model capacity with respect to $N_2O$ emissions. We subsequently review empirical models because they figure prominently in efforts to implement climate change mitigation policies and discuss challenges for improving our modeling tools with respect to SOC dynamics. For forest ecosystems we focus on modeling tools currently applied in policy settings to assess changes in biomass and soil carbon storage. We conclude by summarizing both the lessons learned and outstanding questions for the following three themes: (i) model structure and scale of application, (ii) data limitations and nonlinearity, and (iii) applications to GHG offset projects.

## Institutions that Link Science and Policy to Reduce Net GHG Emissions

### International

Since its adoption in 1992, the United Nations Framework Convention on Climate Change (UNFCCC) has spurred substantial international efforts to quantify annual GHG emissions from various sources, including agriculture and forestry. Most directly, the Kyoto Protocol established the Clean Development Mechanism

(CDM), which enables industrialized nations to meet emission reduction targets by supporting activities in less developed countries. A search of CDM agricultural projects demonstrates an emphasis on reclaiming $CH_4$ from improved manure management or biomass processing. Land-management-based strategies such as REDD (and REDD+)—projects designed to prevent forest clearing or promote C sequestration through reforestation or afforestation—have also received considerable attention.

Similarly, the EU Emissions Trading System (EUETS) has focused on reducing emissions from power stations and manufacturing facilities—sources that can be measured, reported, and verified with a high level of accuracy. Regulated entities can trade allowances with each other, so that facilities that reduce emissions more than required can sell unneeded allowances to companies that have not achieved required targets. While EUETS projects do not focus on land management, these projects can indirectly affect agricultural landscapes. Biomass energy plants initiated as EUETS projects have disrupted traditional agricultural uses for unharvested crop biomass (Gilbertson and Reyes, 2009), with the subsequent reduction in plant residue incorporation into soil potentially reducing SOC stocks.

A number of countries and provinces have developed climate change mitigation programs that include market-based mechanisms. Some of these (e.g., Alberta, Canada) incorporate offsets that include agriculture and forestry management. Others, such as the nascent provincial markets in China (Qiu, 2013), could adopt land management offsets as they evolve, particularly if robust protocols for estimating GHG emission reductions have already been developed. Other nations recognize the need to mitigate climate change and would like to harness the potential of land management to reduce emissions using other policy levers.

All these efforts to develop land-based climate change mitigation policy rely on a strong science base to quantify the relationship between specific land management practices and GHG emissions. Many nations support research on the climate change mitigation potential of agriculture and forestry, and the Global Research Alliance on Agricultural Greenhouse Gases facilitates information sharing among nations to accelerate the design and adoption of land management practices that reduce GHG emissions (http://www.globalresearchalliance.org, accessed 15 Sept. 2015). The United States, with its strength in scientific research and its culture of research collaboration across borders, is a major contributor to this effort. Identifying data needs to improve quantification of management effects, and the resources required to collect and interpret the data, will support wider adoption of agricultural and forest management practices that mitigate climate change. The scientific knowledge and research expertise of US researchers should provide a useful guide on quantifying management effects because of the substantial investment in research that has been made for decades and the associated availability of landscape- and national-scale data.

Major international efforts to promote Climate Smart Agriculture (CSA) include the work by the World Bank and the UN Food and Agriculture Organization (FAO), and the nascent CSA Alliance. Although these organizations have not focused on quantifying emissions, they have played a major role in drawing attention to the potential to reduce GHG emissions via land management. In December 2013, the Climate Change, Agriculture, and Food Security research program of the CGIAR system partnered with Future Earth to convene the third Global Conference on Agriculture, Food and Nutrition Security and Climate Change. The purpose of the conference was to promote a CSA Alliance (http://ccafs.cgiar.org/global-conference-agriculture-food-and-nutrition-security-and-climate-change#.UzniR_ldWgY, accessed 15 Sept. 2015). A number of governments are now working to define the structure and objectives of this nascent organization. At over 550 pages, the FAO's CSA sourcebook defines a framework for linking climate change and agriculture in a landscape context and applies this framework to row crops, livestock, forestry, and fisheries (Food and Agriculture Organization, 2013). Leading institutions with leverage over the trajectory of development initiatives have made the connection between climate change and agriculture a centerpiece of their work. And while the need for adaptation appears to have been the main driver of these initiatives, the role for agriculture in mitigation has also received a great deal of attention.

The US government is also paying attention to CSA and land use in an international context. Secretary of State John Kerry's initial policy guidance statements to the Department of State and the US Agency for International Development. This guidance focused on climate change and included the directive to slow, halt, and reverse emissions from land use.

## United States

Before 2009, in anticipation of federal climate-and-energy legislation in the United States (e.g., HR2454, the American Clean Energy and Security Act), numerous organizations began to develop the analysis necessary for agriculture and forestry to participate in a cap-and-trade program (Broekhoff, 2008). The motivations for considering land use management in policy aimed primarily at reducing GHG emissions included (i) the need to make climate protection policy as affordable (hence feasible) as possible, coupled with analyses suggesting that emissions associated with land use change could be reduced at low cost relative to other emission reduction opportunities (Creyts et al., 2007); (ii) preexisting commitments of some stakeholders to protect forests, and the opportunity to find an additional lever to achieve that end; and (iii) the possibility of finding a revenue stream for farmers linked to climate change mitigation and hence enrolling their support for (or at least reducing opposition to) policies that promote climate change mitigation. Most efforts to take advantage of GHG emission reductions associated with land use change assumed a policy structure in which directly regulated entities such as power plants could purchase carbon credits from growers who reduced emissions by using management practices they would not have employed without this market opportunity. However, for carbon credits to have robust value, there was a critical need for agreed-on, defensible methods to quantify the degree to which various agricultural and forest management practices reduce emissions.

Following the failure of the US Congress to pass this legislation, efforts to take advantage of agriculture's potential role in climate change mitigation continued in several initiatives that also support the need for improved methods to quantify how much changes in agriculture and forest management lead to reduced GHG emissions. Within the United States, prominent public initiatives that include the agriculture and forestry sectors in climate change mitigation efforts are guidance provided to growers for voluntary use by the Climate Change Office of the US Department of Agriculture (Biggar et al., 2013); the Northeast Regional Greenhouse Gas Initiative (RGGI) (Regional Greenhouse Gas Initiative, 2013a, 2013b); and the carbon offsets program administered by the California Air Resources Board, as it implements the California Global Warming Protection Act (Assembly Bill 32 [AB32]) (California Air Resources Board, 2013).

Under AB32, offsets can be used to meet up to 8% of required emission reductions and can originate anywhere in the United States, even though emission reduction requirements apply only to California. To take full advantage of this provision in the law, regulated entities would require approximately 200 million metric tons of $CO_2$ equivalents ($CO_2$e) in offsets; generating that many offsets would require substantial tracts of land across the country (Stevenson et al., 2012). Approved offset protocols grant credits for projects that destroy ozone-depleting substances that have high global warming potential and for projects that sequester carbon in US forests (California Air Resources Board Cap-and-Trade Protocols). However, existing offset protocols do not provide the opportunity to generate 200 million metric tons of $CO_2$e. Project eligibility for offset credits under California's global warming protection law mandates that emission reductions would not have occurred in the absence of the regulation; because a suite of regulations and incentive programs in California are already reducing emissions from many activities, the number of potential offset opportunities is rather limited.

Indeed, whether enough offset opportunities exist to generate 200 million tons of $CO_2e$ is an open question. Agriculture is typically unregulated, making the sector an appealing source of offset projects if the necessary quantification tools can be developed and if risks of emission reductions being temporary can be adequately addressed. Agricultural practices that might qualify as offsets include ones that sequester carbon in soil and ones that reduce emissions of nitrous oxide and methane (California Air Resources Board, 2011a, 2011b). The urgency behind efforts to identify offset opportunities and develop quantification methods and protocols has eclipsed a more fundamental question: When will research on agriculture and GHG emissions be able to credibly support these efforts?

In anticipation of a market-based GHG emissions reduction program, institutions like the Climate Action Reserve (CAR), the American Carbon Registry (ACR), and the Voluntary Carbon Standard (VCS) have produced protocols that define management practices eligible to earn carbon credits and quantify the GHG emissions reductions from specific management practices. Our treatment of carbon registries focuses on CAR. In contrast to the global distribution of CDM, EUETS, and REDD projects, for CAR these protocols primarily outline projects to be implemented within the United States. Exceptions include landfill, livestock, and forestry protocols developed specifically for implementation in Mexico (Climate Action Reserve, 2013). This national scope is consistent with requirements under California's AB32. The California Air Resources Board, which administers AB32, is a potential user of CAR's protocols.

Although research linking land management to GHG emissions and carbon storage in soil and forest biomass has been evolving for decades, only recently have a number of initiatives focused on synthesizing the science in a policy-relevant form. These initiatives are framing how technical understanding is incorporated into policy. For example, with funding from the Packard Foundation, the Technical Working Group on Agriculture and Greenhouse Gases (T-AGG) reviewed the literature to produce a current assessment of the GHG emission reduction potential of dozens of agricultural practices (Eagle et al., 2012). A set of USDA Conservation Innovation Grants (CIGs) focused on climate change mitigation included projects to assess barriers to farmers' use of GHG reduction protocols (USDA-NRCS, 2011). The Coalition for Agriculture and Greenhouse Gases (C-AGG) provides a regular forum for information exchange among project developers, scientists, growers' organizations, investors, and government agencies (www.c-agg.org).

As organizations like C-AGG and USDA-NRCS have recognized, quantification tools can be used in a variety of contexts apart from environmental markets or regulatory arenas. For example, COMET-FARM couples a complex biogeochemical model with a user-friendly interface and incorporates many default values for parameters. It therefore allows growers to estimate "best guesses" about GHG emission reductions that might result from practice changes on their land (http://cometfarm.nrel.colostate.edu/, accessed 15 Sept. 2105).

The limited number of landscapes, climate regimes, and agricultural and forestry practices for which protocols have been approved can be attributed largely to a scarcity of affordable, robust tools that can currently be used to quantify the relationship between changes in land management and GHG emissions. The status of tool development, in turn, can be traced to limited data and understanding. In the following sections we discuss the development and limitations of available GHG quantification tools and subsequently discuss how these limitations have influenced protocol development.

## Quantifying GHG Emissions from Agricultural Land Management

Globally, agriculture is a dominant source of $N_2O$ and $CH_4$, accounting for 82% and 43% of anthropogenic emissions, respectively (USEPA, 2012). In US agricultural systems, soils account for about 70% of total anthropogenic $N_2O$ emissions, The majority of US anthropogenic $CH_4$ emissions result from losses in the natural gas industry (~36%), landfill emissions (~19%), and enteric losses (20%), with manure management (<10%) also contributing significant emissions; in contrast rice cropping systems only account for approximately 1% of US anthropogenic $CH_4$ emissions. Therefore, in the US context, cropping system management decisions impact net $N_2O$ flux, but agricultural impact on $CH_4$ emissions depends mainly on the number of livestock raised and how manure is managed. Because cropping system management is a primary driver of US net $N_2O$ flux, we focus our review of agricultural simulation models on tools that quantify $N_2O$ flux in response to agricultural management decisions.

The contribution of agricultural land management to $CO_2$ emissions results from a net decrease in SOC on existing farms due to management practices; the influence of agricultural expansion on SOC and vegetation C; and fossil fuel combustion for farm management, fertilizer production, pesticide production, and manufacturing agricultural equipment. In the US context, $CO_2$ emissions from agricultural inputs and operations are about 14% of total agricultural GHG sources (Del Grosso and Cavigelli, 2012) and represent a small contribution to net $CO_2$ emissions relative to commercial, industrial, residential, and transportation $CO_2$ emissions. However, changes to agricultural land management could produce a significant sink for atmospheric $CO_2$ via SOC sequestration (e.g., Smith et al., 2008b). We discuss challenges regarding the accurate measurement of SOC accumulation in the section entitled "Managing Agricultural Landscapes to Promote SOC Storage."

Our ability to accurately model agroecosystem biogeochemical dynamics, and therefore net GHG emissions, remains constrained by limited data sets with spatially and temporally concurrent measurements of multiple C and N cycle processes. This challenge applies to both empirical models and mechanistic simulation models, though model performance has improved significantly after decades of model development and validation (e.g., Smith et al., 1997; Shepherd et al., 2011). We recognize that the distinction between mechanistic simulation models and empirical models is imprecise, as empirically derived functions relating ecosystem properties to GHG flux rate have served as the basis of both complex simulation model development (e.g., Parton et al., 1996), as well as simple empirical model development (e.g., Millar et al., 2010). However, we find the categorization of models as complex simulation or simple empirical models useful for understanding how models are being adopted in the policy arena. Below we review (i) agricultural models that explicitly simulate $N_2O$ flux, (ii) simple empirical modeling tools for quantifying $N_2O$ flux in agricultural landscapes, and (iii) broader challenges to accurately quantifying SOC response to agricultural management practices.

## Simulation of Cropping System GHG Flux

Modeling net GHG emissions from agricultural landscapes requires accurate model description of (i) net $CO_2$ flux as the balance between $CO_2$ uptake resulting from C fixation during plant growth, and $CO_2$ emissions from plant respiration and heterotrophic soil respiration due to organic matter decomposition, (ii) $N_2O$ emissions resulting from aerobic pathways during nitrification and anaerobic pathways during denitrification, and (iii) $CH_4$ emissions by explicit modeling of methanogenesis and $CH_4$ uptake through oxidation in non-saturated soils.

All agroecosystem models include the dominant processes of the C cycle, though mechanistic details vary. Models differ in how organic matter (OM) substrate pools are defined, which OM pools are explicitly tracked, and how environmental properties control decomposition rates. Models commonly distinguish between OM substrate quality in determining decomposition rate and OM residence time. Due to our incomplete mechanistic understanding of controls on carbon stabilization and decomposition, many modeled OM pools are conceptual rather than measurable pools. In other words, although models typically recognize that some OM decomposes easily and some resists decay, the division of the total OM pool into different fractions has only limited basis in a mechanistic understanding of why these decay rates vary. Therefore, model ability to project how decay rates will change in response to changes in management practices or environmental conditions is also limited. Without the ability to mechanistically simulate C stabilization, we cannot adequately study a major mechanism of long-term SOC accumulation in agricultural systems. We discuss recent mechanistic work regarding SOC accumulation and stabilization in the section entitled "Managing Agricultural Landscapes to Promote SOC Storage." Models also differ in how they simulate plant growth, with some models explicitly modeling gross primary productivity (GPP) and plant respiration, while other models calculate net primary productivity (NPP).

The complexity of N cycle description differs significantly among agroecosystem models. Models differ in their assumptions regarding controls on net N mineralization during decomposition, as well as subsequent nitrification rates. The most important difference in N cycle description among agricultural models is whether the model explicitly simulates $N_2O$ emissions. We focus our simulation model review on models that explicitly partition $N_2O$ and $N_2$ flux. Since anthropogenic $N_2O$ emissions are largely controlled by agricultural soils, agroecosystem model capacity to track $N_2O$ emissions is a critical model feature for application to GHG accounting questions. We complement the model review by Shepherd et al. (2011) by comparing simulation model commonalities and differences with respect to model description of C, N, and hydrologic cycles.

## Overview of Models

We compare key features of agroecosystem models that are commonly applied to study environmental impacts of agricultural management and that explicitly quantify $N_2O$ emissions. These models have a varied history in terms of original purpose of model development, as well as coordination of model development with management objectives. The disciplinary roots of the development team and the primary objectives of model development influence the complexity or simplicity of process representation. Historically hydrologic models originated with simplified C and N cycle descriptions, while models focused on C and N cycles started with relatively simplified hydrology. However, over the course of model development all ecosystem models have increased the complexity of C, N, and hydrologic cycle descriptions

We review five ecosystem models—three with a historical focus in C and N cycling and two with a historical emphasis on hydrology. DAYCENT (Daily CENTURY) (Parton et al., 1987, 1998, 2001) is a daily time-step model that tracks both $N_2O$ and $N_2$ flux from agricultural systems and is based on the CENTURY model. CENTURY was originally developed as a monthly timescale model to quantify decadal and century-scale processes of SOC dynamics and soil development. The development of the daily time-step DAYCENT model allowed the CENTURY description of soil organic matter (SOM) dynamics to be applied toward agricultural management questions. Model application to address GHG emissions from agricultural landscapes has been a key goal over the course of DAYCENT development. The Denitrification–Decomposition (DNDC) model (Li et al., 1992) is grounded in the explicit representation of microbial populations and environmental constraints that control C and N process rates. Throughout DNDC's history, model developers have focused on describing GHG emissions. As a result, DNDC has been broadly applied to estimate agricultural GHG emissions and model development has been coordinated with various international research teams focused on quantifying GHG emissions for policy compliance, including NitroEurope and Landcare Research in New Zealand. The ECOSYS model (Grant et al., 2001a) uses a highly detailed, mechanistic approach to simulate C and N processes, describing microbial control on C and N dynamics by representing populations of different microbial functional groups. ECOSYS model development has frequently been coordinated with model validation against flux tower measurements, offering a unique opportunity to test and refine modeled $CO_2$ dynamics. Similar to other models reviewed, a fundamental goal of ECOSYS model development has been model application to support improved land management planning with respect to environmental impacts. The Agricultural Policy/Environmental eXtender (APEX) model (Williams et al., 2008) is derived from the Environmental Policy Integrated Climate (EPIC) model (Williams, 1995), which was originally developed to study watershed dynamics and soil erosion. The APEX model combines the hydrologic sophistication of EPIC with a complex N cycle and a CENTURY-like description of SOC decomposition processes. The Root Zone Water Quality Model 2 (RZWQM2) (Ma et al., 2001) is a widely applied model developed by USDA-ARS scientists. This model has been used to study various agricultural management scenarios, including tillage, fertilizer, manure, and pesticide management. The direct involvement of USDA-ARS scientists in APEX and RZWQM2 development has resulted in models that are well suited to addressing major management practices in US agricultural landscapes. Below we summarize key similarities and differences in how these five models describe C, N, and hydrologic processes and agricultural management practices. An overview of model similarities and differences is also summarized in Table 1.

## C and N Process Description

A key distinction among models is whether C and N processes are described as explicit functions of microbial biomass with rates grounded in microbial metabolism or whether descriptions are derived from empirical observations of process

**Table 1. Comparison of ecosystem models that partition $N_2$ and $N_2O$ emissions.†**

| | DAYCENT | DNDC | ECOSYS | APEX-EPIC | RZWQM2 |
|---|---|---|---|---|---|
| **C cycle** | | | | | |
| Vegetation growth | | | | | |
| GPP description‡ | no | radiation use efficiency | Michaelis–Menten enzyme kinetic model | no | radiation use efficiency |
| GPP controls | | plant C, temperature, precipitation, PET, soil moisture, $[CO_2]$, soil N, leaf N, fPAR | LAI, PAR, $[CO_2]$, leaf temperature, leaf nonstructural C:N, leaf structural C:N, leaf water potential, leaf stomatal conductance | | plant C, plant N, light, temperature, soil moisture |
| NPP controls | plant C, temperature, precipitation, PET, soil moisture, soil N, LAI | plant C, temperature, precipitation, PET, soil moisture, soil N, leaf N, fPAR | GPP – autotrophic respiration | light-use efficiency, temperature, soil N, soil moisture | plant C, temperature |
| Soil processes | | | | | |
| SOM pools | 6 | 9 | 9 in each of 5 substrate–microbe complexes | 6 | 8 (including 3 microbial pools) |
| Microbial biomass explicitly controls C, N processes | no | yes | yes | no | yes |
| Decomposition function | first order | first order | Michaelis–Menten (substrate) × active microbial biomass | first order | first order |
| Decomposition controls | substrate, temperature, moisture | substrate, temperature, moisture, clay adsorption | substrate concentration, temperature, moisture, microbial biomass | substrate, temperature, moisture | substrate, temperature, soil moisture, microbial population, $O_2$, pH |
| Methane function | first order | Michaelis–Menten | Michaelis–Menten | | first order |
| Methane controls | substrate, soil moisture, temperature, Eh | substrate, temperature, Eh, pH | DOC, acetate, $[CO_2]$, $[O_2]$ | | |
| Decomposition temperature response | unimodal | unimodal, second-order polynomial | Arrhenius | unimodal | Arrhenius temperature limitation |
| Decomposition moisture response | unimodal | unimodal, second-order polynomial | through effect on microbial access to substrate | unimodal | piecewise step function |

Table continued.

**Table 1. Continued.**

| | DAYCENT | DNDC | ECOSYS | APEX-EPIC | RZWQM2 |
|---|---|---|---|---|---|
| | | | N cycle | | |
| Gross N mineralization | SOC decomposition | SOC decomposition | microbial decomposition | SOC decomposition | SOC decomposition |
| Net N mineralization-immobilization | OM C to N ratio | OM C to N ratio | microbial C:N | OM C to N ratio | explicit microbial growth |
| Nitrification function | empirical fit to observations across a gradient of $NH_4^+$, WFPS, temperature, and pH | Michaelis–Menten | Michaelis–Menten for $NH_3$ | first order | first order for low $[NH_4^+]$, zero order for high $[NH_4^+]$ |
| Nitrification controls | substrate, temperature, texture, WFPS, pH | substrate, temperature, moisture, pH | substrate, $CO_2$, nitrifier biomass | substrate, temperature, WFPS, pH | substrate, temperature, soil moisture, microbial population, $O_2$, pH |
| Plant uptake function | | first order | coupled diffusion and active uptake (Michaelis–Menten) | evapotranspiration | Michaelis–Menten (active), transpiration (passive) |
| Plant uptake controls | root biomass, soil N, plant N need | daily plant growth, soil N, moisture | soil N, water, $O_2$, soil temperature | soil N, moisture | plant N need, soil N |
| $N_2O$ and $N_2$ flux | empirical fit to observations across a gradient of $NO_3^-$, WFPS, and respiration | Michaelis–Menten dentrifier growth rate | Michaelis–Menten | first order | first order |
| $N_2O$ and $N_2$ flux controls | substrate, temperature, texture, WFPS, pH | substrate, temperature, pH, Eh (soil anaerobic volumetric fraction) | substrate, temperature, $O_2$, and nitrifier and denitrifier biomass | substrate, temperature, moisture | substrate, temperature, moisture, pH |
| $N_2/N_2O$ ratio | based on WFPS | clay, WFPS, temperature | from sequential reduction of $NO_3^-$, $NO_2^-$, $N_2O$ | moisture | based on WFPS |
| $N_2$ and $N_2O$ source | nitrification and denitrification | nitrification and denitrification | nitrification and denitrification | denitrification | nitrification and denitrification |
| N fixation | difference between plant N demand and soil available N | fraction of legume N uptake | root-nodule exchange of nonstructural C and N | fraction of legume N uptake | difference between plant N demand and soil available N (generic model); simulated nodule dynamics (DSSAT) |
| N cycle temperature response | nitrification exponential increase with temperature | nitrification monotonically increasing with temperature, fourth-order polynomial | common Arrhenius equation for all biological transformations | | Arrhenius temperature limitation |
| N cycle moisture response | unimodal in nitrification, monotonically increasing in denitrification (shape dependent on soil texture) | unimodal, fourth-order polynomial; soil divided into aerobic and anaerobic fractions on the basis of Eh controlled by moisture and other oxygen consumption processes | through effects on soil $[O_2]$, aqueous microbial concentrations | | piecewise step function |

| | DAYCENT | DNDC | ECOSYS | APEX-EPIC | RZWQM2 |
|---|---|---|---|---|---|
| **Hydrology** | | | | | |
| Soil layers | 10 | 20+ | 15 | 15 | 10 |
| Infiltration and runoff | | SCS curve | | Green–Ampt, SCS curve | Green–Ampt, difference between rainfall intensity and Ksat |
| Drainage and redistribution | gravitational drainage (above field capacity); Darcy's Law (unsaturated soil) | gravitational drainage of soil moisture in excess of field capacity; empirical capillary rise | Richards' equation (unsaturated), Green–Ampt (saturated) for water flux; Fick's Law for vapor flux | gravitational drainage of soil moisture in excess of field capacity | Richards' equation using Brooks–Corey equation for soil water retention curves |
| Lateral flow | no | no | yes, surface and subsurface | yes | yes (for tile flow) |
| **Management** | | | | | |
| Tillage | yes | yes | yes | yes | yes |
| Fertilization method | yes | yes | yes | yes | yes |
| Irrigation | yes | yes | yes | yes | yes |
| Fertilizer type | yes | yes | yes | yes | yes |
| Manure | yes | yes | yes | yes | yes |
| Management schedule | yes | yes | yes | yes | yes |
| Complex rotations | approximated | yes (6 simultaneous crops) | | yes (10 simultaneous crops) | yes |
| **Model history** | | | | | |
| Application | field, regional, national | field, regional, national | field, regional, continental | field, regional, national | field, regional |
| Original goal | long-term SOC dynamics | short-term C and N processes | short-term C and N processes | water quality | water quality |
| References | Del Grosso et al. (2000); Parton et al. (1996) | Zhang et al. (2002); Deng et al. (2011); Giltrap et al. (2010) | Grant et al. (2001a) | Williams et al. (2008); Williams (1995); Gassman et al. (2010) | Ahuja et al. (2000); Ma et al. (2011); Jones et al. (2003) |

† A draft of this table was emailed to model developers, and we incorporated their comments. Blank cells denote processes for which we could not determine descriptions on the basis of model documentation. Abbreviations: APEX, Agricultural Policy/Environmental eXtender; DAYCENT, Daily CENTURY; DNDC, Denitrification–Decomposition; DOC, dissolved organic carbon; DSSAT, Decision Support System for Agrotechnology Transfer; EPIC, Environmental Policy Integrated Climate; fPAR, fractional photosynthetically active radiation; GPP, gross primary productivity; Ksat, saturated hydraulic conductivity; LAI, leaf area index; OM, organic matter; NPP, net primary productivity; PAR, photosynthetically active radiation; PET, potential evapotranspiration; RZWQM2, Root Zone Water Quality Model 2; SCS, Soil Conservation Service runoff curve number; SOM, soil organic matter; WFPS, water-filled pore spaces.

‡ GPP and NPP description from NACP model intercomparison (http://daac.ornl.gov/SURVEY8/survey_results.shtml). RZWQM2 has various plant growth models.

rates across a gradient of environmental controls. Decisions regarding representation of C and N processes affect the method of coupling dynamics across C and N cycles and the parameterization requirements for model use. When microbial biomass controls process rates, model reliability depends on robust parameterization of microbial metabolism rates. When empirical relationships determine process rates, model reliability depends on sufficient observations to parameterize key functional relationships. Ideally empirical functions describe process rate response to environmental gradients relevant to the particular cropping system, climate, and soil type being studied. Currently available data sets are best suited for the parameterization of empirical relationships, because C and N process rates have been studied in the field and lab across a gradient of environmental conditions such as soil texture, soil moisture, soil temperature, and cropping system.

## $CO_2$ Flux

Models have used two approaches to estimate plant uptake of $CO_2$. One group of models bases plant uptake of $CO_2$ on leaf carboxylation rate as a function of leaf traits (leaf area, leaf N or RuBisCO content, and the saturation function limited by leaf $CO_2$ concentration). The other group models $CO_2$ uptake on the basis of photosynthetically active radiation with downregulation controlled by nutrient, water, and temperature stress. Plant release of $CO_2$ through respiration depends on stored C, N content, water balance, and temperature.

Most of the models we reviewed use first-order decay to represent the release of $CO_2$ from soil microbial respiration. An alternative approach to modeling OM decomposition applies Michaelis–Menten kinetics to explicitly simulate microbial population controls on decomposition. When using Michaelis–Menten kinetics, the decomposition rate depends on both substrate concentration and density-dependent microbial activity. Organic matter decay varies depending on the type of OM, and models differ in the categories of OM substrate explicitly represented and the decay rates of these substrates. Finally, models differ in how they represent environmental conditions such as water-filled pore spaces (WFPS), pH, temperature, and $O_2$ and the response of decomposition to these environmental constraints. The description of which environmental properties and how environmental properties affect decomposition rates is a key distinction among models. While a first-order rate equation is the dominant description of OM decomposition, models differ in their simulation of OM decomposition because decomposition rates are dynamically affected by environmental controls. Model assumptions regarding environmental controls and model ability to adequately model gradients in these environmental drivers across a range of soil texture and climates ultimately determines simulated decomposition rates and is a key reason for differences in model outcomes.

## N Gas and Denitrification Description

Models represent denitrification and $N_2$:$N_2O$ partitioning, as well as $N_2O$ lost during nitrification, with varying degrees of complexity. The models describe $N_2O$ flux using first-order kinetics, Michaelis–Menten kinetics, or empirically derived functions. No single functional form dominated model representation of $N_2O$ flux, $N_2$:$N_2O$ partitioning, or $N_2O$ lost during nitrification. As with decomposition, controls on N gas flux differ among models. The activity of denitrifying microbes primarily responds to soil aerobic status, with the observed relationship between WFPS and $N_2O$ flux varying across soil textures (e.g., Parton et al, 1996). Simulated denitrification is controlled by a combination of soil properties including substrate, temperature, moisture or $O_2$ concentration, texture, and pH. While all models calculate $N_2O$ flux on the basis of some measure of soil moisture or $O_2$ state, only DAYCENT and DNDC explicitly include soil texture as a control on $N_2O$ flux.

## C and N Coupling

Model assumptions regarding how C to N ratios constrain decomposition and N mineralization are an important link between simulated C and N dynamics; these assumptions control decomposition as well as soil inorganic N available for plant uptake, or loss as $NO_3^-$ or $N_2O$. For models with explicit simulation of microbial populations, net N immobilization is modeled on the basis of the C:N stoichiometry of the microbial population. In models driven by substrate composition, net immobilization and net mineralization are based on observed threshold C:N relationships across different OM substrate classes.

## $CH_4$ Flux

For models that consider methane flux, model representation of methanogenesis applied either Michaelis–Menten kinetics or first-order kinetics. Methane flux is mainly relevant in flooded rice production systems, waterlogged ecosystems such as wetlands and bogs, or during decomposition of manure stored where anaerobic conditions can develop. In models that explicitly simulate microbial populations, the rate of $CH_4$ production is described using a Michaelis–Menten saturating function limited by substrate, soil acetate, $CO_2$ or $O_2$ concentration, Eh, or pH.

## Water Cycle

Hydrologic complexity varies significantly among the ecosystem models reviewed. Models developed primarily to investigate C and N dynamics commonly limit drainage to vertical flow, while models developed primarily as hydrologic models consider both vertical and lateral flow. The most simplistic representation of drainage is based on gravitational flow, with flow volume determined as soil layer water in excess of water holding capacity or field capacity. More complex representation of drainage includes information about the physical soil matrix by applying Darcy's Law or Richards' equations to describe flux on the basis of soil hydraulic properties of a given soil layer. An increasingly mechanistic modeling approach includes water vapor flux using Fick's Law of diffusion.

A fundamental limitation to modeling denitrification is our inability to accurately model pore-scale soil moisture dynamics. Because it is currently not possible to accurately model the true distribution of pore-scale $O_2$ or moisture conditions, structural model uncertainty is an inevitable component of denitrification model description. However, simulated denitrification dynamics are often able to track observed trends in seasonal flux, implying that an approximation of soil properties using average, aggregate soil conditions is capable of approximating flux patterns that are actually occurring in temporal (hot moment) and spatial (hot spot) peak events.

## Agricultural Management

Management practices that affect $N_2O$ flux include (i) tillage and its effect on soil physical properties, (ii) N fertilization rate, timing and type, and its effect on soil available N, (iii) crop rotation and its effect on N uptake and water balance, and (iv) residue management and its effect on C availability and water and nutrient retention. All the models that we reviewed have been extensively applied in agricultural systems and include common conventional agricultural management practices: tillage, irrigation, fertilization rate and method, inorganic N or P fertilizer type, manure, plant residue management, planting density, planting date, harvest date, harvest amount, and a range of crop types (see Olander et al. [2011] for more details on model management capabilities). Models used in rangeland applications can also simulate grazing and burning. The majority of model applications have focused on conventional rotations.

Model capacity to represent complex management options and ecological complexity is more limited. Techniques such as frost seeding and intercropping are not universal capabilities of agroecosystem models. The lack of capacity to simulate multiple species concurrently limits model applicability for testing ecological management as a means of GHG emission reduction. Agroecosystem models rarely simulate disease or pest effects, which can influence the accuracy of plant productivity modeling as well as soil biogeochemical processes. The effects of flooding on plant productivity or soil biogeochemical processes are also rarely modeled. Such ecological perturbations can significantly alter biogeochemical cycling and may become more frequent under climate change. Therefore, improved model representation of ecological complexity is necessary for assessing climate change scenarios.

While all models reviewed herein simulate common management practices, the sensitivity of modeled ecosystem dynamics to management scenarios differs among models. In models with a hydrologic development history (APEX/EPIC, RZWQM2), simulated changes in tillage affect soil physical properties and water dynamics. In contrast, management decisions in models with a development history emphasizing C and N dynamics (DNDC, ECOSYS) have more resolution in the ecology of different cropping systems. Increased capacity to differentiate soil physical properties resulting from management is useful for testing management impacts on physical processes such as erosion, as well as moisture-driven biogeochemical processes such as denitrification. Enhanced complexity with respect to the cropping system description allows model application to study ecological rotations and more generally the role of litter substrate in controlling net C and N process rates.

## Model Limitations

More accurate simulation of the C cycle in agricultural systems depends on improved simulation of the biophysical impact of agricultural management (especially tillage), improved simulation of ecosystem processes such as decomposition and SOC stabilization, and improved simulation of diverse management strategies such as rotation complexity, nutrient amendment strategy, and residue management (see the section entitled "Managing Agricultural Landscapes to Promote SOC Storage" for further discussion of SOC dynamics). However, even when models represent detailed management practices, there are limited data to validate whether a model mechanistically captures land management effects on SOC accumulation and soil erosion, especially across a wide range of soil types and climatic conditions. Furthermore, the impacts of individual management practices on biogeochemical cycles are often studied independently. In the absence of improved data, a conservative modeling approach would assume an additive relationship for quantifying biogeochemical dynamics in complex landscapes under multiple management practices; this assumption is likely to be erroneous in many cases. Therefore, model validation for diversified cropping systems will require improved access to long-term data sets under different management strategies.

The fundamental challenge to mechanistically simulating $N_2O$ flux is accurately modeling the spatial and temporal variation in soil environmental conditions. A classic study of N flux from structurally intact soil cores by Parkin (1987) demonstrated that 85% of N gas flux occurred in a soil aggregate representing <0.1% of the original core mass and largely resulted from the decay of a piece of pigweed substrate. This study exemplifies the extent to which extreme soil conditions, rather than the mean, control cumulative $N_2O$ flux, and therefore clarifies the challenge to achieving a truly mechanistic model description of N gas flux.

Even though $N_2O$ flux from soils is part of an N cycle that potentially responds to numerous factors, in agricultural lands these emissions respond primarily to changes in soil aerobic state and the availability of labile forms of nitrogen. Simulation models generally describe average system properties, while in the agricultural landscape $N_2O$ flux occurs in microsites, with the majority of emissions occurring during ephemeral "hot moment" periods when high soil N availability coincides with optimal soil moisture (Groffman et al., 2009). Therefore, simulation models have been conceptualized to aggregate the temporal and spatial diversity of soil environmental state to generate an aggregated N flux emission rather than explicitly characterizing diversity. Despite these simplifications in the spatial and temporal resolution of simulated $N_2O$ flux, models have been successfully calibrated to simulate annual N flux observations. They generally produce seasonal patterns where the highest emissions correspond to periods when heavy rainfall coincides with fertilization or with pulses in N mineralization under bare fallow. Both of these patterns are consistent with our mechanistic understanding and observations of N gas flux dynamics.

Global change adds to current limitations on accurate modeling of complex C and N dynamics. Empirical observations of $N_2O$ flux, plant growth, and SOM decomposition across current climate, atmospheric chemical composition, and soil environmental conditions form the basis of both simulation and simple empirical models. Uncertainty about long-term, ecosystem-scale C and N process response to global change (e.g., Norby and Luo, 2004; Reich et al., 2006; Rustad, 2006; Lukac et al., 2009) reduces confidence in ecosystem model accuracy. In particular, it remains challenging to mechanistically model ecosystem process response to moisture, temperature, and anthropogenic forcing of biogeochemical element concentrations for SOM decomposition (e.g., Giardina and Ryan, 2000; Torbert et al., 2000; Norby et al., 2001; Ågren and Bosatta, 2002; Davidson and Janssens, 2006; Davidson et al., 2006; Erhagen et al., 2013; Stockmann et al., 2013; Gabriel and Kellman, 2014), net N mineralization-immobilization (e.g., Kätterer et al., 1998; Thornley and Cannell, 2001; Reichstein et al., 2005; West et al., 2006; Guntiñas et al., 2012), and plant physiology (e.g., Karnosky et al., 2003; Ainsworth and Long, 2004; Li et al., 2007; Ziska and Bunce, 2007; Sun et al., 2009). In addition most ecosystem models do not consider the complexity of plant community response to global change (e.g., Araújo and New, 2007; Thuiller et al., 2008; Beale and Lennon, 2012; Svenning and Sandel, 2013). These responses depend not only on plant growth responses to novel climate conditions but also on climate-driven shifts in the behavior of invasive species (e.g., Whitney and Gabler, 2008; Bradley et al., 2010; Peltzer et al., 2010; Corbin and D'Antonio, 2012; Ibáñez et al., 2014) or pest dynamics (e.g., La Porta et al., 2008; Dukes et al., 2009; Ziska et al., 2011; West et al., 2012; Sutherst, 2014). While modeling remains our best tool for assessing long-term implications of land management, improving the mechanistic description of C and N dynamics as well as community dynamics will make ecosystem model predictions more robust for global change applications.

Making these improvements requires innovation in model development, and adequate model documentation is a barrier to innovation. The USDA-ARS supported model APEX/EPIC provides a stellar example of documenting the current state of a model's scientific framework (Williams et al., 2008). Because comprehensive model documentation is a time-consuming, unrewarded activity, models developed in an academic environment often rely on peer-reviewed publications to document model development. As a result, model documentation may be scattered among various journal articles, white papers, or technical reports. While individual advances in model components may be described in specific publications, it is generally difficult to get a comprehensive understanding of the current functional foundation of a particular model. Further limitations arise when the model code is not open source, which prohibits a researcher from reviewing how ecosystem processes are represented.

If the ecosystem modeling community were to establish documentation standards, benefits would be widespread. Standards could include guidelines for documenting the current conceptual framework of widely applied agricultural management models and for making model descriptions readily available on a website where model executable files are available for downloading. Additionally, open-source code and detailed documentation within the code should be encouraged. A transparent understanding of how processes are described will allow developers to learn which functional forms are best suited to particular simulation scenarios, facilitating the development of new, improved hybrid modeling approaches.

## Empirical Models for Estimating Cropping System GHG Flux

While complex simulation models define interdependencies between multiple ecosystem processes and drivers, simple empirical models determine $N_2O$ flux as a function of observed relationships between gas flux and one or two independent variables, with limited additional environmental information. Simple empirical models for GHG accounting are typically expressed as emission factors (EFs) that define $N_2O$ emissions as a percentage of N addition. The database of $N_2O$ emissions compiled by Stehfest and Bouwman (2006) has served as a basis for estimating EF values across different cropping systems. The Intergovernmental Panel on Climate Change (2006) recognized two categories of empirical $N_2O$ models. Tier 1 models based on globally aggregated flux observations represent the coarsest approach to estimating $N_2O$ in agricultural systems. The Intergovernmental Panel on Climate Change (2006) Tier 1 function for deriving $N_2O$ flux as a function of fertilizer inputs is based on the analysis of Stehfest and Bouwman (2006). Tier 2 approaches can be developed when national or regionally relevant $N_2O$ flux data are available. Because N gas flux is controlled by microbial response to oxygen state, and C and N availability, Tier 2 models define regionally specific functional relationships on the basis of observations relevant to specific cropping systems, soil textures, or climates (e.g., Dalgaard et al., 2011; Leip et al., 2011; Millar et al., 2010; Tonitto et al., 2009). Because empirical models do not mechanistically and dynamically describe environmental state, they cannot capture nuanced patterns of $N_2O$ flux resulting from spatial and temporal variation in environmental properties. The absence of a mechanistic basis for prediction also means that Tier 2 models cannot be applied outside specific conditions and systems that the underlying data represent. However, for environmental conditions representative of the observation site, especially soil texture, climate, and cropping system, empirical model outcomes integrate the effect of environmental complexity on $N_2O$ flux.

Quantifying landscape $N_2O$ flux is difficult because N cycle processes are nonlinear. Nonlinear functions exhibit the property that function evaluation for the mean value of a controlling parameter is not equivalent to taking the mean of the function evaluated over the entire range of a controlling parameter, a property known as Jensen's Inequality. In a study of N losses from Danish agricultural landscapes, Dalgaard et al. (2011) demonstrated the problem of up-scaling nonlinear N cycle processes by calculating that landscape-scale N loss estimates are 30% higher when empirical EF models are applied using farm-specific data compared with estimates using landscape-averaged data. Millar et al. (2010) addressed the nonlinearity of $N_2O$ flux by defining a nonlinear functional response of $N_2O$ flux to N fertilization rate using flux observations from an N addition gradient experiment. Because investigators applied a gradient of fertilizer application rate within a site, the shape of the $N_2O$-N fertilizer rate relationship provides insight into the nonlinear trend in $N_2O$ loss. This data-intensive approach provides a robust method for quantifying $N_2O$ flux in response to changes in fertilizer rate for corn (*Zea mays* L.) grown on coarse textured soils, but few data sets exist for flux across an N addition gradient in other soil textures, cropping systems, or climates. The Millar et al. (2010) analysis serves as the basis of the Tier 2 MSU–EPRI model (developed by Michigan State University and the Electric Power Research Institute) applied in the CAR Nitrogen Management Project Protocol (Climate Action Reserve, 2012a) and the ACR N management protocol (American Carbon Registry, 2012), allowing for increased data support relative to the Tier 1 linear extrapolation of $N_2O$ emissions as a function of N applied. Recent $N_2O$ reviews by Kim et al. (2013) and Van Groenigen et al. (2010) confirmed the applicability of a nonlinear $N_2O$ flux response across an N input gradient, though Kim et al. (2013) reviewed studies where a linear response to N addition is the best fit to the data.

Various approaches have been applied to integrate existing data and quantitative tools into Tier 2 models. Leip et al. (2011) develop a hybrid approach to quantifying $N_2O$ flux by defining EF values using simulation modeling. Leip et al. (2011) implemented spatial simulations using DNDC-Europe to establish a range of EF values for different SOC, fertilizer scenarios, and climate zones in Europe, and they ultimately contrast these simulated EF ranges to Intergovernmental Panel on Climate Change Tier 1 outcomes. They found simulated EF values paralleled Tier 1 results at large scales (national estimates), but for higher-resolution estimates simulation methods allowed for better accounting of the effect of spatial variation in management practices. Dechow and Freibauer (2011) contrasted a fuzzy logic model to empirical $N_2O$ functions (Bouwman, 1996; Freibauer and Kaltschmitt, 2003; Stehfest and Bouwman, 2006; Intergovernmental Panel on Climate Change, 2006). Their MODE model includes uncertainty due to variation in soil properties and climate. Their model often predicts lower $N_2O$ flux than Bouwman (1996) and Stehfest and Bouwman (2006) models. Tonitto et al. (2009) applied Monte Carlo sampling to a subset of the Stehfest and Bouwman (2006) database to define potential $N_2O$ flux ranges for specific grain cropping system and soil texture combinations under rainfed management. They found that observations supported the potential for $N_2O$ flux rates that exceed 5% of available soil inorganic N, a rate significantly larger than Tier 1 estimates; however, these high flux events occurred with low probability.

While agricultural management has great potential for reducing $N_2O$ emissions, the spatial and temporal resolution of emissions data remains poor because of the cost of implementing N trace gas measurements with sufficient sampling intensity. When field observations are available for specific cropping systems on representative soil textures and in relevant climates, empirical models are a robust tool for bounding the range of $N_2O$ flux. Increased funding and improved N gas measurement methods are expanding the spatial coverage and management scenarios included in observations. These data will allow Tier 2 models to be developed for more cropping systems covering a larger geographic range.

Accurately integrating the effect of seasonal flux variation currently remains a challenge for Tier 2 models. Groffman et al. (2000) concluded that multi-year, continuous data are necessary to derive meaningful patterns for establishing ecosystem $N_2O$ flux at regional scales. Their analysis of long-term data sets demonstrated patterns of $N_2O$ flux related to soil N, C, and moisture state for cropland, forest, and grassland ecosystems, but did not find a single unifying description that could be applied broadly. For cropping systems in Ontario, Desjardins et al. (2010) demonstrate good agreement across measurement methods for a transect of tower and aircraft measurements, as well as simulations using DNDC. However, their work demonstrated the importance of spring snow melt in driving high $N_2O$ flux, with variable moisture controlling peak $N_2O$ flux events and thereby controlling annual flux totals. This nonlinear response to weather makes Intergovernmental Panel on Climate Change Tier 2 equations inadequate for capturing flux trends with respect to pulse climatic events, though they may parallel multi-year average flux. Because experiments that measure $N_2O$ flux in the field typically do not continue over many years (or even a full calendar year), observations are unlikely to capture the full climate variation a site can experience. In the absence of measurements focusing on $N_2O$ flux response to extreme weather events, empirical models are unlikely to accurately capture the occurrence of pulse $N_2O$ flux events that often compromise the bulk of emissions.

Model inaccuracy due to the limited ability of empirical models to reflect ecosystem response to extreme weather events may be amplified as ecosystems respond to climate change. Direct limitations will result for systems that experience more frequent extreme weather events. For instance, empirical relationships derived from observations with a uniform distribution of mean annual precipitation may not adequately describe a system in which the same total precipitation largely falls in extreme events. The application of empirical models under climate change is also complicated in systems where climatic change affects plant physiology. In systems experiencing substantial physiological perturbation, historic observations will no longer adequately describe aggregate biogeochemical dynamics.

## GHG Data Availability and Model Validation Studies

Simulation models are often applied to combinations of crops, management practices, and environmental conditions for which there are limited or no GHG flux data. Even if GHG flux data are available, there are very rarely enough auxiliary data to support a mechanistic understanding of environmental drivers on net flux. For example, rarely are there measurements of $N_2O$ fluxes and water and nitrate leaching below the rooting zone throughout an entire year. Simulations of these systems do not focus on improved mechanistic representation of GHG flux. Instead, in these systems model-measurement comparisons are typically conducted for output variables such as SOC or crop yield.

In managed landscapes, long-term SOC data sets have sometimes been available, allowing validation of the C cycle (e.g., Smith et al., 1997) and constraining modeled $CO_2$ emissions resulting from plant and soil respiration, for a given set of assumptions about erosion rates. While some studies have validated models against observed steady state SOC at sites with long-term measurements, SOC response to management is difficult to measure accurately over shorter timescales and is therefore infrequently available for model validation. (We discuss this issue in more detail in the section entitled "Managing Agricultural Landscapes to Promote SOC Storage.") With the exception of the ECOSYS model, validation of modeled $CO_2$ flux against measurements is limited, though the increasing availability of flux tower data is making these comparisons possible (e.g., Schaefer et al., 2012; Dietze et al., 2011; Schwalm et al., 2010). Methane outcome validation exists for rice cropping systems (e.g., Fumoto et al., 2008; Smakgahn et al., 2009; Zhang et al., 2011a), and is not an important GHG source in other cropping systems. However, for croplands managed with manure inputs, manure storage is an important methane source. Methane loss responds to livestock management, especially manure management and diet selection, topics that are outside the scope of this chapter. Crop yield is the easiest C stock to measure and the most frequently validated model outcome. However, validating a model solely against crop yield provides limited constraint on the total C cycle and does not mean that the model delivers sound estimates of GHG flux. As models are increasingly applied in policy design and verification, model users should assess that a particular model has been adequately validated for the system and process of interest.

The full complexity of the N cycle remains a challenge to validate in model studies. Limited validation of modeled N cycle dynamics is largely due to the difficulty and expense of sampling nitrate leaching (Addiscott, 1996) and trace gas emissions (Groffman et al., 2006). Recent multi-scale $N_2O$ emission studies by Desjardins et al. (2010) provide rigorous data for both model validation as well as understanding issues of integrating local-scale chamber measurements to landscape-scale emissions estimates, but high quality $N_2O$ emissions data are unavailable for most agricultural systems.

We conducted a comprehensive survey of model application for the five simulation models reviewed in this chapter. Our survey emphasized the spatial extent of model application, the primary goal of application, the model outcomes reported in the study, and the type and success of model validation studies. Our survey demonstrated that model application was dominated by field-scale studies, though small- and large-scale regional applications were significant for hydrologic-centered applications of APEX, applications of DNDC to well-studied rice systems of China as well as grain and grassland systems in Europe, and DAYCENT application in the United States (Fig. 1a). The primary goal of most studies was to address ecological impacts of agricultural management (Fig. 1b), though some model applications were focused on more agronomic questions such as yield. A few model applications were explicitly concerned with the economic viability of land management scenarios. The models were generally applied to simulate conventional agricultural systems, though model application included ecologically based management scenarios (e.g., cover cropping or buffer strip management) or addressed the sustainability of agricultural management with respect to climate change.

Model development history influenced choices about how and where to apply a model. Studies applying models with a development history focused on hydrology (APEX, RZWQM2) frequently reported hydrologic outcomes, while studies applying models with a development history focused on carbon and nitrogen cycling (DAYCENT, DNDC, ECOSYS) had the most frequent reporting of GHG outcomes (Fig. 1c). Model application frequently included other ecosystem properties such as soil inorganic N status, $NO_3^-$ or P leaching, soil temperature, or system energy balance (Fig. 1c). Models rooted in biogeochemistry (DAYCENT, DNDC, ECOSYS) were used predominantly in ecological studies (studies addressing GHG emissions, C storage, or water pollution). For models rooted in hydrology, APEX was generally used to address water quantity and quality questions, while RZWQM2 was applied to simulate water quantity and crop yield. In many applications the specific metric of the study (water quality or quantity, GHG emissions, C storage, or yield) was framed in a larger context of supporting ecosystem services.

Studies applying DNDC and ECOSYS most frequently validated biogeochemical outcomes, while APEX and RZWQM2 had the highest occurrence of hydrology validation (Fig. 1d). Lack of model validation (applied to manuscripts where no validation data were presented and prior validation work relevant to

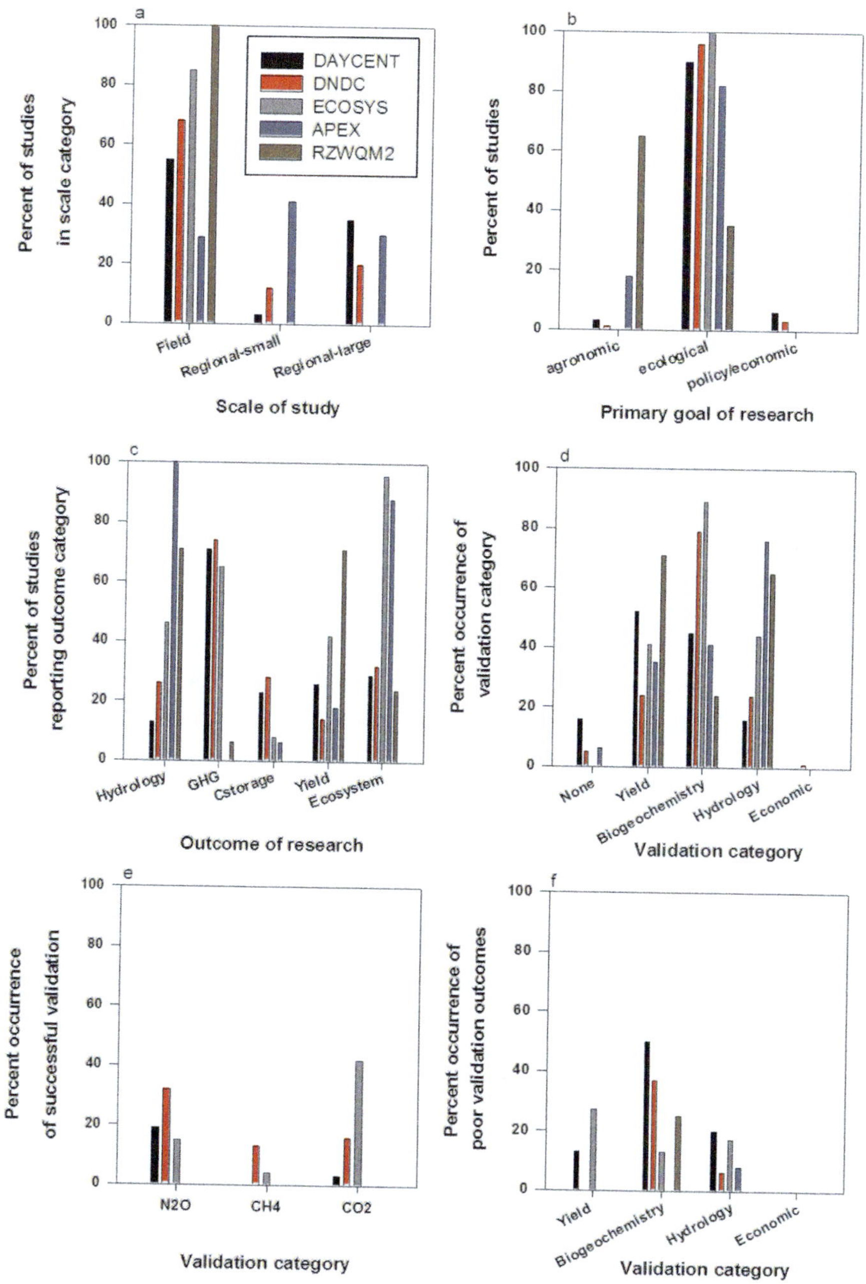


**Fig. 1. Simulation model application and validation.**

the study was not referenced) occurred in 5, 6, and 17% of the studies for DNDC, APEX, and DAYCENT, respectively. While all of the models reviewed can simulate GHG emissions, GHG outcomes have not been broadly validated (Fig. 1e). Nitrous oxide emissions have been most frequently validated for DNDC, but $N_2O$ validation was also conducted in studies using DAYCENT and ECOSYS. Methane emissions have mostly been validated for rice cropping systems in China using the DNDC model. Net ecosystem $CO_2$ emissions have been validated most frequently by comparing ECOSYS outcomes with sites on the basis of flux tower observations. APEX and RZWQM2 have not been validated with respect to GHG flux. Biogeochemical trends remain difficult to model, especially at seasonal timescales (Fig. 1f). At least one instance of poor model–data agreement for biogeochemical trends occurred in 13, 25, 37, and 50% of biogeochemical validation studies for ECOSYS, RZWQM2, DNDC, and DAYCENT, respectively. Few model applications concurrently validated all relevant agroecosystem properties: yield, biogeochemical cycles, and hydrology. While the modeling studies surveyed represent a comprehensive use of available data, limited observations remain a challenge for model validation of biogeochemical outcomes.

Whereas data availability limits development and application of both empirical and process-based models, the availability of dedicated, trained experts is a limitation primarily in the application of process-based models. Accurate application of biophysical models requires significant training to understand how to interpret results or estimate uncertainty. While a rigorously validated process-based model can be expected to generate GHG flux outcomes that reflect the system studied, many model validation studies have demonstrated initial inconsistency in comparisons between model outcomes and measured data, highlighting the need for model improvement (e.g., Beheydt et al., 2007). An individual responsible for model application needs to have a general familiarity with the range of plausible ecosystem fluxes in the system studied, an understanding of the mechanisms that drive ecosystem fluxes, and an understanding of how the applied model describes particular ecosystem processes to judge if model structural error is leading to spurious outcomes. The amount of training required to accurately apply a simulation model is a limitation to the widespread deployment of simulation models as policy verification tools.

## Managing Agricultural Landscapes to Promote SOC Storage

Carbon turnover times have been studied in agricultural soils for well over 100 yr, and biogeochemical models such as Roth-C and CENTURY, developed to test and refine hypotheses about SOM in crop fields, appear to have been well validated for slow turnover SOC pools at steady state (Smith et al., 2012). However, the past decade has witnessed a vibrant debate among competing hypotheses to explain why SOM is sometimes quickly mineralized by microbes and at other times persists for decades to centuries. Inherent biochemical recalcitrance, protection from attack by virtue of position in the soil matrix, and physical stabilization on mineral surfaces have all received support as stabilization mechanisms (Giardina and Ryan, 2000; Kramer et al., 2012; Mikutta et al., 2006; Sollins et al., 2006; Torn et al., 1997, 2005). In contrast, models of SOM decomposition generally assume decay is primarily based on OM substrate chemistry mediated by environmental factors such as soil texture, moisture, and temperature. Because our understanding of SOM dynamics is undergoing a major shift, models cannot incorporate a fully mechanistic description of SOM stabilization and destabilization. It is therefore worth asking whether SOM models capture how management practices influence soil carbon stability on short timescales and how those changes cascade to influence long-term SOC dynamics. If models cannot mechanistically describe the nuanced ways in which management changes the chemical and physical stabilization of SOM, will they provide a sufficiently unbiased comparison of how management techniques influence GHG budgets to develop policy recommendations?

Within the research community, opinions diverge regarding the extent to which the science behind soil carbon storage is robust, and several factors contribute to that split. First, biogeochemical models of soil carbon cycling rely on conceptual definitions of carbon pools (fast-cycling, slow-cycling, and passive) that are based on long laboratory incubations (Paul and Clark, 1996) rather than incubations under field conditions. This is necessary because of the long timescale of SOC decomposition dynamics, though SOC turnover times derived using $^{14}C$ isotope methods have been applied to gauge SOC turnover rate in situ (e.g., Gaudinski et al., 2000) and serve as a foundation for SOM model development (e.g., Tonitto et al., 2014). Because models built using these data can predict changes in total soil carbon that match data from long-term experiments (LTEs), they can be argued to provide a sufficient empirical foundation for model structural development and parameterization. However, as explained above, the underlying mechanisms have not been clearly established. A second school of thought holds that until mechanistic knowledge and understanding advances, it will be challenging to form robust expectations about how long SOM will persist and about how that persistence will change in response to a wide variety of management decisions (Crow et al., 2007) and global change.

Second, experiments comparing SOM content among different land uses and management practices appear to yield clear results, but assumptions and knowledge gaps across most experiments leave potentially important processes unmeasured. Eagle et al. (2012) list 11 agriculture management practices with "significant" research, 8 of which appear to increase soil carbon storage compared with fields managed without that practice. However, empirical knowledge about plant–soil interactions belowground lags behind the understanding of aboveground processes, and sometimes even well-understood interactions are not always incorporated into models. For example, several studies indicate that plant roots and plant shoots do not decompose at the same rate and do not contribute equally to soil carbon storage (Puget and Drinkwater, 2001). Roots may add more carbon to soil than shoots, but we know much less about their growth and decomposition (Freschet et al., 2013; Karlen and Cambardella, 1996; Merrill et al., 2002). In addition, roots can stimulate decomposition of native SOM, including deep SOM that otherwise likely would remain in situ for years or longer (Dijkstra and Cheng, 2007; Fontaine et al., 2007; Fu and Cheng, 2004; Zhang et al., 2013). Despite the complexity of rhizosphere dynamics, most research on soils has focused only on shallow horizons. The dearth of research below 20 to 30 cm in the soil profile means that biophysical processes and different types of tillage that are known to affect deeper parts of the soil profile are not considered in most analyses (Baker et al., 2007). Although models vary in their ability to represent processes such as root growth explicitly, they do not generally capture the type of nuanced result from field experiments discussed above. Similarly, biogeochemical models tend to exclude soils deeper than 1 m from the system boundary, so projections from models largely ignore gains or losses in deep soil. Improved mechanistic representation of management effects on the surface to 1-m depth is the most relevant scale for improving modeled response to management of most grain cropping systems in the United States; however, in cropping systems such as perennial grasses a deep and complex rooting zone can influence C accumulation below 1-m depth. Current efforts to create a community platform for quantitative assessment of agricultural systems (e.g., The Agricultural Model Intercomparison and Improvement Project [AgMIP] or The Global Research Alliance on Agriculture and Greenhouse Gases) may improve our collective ability to systematically address shortcomings in empirical knowledge and model development.

Third, few data sets are available that show changes in soil C in response to agriculture management practices and that enable us to ascertain whether these changes resulted from decomposition or erosion (Van Oost et al., 2007). Where soil carbon stocks have been observed to decline over time, it is often uncertain whether the decline represents a transfer of C from soil to atmosphere or movement of C across the landscape. If the latter, SOM may actually be deposited in streams or ponds, where low oxygen concentrations retard rather than accelerate decomposition. While the mechanism of OM loss is relatively unimportant from the perspective of soil fertility on the field, which declines in either case, it is critical from the perspective of GHG emissions. While simulation models with a hydrologic development history have often been validated with respect to runoff and sediment yield (e.g., Wang et al., 2008b, 2009), erosion processes have often been ignored in models with a development history emphasizing soil C and N dynamics. The assumption that C losses all result from decomposition is a critical weakness in model projections.

Fourth, to establish empirical relationships between a change in practice and soil carbon storage, experiments usually need to run for many years before a response can be detected (Richter et al., 2007). The need for long-term data arises because the soil carbon pool is very large, which means that even if the rate of soil carbon storage increases significantly, the total amount of carbon in the soil will not change enough over 1 to 5 yr for the change to be captured with a feasible number of samples (Paul et al., 1997). Consequently, detecting a change in soil carbon in response to a change in land management requires soil samples taken many years apart, ideally on fields where different management practices have been maintained following initially similar histories (Rasmussen et al., 1998; Richter et al., 2007). These data can come either from experiments that run for at least 15 to 20 yr (Paul et al., 1997) or from working farm fields that have been managed differently but consistently for many years and where we have good reason to believe the soils were similar at the beginning (Reganold et al., 1993). We identified 22 studies focused on agricultural ecosystems that have been running for at least 15 yr in the United States and Canada and that measured soil carbon. The longest-running of these include the Morrow Plots in Illinois, established in 1876, the Sanborn Plots in Missouri, begun in 1888, the Bretton Plots in Canada, initiated in 1929, and the Pendleton Plots in Oregon, started in 1931.

Despite their scientific importance, LTEs in the United States and Canada have included only a few of the many management systems a grower might consider, and they occur in a relatively small subset of the soil, climate, and landscape conditions where agriculture is practiced. For example, only three LTEs—two in the Mid-Atlantic and one in the Midwest—measured soil carbon in organic vs. conventional systems. And although LTEs in the United States and Canada capture large variations in soils and climates, there is a major gap in data showing how soil carbon responds to management change across the significant variation in soil and climate within regions. Similarly, LTEs have tended to focus on commonly used practices—a valuable approach to support choices among familiar options but much less useful for identifying practices with unrecognized promise (Stockmann et al., 2013).

One way to validate agroecosystems models without establishing numerous LTEs—a costly proposition that would not yield the desired answers for many years—might be to periodically resample soil carbon from marked locations at a wide variety of working farms (Spencer et al., 2011). As long as biophysical characteristics are known, and management history has been well-documented, data from farms that vary broadly in crop rotations, fertilizer management, pesticide use (or lack thereof), soils, and climate would enable the research community to calibrate and validate models that project the response of soil carbon storage to changes in agriculture management. The National Resources Inventory (NRI) is a statistical survey of land use and natural resource conditions and trends on non-federal lands in the United States that includes about 200,000 sample points on cultivated cropland. Existing resources already support periodic visits to these plots for sampling purposes, and the data collection for a subset of these plots could readily incorporate soil carbon measurements.

## Reducing Agricultural GHG Emissions through Management of Agricultural Systems

Achieving sustainable, low GHG emitting agricultural landscapes will ultimately include more systemic management changes, such as diversified cropping systems and a more integrated distribution of animal and cropping systems to allow for efficient usage of manure. Current land management GHG mitigation strategies often focus on a single practice change to reduce GHG emissions. For example, the CAR nitrogen management protocol enables growers to earn carbon credits for reducing $N_2O$ flux by reducing N application rate. The structure of C market policy does not promote integrative management approaches, because verifying outcomes is more challenging for these systems. Simulation and empirical modeling tools are most easily applied to quantify how changes in conventional management practices (i.e., tillage, fertilizer method, fertilizer type) lead to net GHG reductions. The recently developed Manure-DNDC model (Li et al., 2012) is one tool for studying more integrated landscapes. Improved capacity for life-cycle analysis (LCA) of complex agricultural systems is needed to accurately quantify net GHG emissions from more diversified landscapes.

There are fundamental inconsistencies between current GHG policy norms and an ecosystem management or LCA approach. Standard GHG accounting methodologies award $CO_2$ reductions due to reduced fertilizer production to the fertilizer factory (Climate Action Reserve, 2013). While factory management can reduce $CO_2$ emissions as a result of improved efficiency (lower $CO_2$ per unit fertilizer), it is the decisions and actions of the land manager that result in lower fertilizer per unit area. In rainfed grain cropping systems of the US Corn Belt, fertilizer accounts for over 30% of fossil fuel energy inputs (Kim et al., 2009). Therefore, agricultural land managers can reduce $CO_2$ emissions by increasing N fixation or improving nutrient retention through rotation diversification, or by applying manure as an N source, all of which reduce the need for fertilizer N application. Under current market approaches to GHG accounting, practices that reduce $CO_2$ emissions by reducing fertilizer use cannot be incentivized.

In the US context, policy focusing on N management is relevant for reducing our GHG emissions from agriculture. The USEPA (2012) estimates direct $N_2O$ emissions are 215.9 Tg $CO_2$-eq $yr^{-1}$ (3.1% of all US GHG emissions), while total GHG emissions from agriculture are about 6% of total US emissions (USDA, 2011). However, in the global context, policies and economic structures that promote building SOC have a greater potential to reduce net GHG emissions while promoting the maintenance of soil fertility. Smith et al. (2008b) estimate the GHG mitigation potential of global agriculture as 5500 to 6000 Tg $CO_2$-eq $yr^{-1}$ by 2030, mostly resulting equally from cropland and grazing land management that promotes SOC accumulation and from the restoration of cultivated organic soils, with each sequestering on the order of 1300 to 1350 Tg $CO_2$-eq $yr^{-1}$. Restoration of degraded lands is also an important SOC accumulation strategy comprising about half of the mitigation potential of cropland and grazing land management scenarios ($\sim$650 Tg $CO_2$-eq $yr^{-1}$), while improved rice and livestock management can reduce $CH_4$ flux by $\sim$250 to 300 Tg $CO_2$-eq $yr^{-1}$. In contrast, improved cropland management is only predicted to decrease $N_2O$ flux by $\sim$200 Tg $CO_2$-eq $yr^{-1}$. However, a reduction in $N_2O$ flux is realized each year best management practices are implemented in contrast to SOC accumulation in response to management, which ultimately achieves a steady state, after which further SOC accumulation is negligible.

Eagle and Olander (2012) reviewed the GHG mitigation potential of 42 agricultural management strategies and determined sufficient data exist to promote 20 strategies for their GHG reduction potential. Significant accumulation of SOC across the agricultural landscape requires management changes to rotation, tillage, and organic amendments in cropland, and improved grazing land management. However, with SOC accumulation, future changes in management could again reduce SOC stocks, so the issue of permanence must be addressed carefully. While small changes to management practices are the easiest policies to initiate, management that promotes building SOC will require different economic incentive structures.

## Quantifying GHG flux from Land Use Change and Forestry

Similar to agricultural systems, in forest ecosystems there are two general modeling approaches used to estimate carbon cycling: (i) simulations of forest stand growth based on climatic and soil conditions and (ii) empirical estimates of stand growth based on repeated measurements of forest carbon stocks. Each approach can be used for plantations and naturally regenerated forests. In this section, we focus on how carbon cycling is modeled in the United States for (i) national-scale reporting under the UNFCCC and (ii) project-scale accounting using tools developed by CAR and RGGI. In these cases, the approaches are primarily empirical and are based on periodic measurements of the key carbon pools, either directly in the parcel of interest for the compliance programs or based on systematic forest surveys from similar forests within the region for voluntary programs and county, state, and national estimates of forest carbon fluxes. Therefore, we focus on empirical models and tools that are based on periodic measurements of forest carbon stocks. Because environmental markets use multiple GHG accounting tools, we briefly discuss how land-based empirical carbon flux estimates compare with those from ecosystem models and atmospheric inverse modeling.

Carbon accumulates in forest ecosystems because of the balance between fixation by photosynthesis and emission from respiration. The sum of these processes is termed net ecosystem productivity (NEP). Both fixation and respiration are strongly affected by forest management. Forested ecosystems store large quantities of carbon in living and dead biomass and soil, and changes in these carbon pools are a significant part of the global carbon budget. Globally, forest ecosystems are major net sinks for $CO_2$ (Pan et al., 2011), and in the US net $CO_2$ uptake by forests is equivalent to approximately 16% of total $CO_2$ emissions from fossil fuel combustion (USEPA, 2013). In many forests, the largest C pools are aboveground live biomass and mineral soils, with lesser amounts in belowground biomass and surface detritus (forest floor). But some of these pools, such as mineral soils, usually change slowly, while vegetation and forest floor pools can change quickly over short timescales in response to natural disturbance or forest management. In the United States, most of the change in forest carbon stocks (carbon sink) is in the aboveground biomass (49%) with an additional 27% in wood products in landfills and the remainder in down dead wood, wood products in use, and forest floor and soil (Woodbury et al., 2007b). However, in some forest ecosystems under some types of management or natural disturbance situations, changes in other pools such as forest floor and coarse woody debris can be very important.

The aboveground biomass of living trees is the most important carbon sink. Tree diameter and height can be readily measured using well-established protocols, and allometric equations then used to estimate aboveground forest carbon stocks. Modeling efforts have focused primarily on forest stands that do not already have measurements and for projecting the effects of changes due to disturbances such as fire and insect pests and management practices such as different types of harvest and pre-commercial thinning.

## National-, State-, and County-Scale Estimates Based on Forest Inventory Measurements

### Flux of $CO_2$ from Living Vegetation

At national, regional, state, and county scales in the United States, the Forest Inventory and Analysis (FIA) program of the USDA Forest Service provides data from many thousands of permanent sampling plots on both private and public lands, and these data are available online (http://fia.fs.fed.us, accessed 15 Sept. 2015). Historically, the main purpose of the FIA program has been to estimate timber supply. However, the measurements taken to estimate timber supply include forest area and the species, diameter, and height of individual trees. These data can be used to estimate the amount of wood in entire trees using allometric equations.

To estimate the change in carbon over time, or net carbon flux based on inventory data, a "stock change approach" is used. In this approach, carbon stocks are estimated at two or more times, and net annual carbon flux is estimated by subtracting one stock estimate from the other and dividing by the number of years between stock estimates. These FIA data have been used for decades to estimate forest carbon stocks and stock changes. Over time the data have generally become available at finer spatial and temporal scales. Recently, additional measurements of carbon pools beyond trees, including down dead wood, forest floor, and soil, have been included (reviewed by Heath, 2013). Forest inventory data are a useful basis for estimating carbon stocks and net fluxes for the sampled area. However, forest carbon pools such as coarse woody debris, soils, and understory have been difficult to estimate on the basis of data collected in inventories until very recently, so there is a need to augment inventory data with data from intensive research sites and models. Using FIA data supplemented with models for carbon pools other than trees, carbon stocks, and stock changes throughout the United States have been estimated for decades, for example, as part of the annual greenhouse gas inventory required under the UNFCCC (e.g., Woodbury et al., 2007a; USEPA, 2013, as reviewed by Heath, 2013). Over time, these estimates have expanded to cover non-forest land to represent effects of afforestation and deforestation, including changes in soil and forest floor carbon stocks, as discussed further in subsequent sections (Woodbury et al., 2007b; Heath, 2013).

## Tools Based on Forest Inventory Data

Publically available tools that use FIA data to calculate forest carbon fluxes include the Carbon Online Estimator (COLE) (Proctor et al., 2005; Van Deusen and Heath, 2010) and the Carbon Calculation Tool (CCT) (Smith et al., 2007, 2010). The COLE was developed to allow easy estimation of forest carbon stocks for a user-specified region of interest (Proctor et al., 2005). During the US voluntary carbon program known as "1605b," the Forest Service adopted COLE as the official web tool for the forest portion of that program (Heath, 2013). Users can select specific forest regions of interest and generate a report that estimates the carbon accumulated in each forest carbon pool over time (Van Deusen and Heath, 2010). The CCT is a program that allows a user to download the latest FIA data and produce estimates at the state scale matching those in the national GHG inventories, and it can be used to develop estimates of state carbon fluxes since 1990 (Smith et al., 2010).

## Flux of $CO_2$ from Coarse Woody Debris

In many forests, large quantities of C accumulate in the form of coarse woody debris also called "down dead wood." Unfortunately, the size of this pool is not closely related to aboveground biomass or forest age, so it is difficult to estimate coarse woody debris on the basis of frequently measured aboveground stand characteristics (Woodall et al., 2008). At state and national scales, estimates have been made on the basis of predictions of residues produced during harvesting. These modeled estimates were recently compared with those based on field measurements. The total stock estimate for the United States differed by only 9% between modeled and measured estimates, with the models underestimating total coarse woody debris in stands with little coarse woody debris and overestimating carbon in stands with high coarse woody debris (Domke et al., 2013). There were also regional differences in model accuracy, with greater accuracy in the Southeast and lower accuracy in the North Central and Northeast. On a regional basis, these discrepancies caused significant differences in coarse woody debris carbon stocks by region and ownership. On the basis of these results the authors suggest that the United States conduct systematic field sampling for the purpose of national inventories.

## Flux of $CO_2$ from Forest Floor, Soil, and Roots

The forest floor is defined broadly as the organic layer above the mineral soil including woody debris smaller than 7.5 cm in diameter. The forest floor contains fine woody debris, a litter layer (designated $O_i$ and $O_e$), and humus ($O_a$). In most forests in the conterminous United States, these layers generally store smaller amounts of C than do mineral soils, but they can change much more rapidly following harvest or fire, highlighting their importance when assessing effects of management. Changes in forest floor carbon for broad forest types and large regions of the United States have been modeled on the basis of available research and inventory data (Smith and Heath, 2002), and these models have been used to estimate national carbon stock changes for UNFCCC reporting. Root biomass is extremely challenging to measure directly, but it is usually closely correlated with aboveground biomass (typically 0.26 of aboveground biomass; Cairns et al., 1997), and thus it can be modeled on the basis of measurements made in forest inventories. Changes in forest soil carbon are discussed in the subsequent section on land use change.

## Land Use Change—Deforestation and Afforestation

When forest regrows immediately following a natural or anthropogenic disturbance, there may not be large, long-term (multi-decadal) changes in forest floor and soil carbon stocks, unless the soil surface is physically disturbed, for example, is burned or plowed. However, if there is a change in land use from forest to plowed agriculture, there can be large and long-term decreases in these carbon stocks. In tropical regions, such deforestation to agriculture causes large emissions of $CO_2$ (Ramankutty et al., 2007). In the United States, afforestation on abandoned agricultural land has acted as a major C sink, especially in the Northeast. This is supported by survey and experimental data for forest floor and soil and an empirical model (Woodbury et al., 2007a) as well as increased tree carbon stocks based on FIA data supplemented with models (Woodbury et al., 2007b). Effects of land use change have also been estimated using ecosystem models rather than empirical survey data (e.g., Houghton and Hackler, 2000; Houghton et al., 2000; Pielke et al., 2002).

It is important to note that carbon in forest floor litter and organic horizons is lost within a few years after deforestation to plowed agriculture, but is gained only over many decades after afforestation (Woodbury et al., 2007a). For this reason, at regional and national scales it is important to model simultaneous afforestation and deforestation on the basis of gross land use changes, not just net land use changes. If only net land use change is modeled, the "lag" effect of afforestation can cause a large bias in the estimates (Woodbury et al., 2007a).

A comparison of tools for estimating the effect of land use change on carbon stocks demonstrates that alternative modeling approaches, which apply different types of input data, result in model outcomes that differ even at the national scale. Estimates based on FIA data supplemented with models (land-based estimates) can be compared with those from ecosystem models. For example in a study of

US forest carbon stocks, Hurtt et al. (2002) estimate 330 Tg of net change in soil, forest floor, and woody debris, which is fivefold greater than a land-based estimate (Woodbury et al., 2007b). Despite these differences, changes in aboveground (tree) carbon still dominate the total carbon stock change from land use change, due to the large stocks in temperate forests and relatively rapid changes compared with forest floor and soil carbon after afforestation (Woodbury et al., 2007b).

### Flux of $CO_2$ and $CH_4$ from Wood Products

Up to now we have focused on changes in carbon stocks within forests. However, after harvest, wood products are usually not instantly decomposed or combusted; instead, wood products may be stored for decades or centuries when incorporated into buildings or other long-lived structures. Eventually, wood products are combusted or decayed to produce $CO_2$ or $CH_4$ (see methane discussion below). When wood products are disposed of in landfills, the carbon may be released many years or decades later, or may be stored almost permanently in the landfill. "Book-keeping" type accounting models have been used for many years to estimate the half-life of various types of wood products and their rate of decay after different types of disposal, including combustion with or without energy recovery (e.g., Skog and Nicholson, 1998; Skog et al., 2004). These approaches have been used for national GHG inventory reporting under the UNFCCC (e.g., USEPA, 2013). In landfills, because of intentionally anaerobic conditions, both $CO_2$ and $CH_4$ may be produced from wood products and other wastes (USEPA, 2013). Because $CH_4$ has such a high global warming potential (circa 23-fold higher than $CO_2$), assumptions about $CH_4$ versus $CO_2$ emission are very important. Surprisingly, landfills are among the fastest growing carbon pools in the United States, where wood products such as paper and construction waste break down only very slowly (Miner and Perez-Garcia, 2007; Skog, 2008). Emissions from landfills are modeled using a book-keeping approach as well as ongoing surveys of waste management practices and of different categories of solid waste, and on measured emissions of $CO_2$ and $CH_4$ from experiments and field studies (USEPA, 2013). A certain proportion of carbon is assumed to be stored indefinitely in landfills. The remainder is assumed to decay over time according to first-order kinetics.

## Forest Stand and Project-Scale Estimates

At the scale of individual forest stands or carbon offset projects, there are limitations to using FIA data and models based on them. First, FIA data are intended to measure forest attributes over wide areas, especially county and larger scales, so they may not represent a small forest parcel. Second, as mentioned above, information on carbon stocks in coarse woody debris, soils, and understory often must be modeled using data from research plots that may not represent the actual site conditions. Third, depending on the location, the sampling interval may be many years or longer and thus may not always represent current conditions. For these reasons, while data from FIA have been used for general planning purposes for estimating the potential value of carbon offsets, they are not generally appropriate for specific offset projects. Instead, such projects generally required repeated sampling of the actual parcel of forest being used as an offset. For both voluntary programs such as the 1605b, and compliance markets such as the former Chicago Climate Exchange (CCX) and the current CAR and RGGI, verifiable estimates of carbon stock changes are required (as reviewed by Fahey et al., 2010).

To obtain carbon credits under compliance programs such as CAR and RGGI, applicants must submit a detailed application that explains how C sequestration will be quantified, monitored, and verified. The project area is divided into relatively homogeneous units on the basis of factors such as forest type, stand age, slope, etc. For each unit, C pools must be sampled to achieve specific statistical confidence, such as >95% confidence that results are within 10% of the true value. Carbon pools include aboveground biomass, belowground biomass, coarse woody debris, soil, and harvested wood products. As discussed in subsequent sections, measuring changes in pools other than aboveground trees can be challenging, and may be very costly to achieve the required precision, but models may be used in some cases to supplement these measurements. Measurements must be repeated periodically, but may be estimated annually using models. Such precision of measurement for multiple pools and requirements for frequent resampling may increase transactions costs so much as to discourage participation in compliance markets.

## Climate Action Reserve Forest Management Protocol

The CAR forest management protocol allows credits for $CO_2$ emission reductions due to improved management associated with US projects supporting reforestation (forest regrowth on previously forested lands) and afforestation (establishment of forests in areas previously under non-forest landcover). Onsite forest carbon pools are broadly grouped into living biomass, dead biomass, and soils. Living biomass includes biomass in live trees and shrubs and herbaceous understory (live non-tree biomass). Onsite dead biomass includes biomass in dead trees, lying dead wood, and litter. Offsite dead biomass includes harvested wood products. A combination of modeling and measurements is used, similar to the forest inventory methods reviewed above, but with forest sampling data collected from the parcel. Projects must be verified by approved third-party verifiers and by the CAR program. A report must be completed each year for 100 yr, although measurements do not need to be made annually. All projects must use the appropriate biomass equations for the project location. The required biomass equations are found on the Reserve's Forest Project Protocol webpage. The calculation of $CO_2e$ for each inventoried tree must be conducted in a manner that provides project estimates for several components, as follows. Whole tree biomass (roots, stump, bark, bole, top, and branches) is used to provide project totals and estimates of emissions associated with harvest activities. Bole biomass is the portion of harvested trees delivered to facilities for processing into wood products. The aboveground portion (stump, bark, bole, top, and branches) is used to compare project data with common practice statistics for improved forest management projects. Further information is available in publications (Climate Action Reserve, 2012b, 2012c) provided on the project website, which should be reviewed for updates (http://www.climateactionreserve.org/how/protocols/adopted/forest/current/, accessed 15 Sept. 2105).

## Regional Greenhouse Gas Initiative Forest Management Protocol

The RGGI protocol allows credits for $CO_2$ emission reductions resulting from reforestation and afforestation projects in RGGI member states in the northeastern United States. The focus is on afforestation with native species to promote restoration and sustainable management of native forests on lands that have not been forest for at least 10 yr before project initiation. Emissions reductions or carbon sequestration must be real, additional, verifiable, enforceable, and permanent. To ensure permanency a relevant state agency must approve a legally binding "permanent conservation easement." This easement must require that all land within the offset project boundary be maintained in a forested state in perpetuity, and the carbon density within the project be maintained at or above that achieved at the end of the $CO_2$ offset crediting period. The net carbon sequestered during the reporting period is also discounted by 10% to account for potential reversals of carbon sequestration, unless the project sponsor retains long-term insurance, approved by the relevant state agency where the offset project is located. Such insurance must guarantee replacement of any lost sequestered carbon for which $CO_2$ offset allowances were awarded. Similar to the CAR protocol, it uses a combination of modeling and measurements including forest sampling data collected from the parcel.

When direct sampling is not possible or cost-effective, empirical models are used to project the results of direct field sampling through simulated forest management activity. Models are also used to assist in updating inventory plots. By modeling tree growth in plots, these plots can represent a reporting year subsequent to the actual sampling date. As of February 2013, RGGI had approved the following growth models for deriving these estimates:

- CACTOS: California Conifer Timber Output Simulator,
- CRYPTOS: Cooperative Redwood Yield Project Timber Output Simulator,
- FVS: Forest Vegetation Simulator,
- SPS: Stand Projection System,
- FPS: Forest Projection System,
- FREIGHTS: Forest Resource Inventory, Growth, and Harvest Tracking System,
- FORESEE: FORESt Ecosystems in a Changing Environment.

Additional models are allowed following approval of a state forestry authority and meeting quality criteria to assure that the model will perform adequately for the forest type and location.

All projects must be verified by approved third-party verifiers. Further information is available in publications (Regional Greenhouse Gas Initiative, 2013a, 2013b) provided on the project website, which should be reviewed for any updates (http://www.rggi.org, accessed 15 Sept. 2105).

## Forest Vegetation Simulator

To project future forest stand growth, carbon sequestration, and the effects of silvicultural treatments, tools such as the Forest Vegetation Simulator (FVS) can be used (Crookston and Dixon, 2005; www.fs.fed.us/fmsc/fvs). The FVS consists of a suite of growth and yield models that have been calibrated using FIA and other data to forests across the United States. The FVS has been approved for use by CAR and RGGI in conjunction with periodic sampling to meet their criteria for providing verifiable estimates of C sequestration for forestry offset projects. We discuss this particular tool because it has been used for decades to model forest stand growth and the effect of silvicultural treatments for many forest types throughout the United States on the basis of FIA data. The FVS is an individual-tree, distance-independent growth and yield model. More recently, the FVS-CarbCalc tool has been added to simulate stand level carbon stocks and changes in forests and in harvested wood products using the approaches discussed above. The methods are consistent with the US 1605b program (USDOE, 2007) calculating and reporting guidelines as well as the Intergovernmental Panel on Climate Change (2003). The following carbon pools are modeled: aboveground live biomass, aboveground dead biomass, belowground carbon, litter and duff (forest floor) biomass, and harvested merchantable biomass. The FVS uses FIA data or project data to describe initial stand conditions. Further information is available at the FVS website (http://www.fs.fed.us/fmsc/fvs/index.shtml, accessed 15 Sept. 2105).

The FVS is a useful tool, but as with any modeling tool, users must be aware of the strengths and limitations of the model and follow guidance to obtain useful results (Hoover and Rebain, 2011). To investigate the utility of the FVS model for carbon offset projects, data from three long-term forest sites were compared with FVS predictions of carbon stock changes (Fahey et al., 2010). For northern hardwood forests in New Hampshire and Allegheny hardwoods in New York State, aboveground tree biomass was predicted accurately over 37-yr and 50-yr measurement intervals, respectively. For a spruce–fir–birch forest in New Hampshire, projections were accurate from 1965 to 1987, but subsequent pollution-induced decline of red spruce (*Picea rubens*) resulted in a substantial overprediction by the model. The response to whole-tree harvest in New Hampshire was also modeled, and FVS greatly underestimated subsequent biomass accumulation, because it did not predict the occurrence of pin cherry (*Prunus pensylvanica*), a fast-growing tree species that can dominate initial biomass accumulation in this region (Fahey et al., 1998). These results suggest that no model simulation can represent events that are not anticipated, such as the decline of a dominant species. But these results also show that even commonly used models may not be correctly parameterized, in this example failing to include a fast-growing tree species.

## Comparing and Evaluating Different Approaches to Modeling Forest Carbon

Estimates of net change in forest carbon stocks often seem to converge when scaling up to regional and national scales. However, this convergence in the overall net change is based on conflicting estimates for particular pools and processes. For example, forest carbon flux estimates of ecosystem models driven by satellite measurements were similar to those derived from land-based surveys supplemented with models, with the exception of the Terrestrial Ecosystem Model, which was only half that of other estimates (Woodbury et al., 2007b). However, estimates of specific carbon pools were extremely variable. For example, the overall estimate of net US forest C change for the ecosystem model of Hurtt et al. (2002) is 330 Tg $yr^{-1}$, which is more than double a land-based estimate of 163 Tg $yr^{-1}$ (Woodbury et al., 2007b). On the basis of an ecosystem growth model, Hurtt et al. (2002) estimate changes in tree carbon stocks similar to the land-based estimate, but they estimate an additional very large change of 100 Tg $yr^{-1}$ due to woody encroachment on non-forest land in the arid Southwest. Furthermore, their estimate of net change in soil, forest floor, and woody debris is fivefold greater than the land-based estimate. As in other ecosystem models, wood products are not included

(Hurtt et al., 2002). At the national, continental, and global scale, improved estimates of forest carbon flux will depend on reconciling estimates derived from different methods such as atmospheric inversion modeling, inventory-based approaches, ecosystem modeling, and land use change modeling.

During recent years, increased spatial and temporal resolution of forest inventory data and increased availability of measurements of additional pools have reduced uncertainties for some carbon pools including coarse woody debris and forest floor (Domke et al., 2013; Woodall et al., 2012). However, there will still be uncertainty in belowground pools that are impractical to measure directly and also because of changes that are not represented in models. For example, the spread of exotic earthworms can greatly reduce the thickness of organic layers, increase the bulk density of soils, and incorporate litter and humus materials into deeper horizons of the soil profile. All of these earthworm impacts could alter carbon stocks in the forest floor and mineral soil, and if this change is not measured or included in a model, it could result in substantial errors in carbon stock changes (Frelich et al., 2006).

At the project or forest stand level, estimates may differ widely depending on the approach and assumptions that are used. The comparison of changes in forest carbon stocks over decades for three sites in the Northeast demonstrated that the FVS modeling system could sometimes predict stock changes well, but in other cases could not because of decline of major species or lack of inclusion of fast-growing species after harvest (Fahey et al., 2010). Discrepancies can occur even when models are used by research scientists familiar with the ecosystems and models themselves, implying that greater deviations can be expected when these models are employed by less experienced individuals or when biases toward economically desirable outcomes might influence modeling procedures. Of course, model simulations cannot represent events that are not anticipated, such as the decline of a dominant species or effects of exotic earthworm invasion. For this reason, compliance programs require third-party verification and periodic sampling of the site. Additionally, none of the models discussed herein include non-GHG effects, such as changes in albedo that might promote local and regional warming (Whitehead, 2011).

For carbon credit markets such as CAR and RGGI, the requirements for high-precision measurement of multiple pools, even those such as soil carbon that change very slowly, and requirements for frequent resampling may increase transactions costs so much as to discourage participation in such markets. In addition, requirements for permanent legal conservation easements and maintaining the land in forest in perpetuity after the project period may also discourage participation.

# Model and Policy Synthesis

## Advantages and Limitations of Mitigation Options

To assess the potential for agriculture and forestry to mitigate climate change, scientists focused first on carbon storage in trees and soil (Campbell et al., 1991; Rasmussen and Collins, 1991; Davidson and Ackerman 1993; Donigian et al., 1994; Hamburg, 2000; Intergovernmental Panel on Climate Change, 2000a,b; Lewandrowski et al., 2004). Carbon storage is a particularly attractive approach because it affords the possibility of reducing current atmospheric $CO_2$ stocks through carbon fixation, whereas strategies available to most sectors of the economy focus on lowering GHG emissions, but do not remove existing atmospheric $CO_2$. In addition, building soil carbon, growing trees, and conserving forests confer numerous benefits to agriculture and society (Magdoff and Van Es, 2010). At first glance, it is easy to imagine strong support for tools that incentivize growers to build soil carbon or that encourage forest managers and governments to reduce rates of deforestation in biologically diverse ecosystems.

### Biomass Carbon Storage

Policy initiatives' early focus on carbon storage in forest biomass (e.g., REDD) reflects the availability of reliable tools for estimating aboveground biomass, and the certainty that tree growth transfers carbon out of the atmosphere into organic matter. Allometric equations have long been used to estimate potential timber productivity (i.e., harvestable tree volume) for specific species on the basis of measurable tree attributes such as diameter at breast height and total tree height. To develop these relationships, researchers collected detailed destructive measurements of individual trees. These equations perform well for species and regions for which they were developed. The challenge is extrapolating the limited set of measurements to all species and regions of interest. Furthermore, data are generally scarce for tropical forests. Though allometric equations generally work well, there can be differences in predicted biomass from different equations (Ahmed et al., 2013). This variation in estimates among equations motivated efforts to develop a standardized set of equations for the conterminous United States (Jenkins et al., 2003). While variation in estimates of biomass using different allometric equations for individual trees can be substantial, a recent study found only modest errors in forest C stocks at the hectare scale of 1 to 5 Mg for two well-studied sites in Massachusetts and Maine (Ahmed et al., 2013).

Forest carbon storage as an approach to climate change mitigation faces a number of challenges. Chief among them has been concern about leakage, i.e., the circumstance under which conservation of a forest in one location simply shifts demand for forest products and associated tree harvest to a different location (Murray et al., 2002). Others include the very long time commitments (typically 100 yr) required of landowners to ensure that climate benefits accrue and persist.

## Soil Carbon Storage

Soil carbon storage as an approach to mitigate climate change also faces several challenges. First, in most cases, soil carbon storage is reversible. Just as soil carbon can accumulate in response to management changes, it can decompose in the absence of continued management or in response to unexpected temperature or precipitation patterns or fire.

Second, because plowing previously uncultivated land initiates a period of rapid decomposition of SOM, some changes in management during that period might only slow the rate of carbon loss rather than result in soil carbon gains (Rasmussen and Collins, 1991). Slowing the rate of soil carbon decomposition confers a climate benefit relative to the business-as-usual case, but soil is still releasing more carbon to the atmosphere than is being stored. Stabilizing the climate will require near-zero emissions (Matthews and Caldeira, 2008). Therefore, using practices that reduce the C loss rate takes us in the right direction, but the remaining emissions ultimately need to be mitigated.

Third, there are limits to the rate and amount of soil carbon storage that a management change can yield. Long-term experiments have revealed that inflows and outflows of carbon in agricultural fields reach equilibrium, often within a few decades (Jenkinson and Rayner, 1977; Paustian et al., 1997). The practical consequence is that changing management to increase carbon inputs (or decrease carbon losses) will lead the amount of soil carbon to increase for a period of time and then to stabilize at a new level. After that time, we expect no additional climate benefit absent *additional* management changes that further increase carbon inputs.

Fourth, estimates of soil carbon stocks and conclusions about the response of SOC to management changes depend on the depth of sampling (Baker et al., 2007; Kravchenko and Robertson, 2011). Most studies of soil carbon have been restricted to the top 20 cm of the soil profile, where SOC content tends to be highest. However, carbon below 20 cm constitutes a very large proportion of total C in the profile and observations suggest that management changes influence carbon down to 1 m. For example, adding deep-rooted perennials to a rotation would be expected to increase SOC to at least 1-m depth (Kell, 2011), and shallow soil samples would fail to capture this change.

The case of reduced tillage illustrates the challenges that field scientists and modelers face as a result of questions about sampling depth. Hundreds of studies support the hypothesis that no-till management leads to higher soil carbon content, but very few of these studies sampled below 20 cm. However, in fields where investigators sampled to greater depths, SOC content tended to be greater in conventionally plowed fields compared with fields managed with conservation tillage (Baker et al., 2007, and references therein). These results suggest that SOC accumulation at depth is greater in conventionally plowed fields than in fields managed with conservation tillage, although the results are not statistically different in part because studies that sample deeply are relatively rare (Baker et al., 2007). One school of thought maintains that because agricultural management influences the soil profile at depth, comparisons among management practices require investigations of the entire profile (Baker et al., 2007). A finding of "no difference" in SOC from 0 to 100 cm between tillage practices precludes assigning climate benefits to conservation tillage. In contrast, Kravchenko and Robertson (2011) argue that no-till practices increase SOC in the top 20 of cm of the profile and not in the 20 to 100 cm segment. The absence of a detectable difference in the lower part of the profile results from the high variability among locations deeper in the profile. An unrealistically large number of samples would be needed to detect a difference given the known variability. According to this view, comparing the entire profile (0–100 cm) between tillage practices masks the clear signal from the upper part of the profile with the high variability in the lower part of the profile, and the correct conclusion would be that conservation tillage increases SOC because that is the finding in the part of the profile where a power analysis shows that we can detect a difference with the number of available samples.

Tillage practices vary across regions and cropping system, raising questions about whether the necessary sampling depth also varies systematically with geography. In a national-scale review of tillage effect on SOC in Canada, VandenBygaart et al. (2003, 2008, 2011) conclude sampling below 30 cm is often necessary to detect SOC change in eastern Canada, because of the extensive use of deeper moldboard plowing. Even for fields with shallower tillage management in western Canada, the authors found that increasing sampling from 15 to 30 cm revealed SOC differences across tilled and no-till fields. These results suggest that in many agroecosystems the sampling depth of currently available measurements comparing SOC accumulation between till and no-till systems is insufficient to accurately describe the effects of tillage in ecosystem models.

Mechanistic understanding about how tillage influences SOC has also played into the debate about sampling depth and whether meta-analyses comprised mainly of studies that used shallow samples provide robust findings that can be used to address climate change mitigation. The finding that no-till management increases soil C in shallow horizons is consistent with the idea that breaking up soil aggregates exposes more surface area to microbial attack. However, considering a more complete picture of how tillage influences the plant–soil ecosystem yields a less conclusive picture. For example, compared with conventional tillage, conservation tillage leaves more plant residue on the surface after harvest, and the insulated soils stay cooler longer into the spring. Because soil temperature controls root growth rate, fields managed with conservation tillage may have slower root growth and therefore add less carbon to the soil (Baker et al., 2007). On the other hand, conservation tillage increases the amount of plant residue on the soil surface, which can increase soil moisture retention and hence root growth, especially in drier soils. Because the relationship between tillage practices and SOC sequestration is mediated by a variety of factors such as these, the net influence of changing tillage practice on SOC is not intuitively obvious.

Despite these limitations, soil carbon storage has been broadly studied as an option in cap-and-trade policies, as occurred in HR2454, the American Clean Energy and Security Act of 2009. Soil carbon storage can, on the basis of some key assumptions, be accomplished fairly cheaply—US$27 or less per ton $CO_2e$—relative to other strategies for climate change mitigation (Creyts et al., 2007). There is a strong need to identify the most appropriate policy tools to take advantage of soil carbon storage, in light of these limitations and the large proportion of the landscape occupied by farms and forests.

Questions about how conservation tillage influences SOC sequestration illuminate the interplay of data availability, model projections, and policy. Although hundreds of studies support the relationship between no-till management and increased SOC, very few of these studies sampled below 20 cm. The mechanistic relationship between tillage and SOC is more complex than it might at first appear, and both tillage practices and the ecology of crop fields vary with geography and climate. A compilation by T-AGG of studies comparing tillage practices and SOC reinforced the "stamp of legitimacy" that no-till agriculture has received from the research community as an effective and quantifiable practice to increase soil carbon storage (Eagle et al., 2012; Eagle and Olander, 2012). Syntheses of available data and statements of consensus about policy-relevant science play a vital role at the science-policy nexus. However, what we can reasonably measure—hence what we have measured—and what we want to know are not necessarily the same thing. And given large investments in data collection, analysis, and interpretation, there is of course a desire to make use of those data to answer key questions (including policy questions) and to drive models. Many of these issues that influence technical recommendations for policy do not receive attention at the science-policy interface even though they bear directly on debates such as whether no-till agriculture reliably increases SOC sequestration.

## Nitrous Oxide Emissions and Nitrogen Management

Issues of reversal and permanence that surround soil carbon storage, and leakage and permanence with forest carbon storage, are generally of less concern for reduced $N_2O$ emissions. It is possible that reduced fertilizer application rates in 1 yr (with associated lower $N_2O$ emissions) would be followed by proportionally higher fertilizer application (and higher $N_2O$ rates) in a following year. But such temporal "catch-up" is not considered likely. In this respect, reduced $N_2O$ emissions are significantly more straightforward to handle in a policy context, compared with soil carbon storage. Nevertheless, there are significant challenges in data availability and data quality related to developing robust policy tools.

Sound interpretation of $N_2O$ flux measurements requires several years of data or more because the response of $N_2O$ to particular management factors often depends on environmental variables such as rainfall and temperature, which vary from year to year (Groffman et al., 2000; Sistani et al., 2010). Data collected for less than one full year is particularly difficult to interpret with confidence because $N_2O$ emissions can occur throughout the year, including under snow, and can exhibit peaks during brief warm and wet periods during winter and early spring (e.g., Johnson et al., 2010; Wagner-Riddle et al., 2007; Wolf et al., 2010; Kim et al., 2012). In addition, the amount of plant growth, which also varies across years, supplies carbon to microbes and therefore affects N cycling and $N_2O$ emissions (Woldendorp, 1962). Therefore, to establish general relationships between management choices and $N_2O$ emissions, it is necessary to capture the $N_2O$ response to management under a range of conditions (Mosier et al., 2006).

In the case of practices such as manure addition, the continued accumulation of organic matter over many years can lead to changes in $N_2O$ emissions (Chang et al., 1998), suggesting that projections of future $N_2O$ emissions need to take into account expectations about how soil conditions will evolve in response to management as well as to changes in climate. So although $N_2O$ emissions do not pose the same challenge for detection as soil carbon storage, we nevertheless need measurements of $N_2O$ emission rates over multiple years following change in management, particularly where management changes influence soil characteristics (Six et al., 2004).

A major challenge to characterizing the response of $N_2O$ emissions to agricultural management is that $N_2O$ emissions are notoriously variable in both space and time, with the majority of emissions occurring in short bursts from particular (and difficult-to-predict) locations (Ambus and Christensen, 1994; Ambus and Robertson, 1998; Smith and Dobbie, 2001; Ellert and Janzen, 2008). For example, Ellert and Janzen routinely observed $N_2O$ emissions at one sampling location 12 to 23 times larger than at 5 other sampling locations within the same treatment. Similarly, $N_2O$ emissions commonly shift by orders of magnitude within short time periods (hours to days) depending on soil moisture (related to rain or irrigation), temperature, and the supply of carbon and nitrogen available to microbes (Parkin, 2008). Therefore, to estimate $N_2O$ emissions accurately, experiments need to use sufficient replication at a site to capture spatial variation and measure emissions at key times of year (e.g., immediately after applying fertilizer) when we expect emissions to be well above background. In addition, to avoid missing the highest emissions—which in the case of $N_2O$ can constitute the majority of total emissions—investigators need to characterize the daily pattern of $N_2O$ fluxes at a site (Parkin, 2008).

Policies designed to reduce $N_2O$ emissions from agricultural soil focus primarily on N management, and this approach raises questions about the appropriate spatial scale of analysis. First, considerable amounts of nitrogen leave the farm field, raising the challenge of how to account for $N_2O$ that may result from its eventual processing. N lost from croplands escapes not only as $N_2O$ but also as $N_2$ gas, $NH_3^+$, or $NO_3^-$. With the exception of $N_2$ gas, all other forms of N remain subject to microbial conversion to $N_2O$, so all management options that influence the rate of N loss from the field could influence rates of $N_2O$ production. Associated accounting methods have so far remained rudimentary, usually using global Intergovernmental Panel on Climate Change default values and failing to differentiate between, for example, fields that use cover crops to trap N during the fallow season and fields that do not.

Available data suggest that the scale of analysis could have dramatic influence on estimates of potential reductions in $N_2O$ loss. Blesh and Drinkwater (2013) compared N budgets for farm fields in the Mississippi River Basin that use common fertilizer management practices with farms that use legumes as the primary N source. Fields using conventional fertilizer typically have an N surplus of 30 to 38 kg $ha^{-1}$ $yr^{-1}$ N while fields managed with a legume N source have an N surplus of less than 4 kg $ha^{-1}$ $yr^{-1}$ N. Similarly, Tonitto et al. (2006) conducted a meta-analysis of nitrate losses from fields managed with and without cover crops and found that using cover crops reduced nitrate loss 40 to 70%. Surplus available N can be lost as $N_2O$ or lost as $NO_3^-$ and denitrified elsewhere in the watershed. Therefore differences in the N saturation of agroecosystem management directly impacts net $N_2O$ loss potential. Models that consider only in-field $N_2O$ emissions and do not represent the difference in N leakiness across management practices would be expected to underestimate total $N_2O$ losses, and consequently to underestimate the benefits of farm management systems that minimize N surplus.

In addition, different sources of fertilizer N have radically different embodied emissions, raising questions about whether "upstream" $CO_2$ emissions should be coupled to policies designed to reduce $N_2O$ emissions. As discussed in the section entitled "Reducing Agricultural GHG Emissions through Management of Agricultural Systems," on the "upstream" end of N management, synthetic N fertilizer production is an energy-intensive process driven by fossil fuel, with associated $CO_2$ emissions. Substituting legume-derived N or manure for synthetic N should drive reduced fertilizer production and hence reduced $CO_2$ emissions. An LCA approach would suggest granting growers credit for these $CO_2$ savings, but in practice who (if anyone) receives this credit is a matter of debate. Broad policy decisions to treat the fertilizer industry as a regulated entity have precluded carbon offset protocols from awarding credit for these emission savings to growers, despite the fact that growers' decisions drive (or diminish) fertilizer production.

Direct measurements of $N_2O$ emissions have shown that nitrification inhibitors (NIs) and slow-release (SL) fertilizers reduce emissions from crop fields by approximately 38 and 21%, respectively, compared with conventional fertilizers that do not incorporate these technologies (Akiyama et al., 2010; Halvorson et al., 2013). The NI estimate represents a meta-analysis of 113 data sets from 35 field studies, and the data represent one of the few cases where we have a substantial number of studies and a consistent response of $N_2O$ flux to a management practice despite the high inherent variability in $N_2O$ emissions from soil. The estimate for SL is based on extensive research on irrigated corn fields at the USDA experiment station in Fort Collins, Colorado. These experiments found strong interactions between fertilizer technology and tillage; SL fertilizer led to reduced $N_2O$ emissions in no-till and strip-till treatments, but not under conventional tillage. These findings underscore the limitations of data that show how $N_2O$ emissions respond to management. Even these unusually robust data represent a narrow geography, and the relationships are restricted to a relatively specific cropping system and management context. In USDA's newly developed method for estimating $N_2O$ emissions from agricultural fields in the United States, the Agency used both of these data sets (NI, SL) as a basis for adjusting estimates of $N_2O$ emissions based on growers' use of NIs (Biggar et al., 2013).

Scientists and managers have devoted considerable attention to questions of scale for managing nitrogen losses from farm fields and their impacts on water quality (Lund et al., 2013, Giri et al., 2012; Jha et al., 2010). David et al. (2010) modeled riverine nitrate losses from counties in the Mississippi River Basin and identified 259 counties with predicted nitrate N yields >7.5 kg $ha^{-1}$ N and 1135 counties with predicted nitrate yields <3 kg $ha^{-1}$ N. They recommended using improved practices in the 259 counties with high nitrate loss rates. Resource conservation professionals have applied targeting at finer scales. For example, riparian zones have long been considered hot spots of nitrogen removal from shallow groundwater and are therefore targeted locations in the landscape to control nitrogen pollution of streams (Gurwick et al., 2008, and references therein; Vidon et al., 2010). Analogous targeting approaches could be applied to reduce $N_2O$ losses from agricultural landscapes. Applying existing knowledge about $N_2O$ losses could help focus the location of management practices, and identification of relevant knowledge gaps could help focus future research.

## Consequences of Science Base for Protocol Development

If offsets are required to be permanent, progress developing offset protocols appears to reflect the relative strengths and limitations of soil carbon storage, forest carbon storage, and $N_2O$ emission reductions. Although no systematic or in-depth study of this area has yet been published to our knowledge, we offer the following observations. In 2011, CAR embarked on development of two agricultural carbon offset protocols: one (termed "cropland management") intended to grant credits for soil carbon storage and the other ("nitrogen management") to grant credits for reduced $N_2O$ emissions. In the first case the Reserve's decision not to adopt a cropland management project protocol (CMPP) highlights the multi-faceted challenge of SOC management. While Eagle et al. (2012) identify agricultural practices that support increased SOC, defining management practices that would qualify for offsets is not straightforward (Climate Action Reserve, 2011a). For instance, the suite of available long-term data sets imposes some limitations on model validation, particularly when the full soil profile is considered. Technical reports considered by the Reserve during the CMPP process identified cover cropping (Gershenson et al., 2011) and grassland management (Diaz et al., 2012) as promising techniques for increasing SOC storage. However, defining these management practices in a policy setting is challenging. Additionality was a major concern for protocol implementation, especially since no-till, reduced-till, and cover crops are supported by other conservation policy initiatives (Climate Action Reserve, 2011b). Permanence was also a major concern (Climate Action Reserve, 2011a). The workgroup was concerned about the ability of land owners to guarantee adherence to a given management practice for a long duration, given both the short timescale nature of agricultural management and the frequency of agricultural land renting.

In the second case ("nitrogen management"), CAR's process for developing a protocol began with a relatively extensive list of practices that might qualify for carbon credits but ended with only one practice—reduced N fertilizer application rates on maize in 12 midwestern states (Climate Action Reserve, 2011c, 2012a). The narrow scope of the N management protocol reflects the guidance Reserve staff received from a panel of expert scientists as well as technical experts on an advisory multistakeholder workgroup, all of whom argued that evaluations of other practices were not yet possible owing to a dearth of necessary data—$N_2O$ emissions collected with appropriate spatial and temporal intensity from side-by-side experiments differing only in management (Climate Action Reserve, 2011d, 2011e). The breadth of the initial proposal indicates the kind of expectations and hopes that have been generated in the policy arena. Subsequent steps taken by the Reserve to refine this protocol, and to establish mechanisms for including additional practices as additional data become available, further demonstrate the extent to which data availability and advice from the research community have shaped this protocol.

## Summary

Here we highlight the most important lessons learned for each topic in this chapter as bullet points under the following three themes: (i) model structure and scale of application, (ii) data limitations and nonlinearity, and (iii) applications to GHG offset projects.

### Model Structure and Scale of Application

Even though models may perform well across large regions over long time periods, there are almost always major uncertainties in estimates for specific locations and management practices.

- Process models share basic approaches to representing ecosystem processes (e.g., controls on decomposition; which environmental conditions affect C and N cycling). However, apparently subtle differences in the way models represent ecosystem complexity lead to variation in strengths and weaknesses of different models.
- Our ability to test the mechanisms that underlie model estimates of net ecosystem GHG fluxes is constrained by the availability of data for specific processes such as shoot, root, or heterotrophic soil respiration. Therefore, even if model estimates of an integrative variable like soil carbon storage or net $CO_2$ flux reasonably match results of several experiments, there remains substantial uncertainty about whether the model will provide accurate estimates under different environmental conditions.
- Both process and empirical models tend to smooth out variation in GHG fluxes and estimate average values over time and space. For example, models of $N_2O$ flux perform much better at the annual scale than at the daily scale. Similarly, at very large spatial and temporal scales such as the conterminous

United States over a decade, estimates of total GHG flux from complex process models and simple empirical models often converge. The utility of modeling average values for integrative variables like $N_2O$ flux is limited if management recommendations require estimates that apply to specific localities.

- Even at large scales, where values of ecosystem-scale variables converge among models, the extent to which different processes contribute to these values can vary substantially among models. For example, $CO_2$ fluxes can originate mainly from litter decomposition in one model and more from woody debris and soil respiration in another. The source of GHG emissions is relevant for designing best management practices; therefore, discrepancies in model outcomes regarding GHG source affects the suitability of models for broad application to management questions.
- Because different models identify different underlying mechanisms for overall GHG flux, modelers and policymakers should exercise caution in concluding that any particular model accurately estimates GHG emissions.

## Outstanding Questions

- Why do models converge and diverge?
- How important are the differences in representation of processes that control integrative variables?
- How much do different model representations of factors such as plant growth and soil water content account for variation in model performance under different environmental conditions?
- What should the next set of model comparisons look like to maximize understanding of underlying ecosystem processes as well as to accelerate our ability to apply models for management at more local and regional scales?

## Data Limitations and Nonlinearity

Models are only as good as the data used to develop, parameterize, and validate them, and relevant data are scarce for many agricultural and forest ecosystems. Data scarcity is a major limitation for modeling all but the most common agricultural systems in the industrialized world and for nearly all ecosystems in less-developed countries. Thus, it is important for model developers and model users to understand and communicate clearly about which model estimates are most robust and which require further validation for specific processes, pools, management practices, soil types, etc.

- Agriculture and forest ecosystems are responding to shifting environmental and ecological conditions, and models often do not reflect such changing conditions. Changes include species invasions, changes in fire regimes, use of new management practices, and climate change. Models often omit such processes because starting with a simpler set of conditions is tractable and because data for many topics is very limited. Simple models are quite useful for research because they teach us how much variability in ecosystem processes can be explained by a small number of key factors. But when processes or practices not included in a model are influential, estimates of GHG emissions are unlikely to be correct.

- Despite increases in data availability, few data sets include concurrent measurements of key processes, pools, and GHGs (e.g., plant growth, $CO_2$, $N_2O$, $NO_3^-$, SOC, soil nutrient status). For example, there are hundreds of data sets that report $N_2O$ measurements in agricultural systems, but few are adequate to support robust understanding of how management practices influence $N_2O$ emissions. To assess the influence of different management practices on $N_2O$ emissions, we need side-by-side trials of different management systems. We also need measurements of fluxes from these side-by-side comparisons over entire years, not just growing seasons; over multiple years; and with sufficient temporal and spatial frequency to capture both hot spots and hot moments.
- Estimating $N_2O$ flux is inherently challenging as flux often exhibits a nonlinear response to controlling factors. Even when a model performs well after calibration at an annual scale, the seasonal pattern of flux may not be accurate, suggesting that simpler empirical models operating at an annual scale may be just as useful for some purposes.

## Outstanding Questions

- Are data needs for model development and testing (to further our understanding of ecosystem dynamics) different from needs driven by existing policy and management frameworks? If so, what would be an appropriate process for balancing investments in data collection and model research to meet both objectives?

## Applications to GHG Offset Projects

The application of models to estimate GHG emission reductions raises several issues that might not have been immediately apparent when agricultural offsets were first proposed for inclusion in environmental markets. These include (i) trade-offs between the cost of implementing an offset project—and hence the cost of carbon credits—and the precision and accuracy of estimates of GHG fluxes; (ii) the availability of tools to estimate agriculture-related GHG emissions at different scales; (iii) the likelihood of different kinds of crop fields being enrolled in carbon offsets programs; and (iv) the level of understanding about models within different policy institutions.

- Models have increasingly been used to quantify GHG budgets associated with carbon offset projects, supplementing measurements to estimate carbon in aboveground biomass of trees, or predicting $N_2O$ fluxes that are too difficult and expensive to measure directly. But even well-validated models are unlikely to make accurate predictions across all combinations of soils, climate, species, and management practices, especially with limited site-specific measurements. Collecting sufficient site-specific data to accurately parameterize a model can be very costly and may still miss factors or practices that are not included in the model.

- During the last decade, there have been many applications and modifications of some models to enhance their use for GHG offset projects. However, there will always be a tension between the need for models to evolve and improve and the need for tractable and stable rules for assessing compliance with GHG offset projects.
- We can be confident that there is huge variation in the potential for $N_2O$ emission reductions and soil carbon storage among agricultural fields. This variation has implications that merit more attention than they sometimes receive in the agricultural GHG offset arena.
- The understanding that ecosystem models do not produce robust estimates of GHG budgets at individual sites has led to a focus on average performance over large spatial scales. This approach assumes that the fields over which a protocol is applied are representative of the population of fields for which the model has been validated. If, for example, fields enrolled under a GHG offset protocol tend to have high baseline soil C or low baseline $N_2O$ emissions compared with most fields, then they are not "average," and the average GHG benefit estimated by the model may not be realized. Institutions that apply an "average" approach should provide some assurance that the distribution of enrolled fields represents average conditions.
- In the context of water quality protection, recognition of variability in N loss from crop fields has resulted in efforts to target areas of the landscape with the highest losses and the greatest potential for reducing N pollution. This targeting approach has not yet been applied in the context of agricultural GHG offsets. Such targeting could potentially enable much greater returns on investment in agricultural GHG emission reductions, underscoring the benefits of developing the research base that enables site-specific estimation.
- For carbon offset projects, the desired accuracy often exceeds scientific capacity. For example, the effect of reduced tillage in sequestering carbon was modeled for many years, but then questions were raised about whether estimates were at all accurate when deeper soil horizons were included (Baker et al., 2007). Additionally, only a single agricultural practice was included in the CAR N management protocol- reduced N fertilizer rate on corn in 12 states because scientists said tools were not adequate to quantify the benefit of other management practices.
- Institutions like CAR that are charged with policy development have demonstrated an impressive capacity to assimilate relevant technical knowledge and craft policies that reflect an appreciation for the abilities and limitations of models. In the case of CAR's nitrogen management protocol, the

perspective of CAR staff shifted radically over the course of several months, illustrating the initial absence of a clear appreciation for many of the issues raised in this paper and also the Reserve's ability to learn quickly.

- When a model has become visible to the policy and management community, the interaction with the modeling community can be critical in establishing realistic expectations about the model's capability and limitations.
- Regardless of model accuracy, there remain issues with permanence for carbon sequestration in soil and vegetation.

## Outstanding Questions

- Once a model has been published and accepted for certain uses in certain places, does it acquire legitimacy and credibility for application beyond those situations or processes for which it has been validated?
- How many policy institutions related to agriculture and greenhouse gas emissions have a sufficiently strong understanding of model capabilities and limitations to instill confidence that these institutions will use them appropriately and encourage investment in model development in the most useful ways?

## Acknowledgments

The authors thank Steve Del Grosso, Laj Ahuja, and Bill Parton for their rigorous coordination of this book. We sincerely thank Steve Del Grosso and anonymous reviewers for very thorough and thoughtful comments on an earlier draft of this manuscript. We also thank Robert Grant, Changshang Li, and Liwing Ma for their review of Table 1. Partial funding support for P. Woodbury was provided by USDA-NIFA Project "New Tools and Incentives for Carbon, Nitrogen, and Greenhouse Gas Accounting and Management in Corn Cropping Systems" (2011-67003-30205) and USDA-NIFA (MacIntyre-Stennis) Project "Managing Forests for Mitigating Climate Change: Carbon Sequestration, Bioenergy, and Wood Products" (1000999).

# References

Abdalla, M., S. Kumar, M. Jones, J. Burke, and M. Williams. 2011. Testing DNDC model for simulating soil respiration and assessing the effects of climate change on the $CO_2$ gas flux from Irish agriculture. Global Planet. Change. 78(3–4):106–115.

Abdalla, M., K. Rueangritsarakul, M. Jones, B. Osborne, M. Helmy, B. Roth, J. Burke, P. Nolan, P. Smith, and M. Williams. 2012. How effective is reduced tillage–cover crop management in reducing $N_2O$ fluxes from arable crop soils? Water Air Soil Pollut. 223(8):5155–5174.

Addiscott, T.M. 1996. Measuring and modeling nitrogen leaching: Parallel problems. Plant Soil 181:1–6. doi:10.1007/BF00011284

Adler, P.R., S.J. Del Grosso, and W.J. Parton. 2007. Life-cycle assessment of net greenhouse-gas flux for bioenergy cropping systems. Ecol. Appl. 17(3):675–691.

Ågren, G., and E. Bosatta. 2002. Reconciling differences in predictions of temperature response of soil organic matter. Soil Biol. Biochem. 34:129–132. doi:10.1016/S0038-0717(01)00156-0

Ahmed, R., P. Siqueira, S. Hensley, and K. Bergen. 2013. Uncertainty of forest biomass estimates in north temperate forests due to allometry: Implications for remote sensing. Remote Sens. 5:3007–3036. doi:10.3390/rs5063007

Ahuja, L.R., L. Ma, and T.R. Green. 2010. Effective soil properties of heterogeneous areas for modeling infiltration and redistribution. Soil Sci. Soc. Am. J. 74(5):1469–1482.

Ahuja, L.R., K.W. Rojas, J.D. Hanson, M.J. Shaffer, and L. Ma. 2000. The Root Zone Water Quality Model. Water Resour. Publ., Highlands Ranch, CO.

Ainsworth, E.A., and S.P. Long. 2004. What have we learned from 15 years of free-air $CO_2$ enrichment (FACE)? A meta-analytic review of the responses of photosynthesis, canopy properties and plant production to rising $CO_2$. New Phytol. 165:351–372. doi:10.1111/j.1469-8137.2004.01224.x

Akiyama, H., X. Yan, and K. Yagi. 2010. Evaluation of effectiveness of enhanced-efficiency fertilizers as mitigation options for $N_2O$ and NO emissions from agricultural soils: Meta-analysis. Global Change Biol. 16(6):1837–1846. doi:10.1111/j.1365-2486.2009.02031.x

Ambus, P., and S. Christensen. 1994. Measurement of $N_2O$ emission from a fertilized grassland: An analysis of spatial variability. J. Geophys. Res. 99:16549–16555. doi:10.1029/94JD00267

Ambus, P., and G.P. Robertson. 1998. Automated near-continuous measurement of carbon dioxide and nitrous oxide fluxes from soil. Soil Sci. Soc. Am. J. 62:394–400. doi:10.2136/sssaj1998.03615995006200020015x

American Carbon Registry. 2012. Methodology for quantifying nitrous oxide ($N_2O$) emissions reductions through reduced use of nitrogen fertilizer on agricultural crops. American Carbon Registry, Arlington, VA. http://americancarbonregistry.org/carbon-accounting/standards-methodologies/emissions-reductions-through-reduced-use-of-nitrogen-fertilizer-on-agricultural-crops/msu-epri-methodology-acr-v1-0_final.pdf (accessed 2 Aug. 2015).

Anomaa Senaviratne, G.M.M.M., R.P. Udawatta, C. Baffaut, and S.H. Anderson. 2013. Agricultural Policy Environmental eXtender simulation of three adjacent row-crop watersheds in the claypan region. J. Environ. Qual. 42:726–736.

Araújo, M.B., and M. New. 2007. Ensemble forecasting of species distribution. Trends Ecol. Evol. 22:42–47. doi:10.1016/j.tree.2006.09.010

Baker, J.M., T.E. Ochsner, R.T. Venterea, and T.J. Griffis. 2007. Tillage and soil carbon sequestration—What do we really know? Agric. Ecosyst. Environ. 118:1–5. doi:10.1016/j.agee.2006.05.014

Beale, C.M., and J.J. Lennon. 2012. Incorporating uncertainty in predictive species distribution modelling. Philos. Trans. R. Soc. London, Ser. B 367:247–258. doi:10.1098/rstb.2011.0178

Beheydt, D., P. Boeckx, S. Sleutel, C.S. Li, and O. Van Cleemput. 2007. Validation of DNDC for 22 long-term $N_2O$ field emission measurements. Atmos. Environ. 41:6196–6211. doi:10.1016/j.atmosenv.2007.04.003

Biggar, S., D. Man, K. Moffroid, D. Pape, M. Riley-Gilbert, R. Steele, and V. Thompson. 2013. Greenhouse gas mitigation options and costs for agricultural land and animal production within the United States. USDA, Washington, DC.

Blesh, J., and L.E. Drinkwater. 2013. The impact of nitrogen source and crop rotation on nitrogen mass balances in the Mississippi river basin. Ecol. Appl. 23:1017–1035. doi:10.1890/12-0132.1

Bouwman, A.F. 1996. Direct emission of nitrous oxide from agricultural soils. Nutr. Cycl. Agroecosyst. 46:53–70. doi:10.1007/BF00210224

Bradley, B.A., D.M. Blumenthal, D.S. Wilcove, and L.H. Ziska. 2010. Predicting plant invasions in an era of global change. Trends Ecol. Evol. 25:310–318. doi:10.1016/j.tree.2009.12.003

Britz, W., and A. Leip. 2009. Development of marginal emission factors for N losses from agricultural soils with the DNDC–CAPRI meta-model. Agric. Ecosyst. Environ. 133(3–4):267–279.

Broekhoff, D. 2008. Creating jobs with climate solutions: How agriculture and forestry can help lower costs in a low-carbon economy. Senate Subcommittee on Rural Revitalization, Conservation, Forestry, and Credit of the US Senate Committee on Agriculture, Nutrition, and Forestry, Washington, DC.

Cairns, M.A., S. Brown, E.H. Helmer, and G.A. Baumgardner. 1997. Root biomass allocation in the world's upland forests. Oecologia 111:1–11. doi:10.1007/s004420050201

California Air Resources Board. 2011a. Minutes of the Research Screening Committee, December 2, 2011, Sacramento, CA. http://www.arb.ca.gov/research/rsc/12-2-11/dec11book.pdf (accessed 31 July 2015).

California Air Resources Board. 2011b. Compliance Offset Protocol, livestock projects. Capturing and destroying methane from manure management systems. October 20, 2011. http://www.arb.ca.gov/regact/2010/capandtrade10/compoffprotfin.pdf (accessed 31 July 2015).

California Air Resources Board. 2013. Compliance Offset Program. California Air Resources Board, Sacramento, CA. http://www.arb.ca.gov/cc/capandtrade/offsets/offsets.htm (accessed 31 July 2015).

Campbell, C.A., K.E. Bowren, M. Schnitzer, R.P. Zentner, and L. Townley-Smith. 1991. Effect of crop rotations and fertilization on soil organic matter and some biochemical properties of a thick Black Chernozem. Can. J. Soil Sci. 71:377–387. doi:10.4141/cjss91-036

Cavero, J., R. Barros, F. Sellam, S. Topcu, D. Isidoro, T. Hartani, A. Lounis, H. Ibrikci, M. Cetin, J.R. Williams, and R. Aragüés. 2012. APEX simulation of best irrigation and N management strategies for off-site N pollution control in three Mediterranean irrigated watersheds. Agric. Water Manage. 103:88–99.

Chamberlain, J.F., S.A. Miller, and J.R. Frederick. 2011. Using DAYCENT to quantify on-farm GHG emissions and N dynamics of land use conversion to N-managed switchgrass in the Southern U.S. Agric., Ecosyst. Environ. 141(3–4):332–341.

Chang, C., C.M. Cho, and H.H. Janzen. 1998. Nitrous oxide emission from long-term manured soils. Soil Sci. Soc. Am. J. 62:677–682. doi:10.2136/sssaj1998.03615995006200030019x

Chianese, D.S., C.A. Rotz, and T.L. Richard. 2009. Simulation of methane emissions from dairy farms to assess greenhouse gas reduction strategies. Trans. ASABE 52(4):1325–1335. doi:10.13031/2013.27782

Chirinda, N., D. Kracher, M. Lægdsmand, J.R. Porter, J.E. Olesen, B.M. Petersen, J. Doltra, R. Kiese, and K. Butterbach-Bahl. 2011. Simulating soil $N_2O$ emissions and heterotrophic $CO_2$ respiration in arable systems using FASSET and MoBiLE-DNDC. Plant Soil. 343:139–160.

Climate Action Reserve. 2011a. Methodology synthesis to supplement cropland management protocol development. March 10, 2011. Climate Action Reserve, Los Angeles, CA.

Climate Action Reserve. 2011b. Summary of background research on additionality for cropland management project protocol (draft). April 2011. Climate Action Reserve, Los Angeles, CA.

Climate Action Reserve. 2011c. Nutrient Management Project Protocol Workgroup Draft Version 1.0. Climate Action Reserve, Los Angeles, CA. http://www.climateactionreserve.org/how/protocols/nitrogen-management/dev/ (accessed 31 July 2015).

Climate Action Reserve. 2011d. Nitrogen Management Project Protocol Science Advisory Committee Meeting Report. September 7, 2011. Climate Action Reserve, Los Angeles, CA.

Climate Action Reserve 2011e. Multistakeholder workgroup meetings. Climate Action Reserve, Los Angeles, CA. http://www.climateactionreserve.org/how/protocols/nitrogen-management/dev/ (accessed 31 July 2015).

Climate Action Reserve 2012a. Nutrient Management Project Protocol Version 1.0. Climate Action Reserve, Los Angeles, CA. http://www.climateactionreserve.org/how/protocols/nitrogen-management/ (accessed 31 July 2015).

Climate Action Reserve. 2012b. Forest Project Protocol. Climate Action Reserve, Los Angeles, CA. http://www.climateactionreserve.org/how/protocols/forest/ (accessed 31 July 2015).

Climate Action Reserve. 2012c. Quantification guidance for use with forest carbon projects. Climate Action Reserve, Los Angeles. http://www.climateactionreserve.org/wp-content/uploads/2012/08/FPP_Quantification_Guidance.pdf (accessed 31 July 2015).

Climate Action Reserve. 2013. Protocols. Climate Action Reserve, Los Angeles, CA. http://www.climateactionreserve.org/how/protocols/ (accessed 31 July 2015).

Corbin, J.D., and C.M. D'Antonio. 2012. Gone but not forgotten? Invasive plants' legacies on community and ecosystem properties. Invasive Plant Sci. Manage. 5:117–124. doi:10.1614/IPSM-D-11-00005.1

Creyts, J., A. Derkach, S. Nyquist, K. Ostrowski, and J. Stephenson. 2007. Reducing US greenhouse gas emissions: How much at what cost? McKinsey & Company, New York.

Crookston, N.L., and G.E. Dixon. 2005. The forest vegetation simulator: A review of its structure, content, and applications. Comput. Electron. Agric. 49:60–80. doi:10.1016/j.compag.2005.02.003

Crow, S.E., C.W. Swanston, K. Lajtha, J.R. Brooks, and H. Keirstead. 2007. Density fractionation of forest soils: Methodological questions and interpretation of incubation results and turnover time in an ecosystem context. Biogeochemistry 85:69–90.

Dalgaard, T., N. Hutchings, U. Dragosits, J.E. Olesen, C. Kjeldsen, J.L. Drouet, and P. Cellier. 2011. Effects of farm heterogeneity and methods for upscaling on modelled nitrogen losses in agricultural landscapes. Environ. Pollut. 159:3183–3192.

David, M.B., S.J. Del Grosso, X. Hu, E.P. Marshall, G.F. McIsaac, W.J. Parton, C. Tonitto, and M.A. Youssef. 2009. Modeling denitrification in a tile-drained, corn and soybean agroecosystem of Illinois, USA. Biogeochemistry 93(1–2):7–30.

David, M.B., L.E. Drinkwater, and G.F. McIsaac. 2010. Sources of nitrate yields in the Mississippi River basin. J. Environ. Qual. 39:1657–1667. doi:10.2134/jeq2010.0115

Davidson, E.A., and I.L. Ackerman. 1993. Changes in soil carbon inventories following cultivation of previously untilled soils. Biogeochemistry 20:161–193. doi:10.1007/BF00000786

Davidson, E.A., and I.A. Janssens. 2006. Temperature sensitivity of soil carbon decomposition and feedbacks to climate change. Nature 440:165–173. doi:10.1038/nature04514

Davidson, E.A., I.A. Janssens, and Y. Luo. 2006. On the variability of respiration in terrestrial ecosystems: Moving beyond $Q_{10}$. Global Change Biol. 12:154–164. doi:10.1111/j.1365-2486.2005.01065.x

Davis, S.C., W.J. Parton, F.G. Dohleman, C.M. Smith, S. Del Grosso, A.D. Kent, and E.H. DeLucia. 2009. Comparative biogeochemical cycles of bioenergy crops reveal nitrogen-fixation and low greenhouse gas emissions in a *Miscanthus × giganteus* agro-ecosystem. Ecosystems 13(1):144–156.

Davis, S.C., W.J. Parton, S.J. Del Grosso, C. Keough, E. Marx, P.R. Adler, and E.H DeLucia. 2012. Impact of second-generation biofuel agriculture on greenhouse-gas emissions in the corn-growing regions of the US. Front. Ecol. Environ. 10(2):69–74.

Deb, S.K., M.K. Shukla, and J.G. Mexal. 2012. Simulating deep percolation in flood-irrigated mature pecan orchards with RZWQM2. Trans. ASABE 55(6):2089–2100.

Dechow, R., and A. Freibauer. 2011. Assessment of German nitrous oxide emissions using empirical modelling approaches. Nutr. Cycl. Agroecosyst. 91:235–254. doi:10.1007/s10705-011-9458-9

De Gryze, S. J. Lee, S. Ogle, K. Paustian, and J. Six. 2011. Assessing the potential for greenhouse gas mitigation in intensively managed annual cropping systems at the regional scale. Agric. Ecosyst. Environ. 144:150–158.

De Gryze, S., A. Wolf, S.R. Kaffka, J. Mitchell, D.E. Rolston, S.R. Temple, J. Lee, and J. Six. 2010. Simulating greenhouse gas budgets of four California cropping systems under conventional and alternative management. Ecol. Appl. 20(7):1805–1819.

Delgado, J.A., S.J. Del Grosso, and S.M. Ogle. 2009. N isotopic crop residue cycling studies and modeling suggest that IPCC methodologies to assess residue contributions to $N_2O$–N emissions should be reevaluated. Nutr. Cycling Agroecosyst. 86(3):383–390.

Del Grosso, S.J., and M.A. Cavigelli. 2012. Climate stabilization wedges revisited: Can agricultural production and greenhouse-gas reduction goals be accomplished? Front. Ecol. Environ 10:571–578. doi:10.1890/120058

Del Grosso, S.J., A.D. Halvorson, and W.J. Parton. 2008. Testing DAYCENT model simulations of corn yields and nitrous oxide emissions in irrigated tillage systems in Colorado. J. Environ. Qual. 37(4):1383–1389.

Del Grosso, S.J., A.R. Mosier, W.J. Parton, and D.S. Ojima. 2005. DAYCENT model analysis of past and contemporary soil $N_2O$ and net greenhouse gas flux for major crops in the USA. Soil Tillage Res. 83(1):9–24.

Del Grosso, S.J., S.M. Ogle, W.J. Parton, and F.J. Breidt. 2009a. Estimating uncertainty in $N_2O$ emissions from U.S. cropland soils. Global Biogeochem. Cycles. 24:GB1009.

Del Grosso, S.J., D.S. Ojima, W.J. Parton, E. Stehfest, M. Heistemann, B. DeAngelo, and S. Rose. 2009b. Global scale DAYCENT model analysis of greenhouse gas emissions and mitigation strategies for cropped soils. Global Planet. Change. 67(1–2):44–50.

Del Grosso, S.J., W.J. Parton, A.R. Mosier, D.S. Ojima, A.E. Kulmala, and S. Phongpan. 2000. General model for $N_2O$ and $N_2$ gas emissions from soils due to dentrification. Global Biogeochem. Cycles 14:1045–1060. doi:10.1029/1999GB001225

Del Grosso, S.J., W.J. Parton, A.R. Mosier, M.K. Walsh, D.S. Ojima, and P.E. Thornton. 2006. DAYCENT national-scale simulations of nitrous oxide emissions from cropped soils in the United States. J. Environ. Qual. 35(4):1451–1460.

Deng, J., Z. Zhou, B. Zhu, X. Zheng, C. Li, X. Wang, and Z. Jian. 2011. Modeling nitrogen loading in a small watershed in southwest China using a DNDC model with hydrological enhancements. Biogeosciences 8:2999–3009.

Deng, J., B. Zhu, Z.X. Zhou, X.H. Zheng, C.S. Li, T. Wang, and J.L. Tang. 2011. Modeling nitrogen loadings from agricultural soils in southwest China with modified DNDC. J. Geophys. Res. 116:G02020. doi:10.1029/2010JG001609

Desjardins, R.L., E. Pattey, W.N. Smith, D. Worth, B. Grant, R. Srinivasan, J.I. MacPherson, and M. Mauder. 2010. Multiscale estimates of $N_2O$ emissions from agricultural lands. Agric. For. Meteorol. 150:817–824. doi:10.1016/j.agrformet.2009.09.001

Desjardins, R.L., W. Smith, B. Grant, C. Campbell, and R. Riznek. 2005. Management strategies to sequester carbon in agricultural soils and to mitigate greenhouse gas emissions. Clim. Change. 70(1–2):283–297.

Diaz, D., B. Rashford, S. DeGryze, S. Zakreski, R. Dell, and M. Niles. 2012. Evaluation of avoided conversion and cropland conversion to grassland as potential carbon offset project types. An issue paper prepared for the Climate Action Reserve. The Climate Trust, Portland, OR.

Dietiker, D., N. Buchmann, and W. Eugster. 2010. Testing the ability of the DNDC model to predict $CO_2$ and water vapour fluxes of a Swiss cropland site. Agric. Ecosyst. Environ. 139(3):396–401.

Dietze, M.C., R. Vargas, A.D. Richardson, P.C. Stoy, A.G. Barr, R.S. Anderson, M.A. Arain, I.T. Baker, T.A. Black, J.M. Chen, P. Ciais, L.B. Flanagan, C.M. Gough, R.F. Grant, D. Hollinger, R.C. Izaurralde, C.J. Kucharik, P. Lafleur, S.G. Liu, E. Lokupitiya, Y.Q. Luo, J.W. Munger, C.H. Peng, B. Poulter, D.T. Price, D.M. Ricciuto, W.J. Riley, A.K. Sahoo, K. Schaefer, A.E. Suyker, H.Q. Tian, C. Tonitto, H. Verbeeck, S.B. Verma, W.F. Wang, and E.S. Weng. 2011. Characterizing the performance of ecosystem models across time scales: A spectral analysis of the North American Carbon Program site-level synthesis. J. Geophys. Res. 116:G04029. doi:10.1029/2011JG001661

Dijkstra, F.A., and W. Cheng. 2007. Moisture modulates rhizosphere effects on C decomposition in two different soil types. Soil Biol. Biochem. 39:2264–2274. doi:10.1016/j.soilbio.2007.03.026

Domke, G.M., C.W. Woodall, B.F. Walters, and J.E. Smith. 2013. From models to measurements: Comparing downed dead wood carbon stock estimates in the US forest inventory. PLoS ONE 8(3):e59949. doi:10.1371/journal.pone.0059949

Donigian, A.S., Jr., J.T.O. Barnwell, I.R.B. Jackson, A.S. Patwardhan, K.B. Weinrich, A.L. Rowell, R.V. Chinnaswamy, and C.V. Cole. 1994. Assessment of alternative management practices and policies affecting soil carbon in agroecosystems of the Central United States. USEPA, Athens, GA.

Dukes, J.S., J. Pontius, D. Orwig, J.R. Garnas, V.L. Rodgers, N. Brazee, B. Cooke, K.A. Theoharides, E.E. Stange, R. Harrington, J. Ehrenfeld, J. Gurevitch, M. Lerdau, K. Stinson, R. Wick, and M. Ayres. 2009. Responses of insect pests, pathogens, and invasive plant species to climate change in the forests of northeastern North America: What can we predict? Can. J. Res. 39:231–248. doi:10.1139/X08-171

Eagle, A.J., L.R. Henry, L.P. Olander, K. Haugen-Kozyra, N. Millar, and G.P. Robertson. 2012. Greenhouse gas mitigation potential of agricultural land management in the United States. A synthesis of the literature. Nicholas Institute for Environmental Policy Solutions, Durham, NC.

Eagle, A.J., and L.P. Olander. 2012. Greenhouse gas mitigation with agricultural land management activities in the United States—A side-by-side comparison of biophysical potential. Adv. Agron. 115:.79–179.

Ellert, B.H., and H.H. Janzen. 2008. Nitrous oxide, carbon dioxide and methane emissions from irrigated cropping systems as influenced by legumes, manure and fertilizer. Can. J. Soil Sci. 88:207–217. doi:10.4141/CJSS06036

Erhagen, B., M. Öquist, T. Sparrman, M. Haei, U. Ilstedt, M. Hedenström, J. Schleucher, and M.B. Nilsson. 2013. Temperature response of litter and soil organic matter decomposition is determined by chemical composition of organic material. Global Change Biol. 19:3858–3871. doi:10.1111/gcb.12342

Fahey, T.J., J.J. Battles, and G.F. Wilson. 1998. Responses of early successional northern hardwood forests to changes in nutrient availability. Ecol. Monogr. 68:183–212. doi:10.1890/0012-9615(1998)068[0183:ROESNH]2.0.CO;2

Fahey, T.J., P.B. Woodbury, J.J. Battles, C.L. Goodale, S.P. Hamburg, S.V. Ollinger, and C.W. Woodall. 2010. Forest carbon storage: Ecology, management, and policy. Front. Ecol. Environ 8:245–252. doi:10.1890/080169

Fang, Q., L. Ma, Q. Yu, L.R. Ahuja, R.W. Malone, and G. Hoogenboom. 2010. Irrigation strategies to improve the water use efficiency of wheat–maize double cropping systems in North China Plain. Agric. Water Manage. 97(8):1165–1174.

Fang, Q.X., R.W. Malone, L. Ma, D.B. Jaynes, K.R. Thorp, T.R. Green, and L.R. Ahuja. 2012. Modeling the effects of controlled drainage, N rate and weather on nitrate loss to subsurface drainage. Agric. Water Manage. 103:150–161.

Farahbakhshazad, N., D.L. Dinnes, C. Li, D.B. Jaynes, and W. Salas. 2008. Modeling biogeochemical impacts of alternative management practices for a row-crop field in Iowa. Agric. Ecosyst. Environ. 123(1–3):30–48.

Foereid, B., P.H. Bellamy, A. Holden, and G.J.D. Kirk. 2012. On the initialization of soil carbon models and its effects on model predictions for England and Wales. Eur. J. Soil Sci. 63(1):32–41.

Fontaine, S., S. Barot, P. Barre, N. Bdioui, B. Mary, and C. Rumpel. 2007. Stability of organic carbon in deep soil layers controlled by fresh carbon supply. Nature 450:277–281. doi:10.1038/nature06275

Food and Agriculture Organization. 2013. Climate-smart agriculture sourcebook. FAO, Rome. http://www.fao.org/docrep/018/i3325e/i3325e00.htm

Freibauer, A., and M. Kaltschmitt. 2003. Controls and models for estimating direct nitrous oxide emissions from temperate and sub-boreal agricultural mineral soils in Europe. Biogeochemistry 63:93–115. doi:10.1023/A:1023398108860

Frelich, L.E., C.M. Hale, S. Scheu, A.R. Holdsworth, L. Heneghan, P.J. Bohlen, and P.B. Reich. 2006. Earthworm invasion into previously earthworm-free temperate and boreal forests. Biol. Invasions 8:1235–1245. doi:10.1007/s10530-006-9019-3

Freschet, G.T., W.K. Cornwell, D.A. Wardle, T.G. Elumeeva, W. Liu, B.G. Jackson, V.G. Onipchenko, N.A. Soudzilovskaia, J. Tao, and J.H.C. Cornelissen. 2013. Linking litter decomposition of above- and below-ground organs to plant–soil feedbacks worldwide. J. Ecol. 101:943–952. doi:10.1111/1365-2745.12092

Fu, S., and W. Cheng. 2004. Defoliation affects rhizosphere respiration and rhizosphere priming effect on decomposition of soil organic matter under a sunflower species: *Helianthus annuus*. Plant Soil 263:345–352. doi:10.1023/B:PLSO.0000047745.30929.32

Fumoto, T., K. Kobayashi, C. Li, K. Yagi, and T. Hasegawa. 2008. Revising a process-based biogeochemistry model (DNDC) to simulate methane emission from rice paddy fields under various residue management and fertilizer regimes. Global Change Biol. 14:382–402. doi:10.1111/j.1365-2486.2007.01475.x

Gabriel, C.E., and L. Kellman. 2014. Investigating the role of moisture as an environmental constraint in the decomposition of shallow and deep mineral soil organic matter of a temperate coniferous soil. Soil Biol. Biochem. 68:373–384.

Gao, J., K.D. Thelen, and X. Hao. 2013. Life cycle analysis of corn harvest strategies for bioethanol production. Agron. J. 105(3):705–712.

Gassman, P.W., J.R. Williams, X. Wang, A. Saleh, E. Osei, L.M. Hauck, R.C. Izaurralde, and J.D. Flowers. 2010. The Agricultural Policy/Environmental eXtender (APEX) Model: An emerging tool for landscape and watershed environmental analysis. Trans. ASABE 53:711–740. doi:10.13031/2013.30078

Gaudinski, J.B., S.E. Trumbore, E.A. Davidson, and S.H. Zheng. 2000. Soil carbon cycling in a temperate forest: Radiocarbon-based estimates of residence times, sequestration rates and partitioning of fluxes. Biogeochemistry 51:33–69. doi:10.1023/A:1006301010014

Gershenson, A., J. Barsimantov, and D. Mulvaney. 2011. Climate Action Reserve background paper on greenhouse gas assessment boundaries and leakage for the cropland management project protocol. EcoShift Consult., Santa Cruz, CA.

Giardina, C.P., and M.G. Ryan. 2000. Evidence that decomposition rates of organic carbon in mineral soil do not vary with temperature. Nature 404:858–861. doi:10.1038/35009076

Gilbertson, T., and O. Reyes. 2009. Carbon trading: How it works and why it fails. Crit. Curr. Occas. Pap. Ser. No. 7. Dag Hammarskjold Foundation, Uppsala, Sweden.

Giltrap, D.L., C. Li, and S. Saggar. 2010. DNDC: A process-based model of greenhouse gas fluxes from agricultural soils. Agric. Ecosyst. Environ. 136:292–300. doi:10.1016/j.agee.2009.06.014

Giltrap, D.L., S. Saggar, J. Singh, M. Harvey, A. McMillan, and J. Laubach. 2011. Field-scale verification of nitrous oxide emission reduction with DCD in dairy-grazed pasture using measurements and modeling. Soil Res. 49:696–702.

Giri, S., A.P. Nejadhashemi, and S.A. Woznicki. 2012. Evaluation of targeting methods for implementation of best management practices in the Saginaw river watershed. J. Environ. Manage. 103(0):24–40. doi:10.1016/j.jenvman.2012.02.033

Gopalakrishnan, G., M. Cristina Negri, and W. Salas. 2012. Modeling biogeochemical impacts of bioenergy buffers with perennial grasses for a row-crop field in Illinois. Bioenergy 4(6):739–750.

Gramig, B.M., C.J. Reeling, R. Cibin, and I. Chaubey. 2013. Environmental and economic trade-offs in a watershed when using corn stover for bioenergy. Environ. Sci. Technol. 47(4):1784–1791.

Grant, R.F. 1994a. Salinity, water use and yield of maize: Testing of the mathematical model ecosys. Plant Soil 172(2):309–322.

Grant, R.F. 1994b. Dynamics of energy, water, carbon and nitrogen in agricultural ecosystems: Simulation and experimental validation. Ecol. Modell. 81:169–181.

Grant, R.F. 1995. Mathematical modelling of nitrous oxide evolution during nitrification. Soil Biol. Biochem. 27(9):1117–1125.

Grant, R.F. 1997. Changes in soil organic matter under different tillage and rotation: Mathematical modeling in ecosys. Soil Sci. Soc. Am. J. 61(4):1159–1175.

Grant, R.F. 1998. Simulation of methanogenesis in the mathematical model ecosys. Soil Biol. Biochem. 30(7):883–896.

Grant, R.F., M. Amrani, D.J. Heaney, R. Wright, and M. Zhang. 2004a. Mathematical modeling of phosphorus losses from land application of hog and cattle manure. J. Environ. Qual. 33(1):210–231.

Grant, R.F., T.J. Arkebauer, A. Dobermann, K.G. Hubbard, T.T. Schimelfenig, A.E. Suyker, S.B. Verma, and D.T. Walters. 2007. Net biome productivity of irrigated and rainfed maize–soybean rotations: Modeling vs. measurements. Agron. J. 99:1404–1423.

Grant, R.F., D.D. Baldocchi, and S. Ma. 2012a. Ecological controls on net ecosystem productivity of a seasonally dry annual grassland under current and future climates: Modelling with ecosys. Agric. For. Meteorol. 152:189–200.

Grant, R.F., A.R. Desai, and B.N. Sulman. 2012b. Modelling contrasting responses of wetland productivity to changes in water table depth. Biogeosciences 9:4215–4231.

Grant, R.F., and L.B. Flanagan. 2007. Modeling stomatal and nonstomatal effects of water deficits on $CO_2$ fixation in a semiarid grassland. J. Geophys. Res. 112(G3):G03011.

Grant, R.F., and D.J. Heaney. 1997. Inorganic phosphorus transformation and transport in soils: Mathematical modeling in ecosys. Soil Sci. Soc. Am. J. 61(3):752–764.

Grant, R.F., N.G. Juma, J.A. Robertson, R.C. Izaurralde, and W.B. McGill. 2001a. Long-term changes in soil carbon under different fertilizer, manure, and rotation: Testing the mathematical model ecosys with data from the Breton plots. Soil Sci. Soc. Am. J. 65:205–214. doi:10.2136/sssaj2001.651205x

Grant, R.F., B.A. Kimball, T.J. Brooks, G.W. Wall, P.J. Pinter, D.J. Hunsaker, F.J. Adamsen, R.L. Lamorte, S.W. Leavitt, T.L. Thompson, and A.D. Matthias. 2001b. Modeling interactions among carbon dioxide, nitrogen, and climate on energy exchange of wheat in a free air carbon dioxide experiment. Agron. J. 93(3):638–649.

Grant, R.F., B.A. Kimball, M.M. Conley, J.W. White, G.W. Wall, and M.J. Ottman. 2011. Controlled warming effects on wheat growth and yield: Field measurements and modeling. Agron. J. 103(6):1742–1754.

Grant, R.F., B.A. Kimball, P.J. Pinter, G.W. Wall, R.L. Garcia, R.L. La Morte, and D.J. Hunsaker. 1995. Carbon dioxide effects on crop energy balance: Testing ecosys with a free-air $CO_2$ enrichment (FACE) experiment. Agron. J. 87(3):446–457.

Grant, R.F., B.A. Kimball, G.W. Wall, J.M. Triggs, T.J. Brooks, P.J. Pinter, M.M. Conley, M.J. Ottman, R.L. Lamorte, S.W. Leavitt, T.L. Thompson, and A.D. Matthias. 2004b. Modeling elevated carbon dioxide effects on water relations, water use, and growth of irrigated sorghum. Agron. J. 96(6):1693–1705.

Grant, R.F., and E. Pattey. 1999. Mathematical modeling of nitrous oxide emissions from an agricultural field during spring thaw. Global Biogeochem. Cycles 13(2):679–694.

Grant, R.F., and E. Pattey. 2003. Modelling variability in $N_2O$ emissions from fertilized agricultural fields. Soil Biol. Biochem. 35(2):225–243.

Grant, R.F., and E. Pattey. 2008. Temperature sensitivity of $N_2O$ emissions from fertilized agricultural soils: Mathematical modeling in ecosys. Global Biogeochem. Cycles 22(4):GB4019.

Grant, R.F., E. Pattey, T.W. Goddard, L.M. Kryzanowski, and H. Puurveen. 2006. Modeling the effects of fertilizer application rate on nitrous oxide emissions. Soil Sci. Soc. Am. J. 70(1):235–248.

Grant, R.F., and P. Rochette. 1994. Soil microbial respiration at different water potentials and temperatures: Theory and mathematical modeling. Soil Sci. Soc. Am. J. 58(6):1681–1690.

Grant, R.F., G.W. Wall, B.A. Kimball, K.F.A. Frumau, P.J. Pinter Jr., D.J. Hunsaker, and R.L. Lamorte. 1999. Crop water relations under different $CO_2$ and irrigation: Testing of ecosys with the free air $CO_2$ enrichment (FACE) experiment. Agric. For. Meterol. 95(1):27–51.

Groffman, P.M., M.A. Altabet, J.K. Bohlke, K. Butterbach-Bahl, M.B. David, M.K. Firestone, A.E. Giblin, T.M. Kana, L.P. Nielsen, and M.A. Voytek. 2006. Methods for measuring denitrification: Diverse approaches to a difficult problem. Ecol. Appl. 16:2091–2122.

Groffman, P.M., R. Brumme, K. Butterbach-Bahl, K.E. Dobbie, A.R. Mosier, D. Ojima, H. Papen, W.J. Parton, K.A. Smith, and C. Wagner-Riddle. 2000. Evaluating annual nitrous oxide fluxes at the ecosystem scale. Global Biogeochem. Cycles 14:1061–1070. doi:10.1029/1999GB001227

Groffman, P.M., K. Butterbach-Bahl, R.W. Fulweiler, A.J. Gold, J.L. Morse, E.K. Stander, C. Tague, C. Tonitto, and P. Vidon. 2009. Challenges to incorporating spatially and temporally explicit phenomena (hotspots and hot moments) in denitrification models. Biogeochemistry 93:49–77. doi:10.1007/s10533-008-9277-5

Guggenberger, G., and K. Kaiser. 2003. Dissolved organic matter in soil: Challenging the paradigm of sorptive preservation. Geoderma 113:293–310. doi:10.1016/S0016-7061(02)00366-X

Guntiñas, M.E., M.C. Leirós, C. Trasar-Cepda, and F. Gil-Sotres. 2012. Effects of moisture and temperature on net soil nitrogen mineralization: A laboratory study. Eur. J. Soil Biol. 48:73–80. doi:10.1016/j.ejsobi.2011.07.015

Guo, M., C. Li, J. Nigel B. Bell, and R.J. Murphy. 2012. Influence of agro-ecosystem modeling approach on the greenhouse gas profiles of wheat-derived biopolymer products. Environ. Sci. Technol. 46(1):320–330.

Gurwick, N.P., D.C. McCorkle, P.M. Groffman, A.J. Gold, D.Q. Kellogg, and P. Seitz-Rundlett. 2008. Mineralization of ancient carbon in the subsurface of riparian forests. J. Geophys. Res. 113:G02021. doi:10.1029/2007JG000482

Halvorson, A.D., C.S. Snyder, A.D. Blaylock, and S.J. Del Grosso. 2013. Enhanced-efficiency nitrogen fertilizers: Potential role in nitrous oxide emission mitigation. Agron. J. 106:715–722. doi:10.2134/agronj2013.0081

Hamburg, S.P. 2000. Simple rules for measuring changes in ecosystem carbon in forestry-offset projects. Mitig. Adapt. Strategies Global Change 5:25–37. doi:10.1023/A:1009692114618

Hartman, M.D., E.R. Merchant, W.J. Parton, M.P. Gutmann, S.M. Lutz, and S.A. Williams. 2011. Impact of historical land-use changes on greenhouse gas exchange in the U.S. Great Plains, 1883–2003. Ecol. Appl. 21(4):1105–1119.

Hastings, A.F., M. Wattenbach, W. Eugster, C. Li, N. Buchmann, and P. Smith. 2010. Uncertainty propagation in soil greenhouse gas emission models: An experiment using the DNDC model and at the Oensingen cropland site. Agric. Ecosyst. Environ. 136(1–2):97–110.

Heath, L.S. 2013. Using FIA data to inform United States forest carbon national-level accounting needs: 1990–2010. In: A.E. Camp, L.C. Irland, and C.J.W. Carroll, editors, Long-term silvicultural & ecological Studies: Results for science and management. Vol. 2. Yale University School of Forestry & Environmental Studies, Global Institute of Sustainable Forestry, New Haven, CT.

Hoover, C.M., and S.A. Rebain. 2011. Forest carbon estimation using the forest vegetation simulator: Seven things you need to know. Gen. Tech. Rep. NRS-77. USDA Forest Service, Northern Research Station, Newtown Square, PA.

Houghton, R.A., and J.L. Hackler. 2000. Changes in terrestrial carbon storage in the United States. 1: The roles of agriculture and forestry. Global Ecol. Biogeogr. 9:125–144. doi:10.1046/j.1365-2699.2000.00166.x

Houghton, R.A., J.L. Hackler, and K.T. Lawrence. 2000. Changes in terrestrial carbon storage in the United States. 2: The role of fire and fire management. Global Ecol. Biogeogr. 9:145–170. doi:10.1046/j.1365-2699.2000.00164.x

Hurtt, G.C., S.W. Pacala, P.R. Moorcroft, J. Caspersen, E. Shevliakova, R.A. Houghton, and B. Moore. 2002. Projecting the future of the US carbon sink. Proc. Natl. Acad. Sci. USA 99:1389–1394. doi:10.1073/pnas.012249999

Ibáñez, I., J.M. Diez, L.P. Miller, J.D. Olden, C.J.B. Sorte, D.M. Blumenthal, B.A. Bradley, C.M. D'Antonio, J.S. Dukes, R.I. Early, E.D. Grosholz, and J.J. Lawler. 2014. Integrated assessment of biological invastions. Ecol. Appl. 24:25–37. doi:10.1890/13-0776.1

Intergovernmental Panel on Climate Change. 2000a. Special report on land use, land use change, and forestry, summary for policymakers. Cambridge Univ. Press., Cambridge, UK.

Intergovernmental Panel on Climate Change. 2000b. Good practice guidance and uncertainty management in national greenhouse gas inventories. Inst. for Global Environ. Strategies, Hayama, Japan.

Intergovernmental Panel on Climate Change. 2003. Good practice guidance for land use, land-use change and forestry. Inst. for Global Environ. Strategies, Hayama, Japan.

Intergovernmental Panel on Climate Change. 2006. $N_2O$ emissions from managed soils, and $CO_2$ emissions from lime and urea application. In: 2006 IPCC guidelines for national greenhouse gas inventories. Inst. for Global Environ. Strategies, Hayama, Japan. p. 11.1–11.54.

Islam, A., L.R. Ahuja, L.A. Garcia, L. Ma, A.S. Saseendran, and T.J. Trout. 2012. Modeling the impacts of climate change on irrigated corn production in the Central Great Plains. Agric. Water Manage. 110:94–108.

Jagadeesh Babu, Y., C. Li, S. Frolking, D.R. Nayak, and T.K. Adhya. 2006. Field validation of DNDC model for methane and nitrous oxide emissions from rice-based production systems of India. Nutr. Cycl. Agroecosyst. 74(2):157–174.

Jenkins, J.C., D.C. Chojnacky, L.S. Heath, and R.A. Birdsey. 2003. National-scale biomass estimators for United States tree species. For. Sci. 49:12–35.

Jenkinson, D.S., and J.H. Rayner. 1977. The turnover of soil organic matter in some of the Rothamsted classical experiments. Soil Sci. 123:298–305. doi:10.1097/00010694-197705000-00005

Jha, M.K., K.E. Schilling, P.W. Gassman, and C.F. Wolter. 2010. Targeting land-use change for nitrate-nitrogen load reductions in an agricultural watershed. J. Soil Water Conserv. 65:342–352. doi:10.2489/jswc.65.6.342

Johnson, J.M.F., D. Archer, and N. Barbour. 2010. Greenhouse gas emission from contrasting management scenarios in the northern Corn Belt. Soil Sci. Soc. Am. J. 74:396–406. doi:10.2136/sssaj2009.0008

Jones, J.W., G. Hoogenboom, C.H. Porter, K.J. Boote, W.D. Batchelor, L.A. Hunt, P.W. Wilkens, U. Singh, A.J. Gijsman, and J.T. Ritchie. 2003. The DSSAT cropping system model. Eur. J. Agron. 18:235–265. doi:10.1016/S1161-0301(02)00107-7

Kang, X., Y. Hao, C. Li, X. Cui, J. Wang, Y. Rui, H. Niu, and Y. Wang. 2011. Modeling impacts of climate change on carbon dynamics in a steppe ecosystem in Inner Mongolia, China. J. Soils Sediments. 11(4):562–576.

Kariyapperuma, K.A., C. Wagner-Riddle, A.C. Furon, and C. Li. 2011. Assessing spring thaw nitrous oxide fluxes simulated by the DNDC model for agricultural soils. Soil Sci. Soc. Am. J. 75(2):678–690.

Karlen, D.L., and C.A. Cambardella. 1996. Conservation strategies for improving soil quality and organic matter storage. In: Structure and organic matter storage in agricultural soils. Advances in soil science. CRC Press, New York. p. 395–420.

Karnosky, D.F., D.R. Zak, K.S. Pregitzer, C.S. Awmack, J.G. Bockheim, R.E. Dickson, G.R. Hendrey, G.E. Host, J.S. King, B.J. Kopper, E.L. Kruger, M.E. Kubiske, R.L. Lindroth, W.J. Mattson, E.P. McDonald, A. Noormets, E. Oksanen, W.F.J. Parsons, K.E. Percy, G.K. Podila, D.E. Riemenschneider, P. Sharma, R. Thakur, A. Sôber, J. Sôber, W.S. Jones, S. Anttonen, E. Vapaavuori, B. Mankovska, W. Heilman, and J.G. Isebrands. 2003. Tropospheric $O_3$ moderates responses of temperate hardwood forests to elevated $CO_2$: A synthesis of molecular to ecosystem results from the Aspen FACE project. Funct. Ecol. 17:289–304. doi:10.1046/j.1365-2435.2003.00733.x

Katayanagi, N., Y. Furukawa, T. Fumoto, and Y. Hosen. 2012.Validation of the DNDC-Rice model by using $CH_4$ and $N_2O$ flux data from rice cultivated in pots under alternate wetting and drying irrigation management. Soil Sci. Plant Nutr. 58(3):360–372.

Kätterer, T., M. Reichstein, A. Andrén, and A. Lomander. 1998. Temperature dependence of organic matter decomposition: A critical review using literature data analyzed with different models. Biol. Fertil. Soils 27:258–262. doi:10.1007/s003740050430

Kell, D.B. 2011. Breeding crop plants with deep roots: Their role in sustainable carbon, nutrient and water sequestration. Ann. Bot. 108(3): 407–418.

Kim, D.-G., G. Hernandez-Ramirez, and D. Giltrap. 2013. Linear and nonlinear dependency of direct nitrous oxide emissions on fertilizer nitrogen input: A meta-analysis. Agric. Ecosyst. Environ. 168:53–65. doi:10.1016/j.agee.2012.02.021

Kim, D.G., R. Vargas, B. Bond-Lamberty, and M.R. Turetsky. 2012. Effects of soil rewetting and thawing on soil gas fluxes: A review of current literature and suggestions for future research. Biogeosciences 9:2459–2483. doi:10.5194/bg-9-2459-2012

Kim, S., B.E. Dale, and R. Jenkins. 2009. Life cycle assessment of corn grain and corn stover in the United States. Int. J. Life Cycle Assess. 14:160–174. doi:10.1007/s11367-008-0054-4

Kleber, M., and M.G. Johnson. 2010. Advances in understanding the molecular structure of soil organic matter: Implications for interactions in the environment. Adv. Agron. 106:77–142.

Ko, J., L. Ahuja, B. Kimball, S. Anapalli, L. Ma, T.R. Green, A.C. Ruane, G.W. Wall, P. Pinter, and D.A. Bader. 2010. Simulation of free air $CO_2$ enriched wheat growth and interactions with water, nitrogen, and temperature. Agric. For. Meteorol. 150(10):1331–1346.

Ko, J., L.R. Ahuja, S.A. Saseendran, T.R. Green, L. Ma, D.C. Nielsen, and C.L. Walthall. 2012. Climate change impacts on dryland cropping systems in the Central Great Plains, USA. Clim. Change. 111(2):445–472.

Kramer, M.G., J. Sanderman, O.A. Chadwick, J. Chorover, and P.M. Vitousek. 2012. Long-term carbon storage through retention of dissolved aromatic acids by reactive particles in soil. Global Change Biol. 18:2594–2605. doi:10.1111/j.1365-2486.2012.02681.x

Kravchenko, A.N., and G.P. Robertson. 2011. Whole-profile soil carbon stocks: The danger of assuming too much from analyses of too little. Soil Sci. Soc. Am. J. 75:235–240. doi:10.2136/sssaj2010.0076

Kröbel, R., W. Smith, B. Grant, R. Desjardins, C. Campbell, N. Tremblay, C. Li, R. Zentner, and B. McConkey. 2011. Development and evaluation of a new Canadian spring wheat sub-model for DNDC. Can. J. Soil Sci. 91(4):503–520.

Kumar, S., R.P. Udawatta, S.H. Anderson, and A. Mudgal. 2011. APEX model simulation of runoff and sediment losses for grazed pasture watersheds with agroforestry buffers. Agroforestry Syst. 83:51–62.

La Porta, N., P. Capretti, I.M. Thomsen, R. Kasanen, A.M. Hietala, and K. von Weissenberg. 2008. Forest pathogens with higher damage potential due to climate change in Europe. Can. J. Plant Pathol. 30:177–195. doi:10.1080/07060661.2008.10540534

Lee, J., S. De Gryze, and J. Six. 2011. Effect of climate change on field crop production in California's Central Valley. Clim. Change 109(Suppl 1):S355–S353.

Lee, J., G. Pedroso, B.A. Linquist, D. Putnam, C. van Kessel, and J. Six. 2012. Simulating switchgrass biomass production across ecoregions using the DAYCENT model. GCB Bioenergy, 4:521–533.

Leip, A., M. Busto, and W. Winiwarter. 2011. Developing spatially stratified $N_2O$ emission factors for Europe. Environ. Pollut. 159:3223–3232. doi:10.1016/j.envpol.2010.11.024

Lewandrowski, J., M. Peters, C. Jones, R. House, M. Sperow, M. Eve, and K. Paustian. 2004. Economics of sequestering carbon in the US agricultural sector. Tech. Bull. 1909. USDA Economic Research Service, Washington, DC.

Li, C., N. Farahbakhshazad, D.B. Jaynes, D.L. Dinnes, W. Salas, and D. McLaughlin. 2006. Modeling nitrate leaching with a biogeochemical model modified based on observations in a row-crop field in Iowa. Ecol. Modell. 196(1–2):116–130.

Li, C.S., S. Frolking, and T.A. Frolking. 1992. A model of nitrous-oxide evolution from soil driven by rainfall events.1. Model structure and sensitivity. J. Geophys. Res. 97:9759–9776. doi:10.1029/92JD00509

Li, C.S., W. Salas, R.H. Zhang, C. Krauter, A. Rotz, and F. Mitloehner. 2012. Manure-DNDC: A biogeochemical process model for quantifying greenhouse gas and ammonia emissions from livestock manure systems. Nutr. Cycl. Agroecosyst. 93:163–200. doi:10.1007/s10705-012-9507-z

Li, D., G. Lanigan, and J. Humphreys. 2011. Measured and simulated nitrous oxide emissions from ryegrass- and ryegrass/white clover-based grasslands in a moist temperate climate. PLoS One 6(10):e26176.

Li, H., J. Qiu, L. Wang, H. Tang, C. Li, and E. Van Ranst. 2010. Modelling impacts of alternative farming management practices on greenhouse gas emissions from a winter wheat–maize rotation system in China. Agric. Ecosyst. Environ. 135(1–2):24–33.

Li, H., J. Qiu, L. Wang, M. Xu, Z. Liu, and W. Wang. 2012. Estimates of $N_2O$ emissions and mitigation potential from a spring maize field based on DNDC model. J. Integr. Agric. 11(12):2067–2078.

Li, P., H.J. Bohnert, and R. Grene. 2007. All about FACE—Plants in a high-[$CO_2$] world. Trends Plant Sci. 12:87–89. doi:10.1016/j.tplants.2007.01.003

Li, T., R.F. Grant, and L.B. Flanagan. 2004. Climate impact on net ecosystem productivity of a semi-arid natural grassland: Modeling and measurement. Agric. For. Meteorol. 126(1–2):99–116.

Li, Y., D. Chen, Y. Zhang, R. Edis, and H. Ding. 2005. Comparison of three modeling approaches for simulating denitrification and nitrous oxide emissions from loam-textured arable soils. Global Biogeochem. Cycles. 19(3):GB3002.

Liu, Y., Z. Yu, J. Chen, F. Zhang, R. Doluschitz, and J.C. Axmacher. 2006. Changes of soil organic carbon in an intensively cultivated agricultural region: A denitrification–decomposition (DNDC) modelling approach. Sci. Total Environ. 372(1):203–214.

Ludwig, B., A. Bergstermann, E. Priesack, and H. Flessa. 2011a. Modelling of crop yields and $N_2O$ emissions from silty arable soils with differing tillage in two long-term experiments. Soil Tillage Res. 112(2):114–121.

Ludwig, B., N. Jäger, E. Priesack, and H. Flessa. 2011b. Application of the DNDC model to predict $N_2O$ emissions from sandy arable soils with differing fertilization in a long-term experiment. J. Plant Nutr. Soil Sci. 174(3):350–358.

Lukac, M., A. Lagomarsino, M.C. Moscatelli, P. DeAngelis, M.F. Cotrufo, and D.L. Godbold. 2009. Forest soil carbon cycle under elevated $CO_2$—A case of increased throughput? Forestry 82:75–86. doi:10.1093/forestry/cpn041

Lund, D., J.D. Atwood, J.K. Bagdon, J. Benson, J. Goebel, K. Ingram, M.-V.V. Johnson, R.L. Kellogg, J. Lemunyon, L. Norfleet, E. Steglich, J. Arnold, M. White, T. Gerik, S. Chinnasamy, M.D. Luzio, A. King, D.C. Moffitt, K. Narayanan, T. Pitts, X.S. Wang, J. Williams, and S. Plotkin. 2013. Assessment of the effects of conservation practices on cultivated cropland in the lower Mississippi River Basin. Conservation Effects Assessment Project (CEAP), USDA NRCS, Washington, DC.

Ma, L., L.R. Ahuja, S.A. Saseendran, R.W. Malone, T.R. Green, B.T. Nolan, P.N.S. Bartling, G.N. Flerchinger, K.J. Boote, and G. Hoogenboom. 2011. A protocol for parameterization and calibration of RZWQM2 in field research. In: L.R. Ahuja and L. Ma, editors, Advances in agricultural systems modeling. Ser. 2. ASA, CSSA, and SSSA, Madison, WI. p. 1–64.

Ma, L., G. Hoogenboom, S.A. Saseendran, P.N.S. Bartling, L.R. Ahuja, and T.R. Green. 2009. Effects of estimating soil hydraulic properties and root growth factor on soil water balance and crop production. Agron. J. 101(3):572–583.

Ma, L., M.J. Shaffer, and L.R. Ahuja. 2001. Application of RZWQM for soil nitrogen management. In: M.J. Shaffer, L. Ma, and S. Hansen, editors, Modeling carbon and nitrogen dynamics for soil management. Lewis Publ., Boca Raton, FL. p. 265–301.

Ma, L., T.J. Trout, L.R. Ahuja, W.C. Bausch, S.A. Saseendran, R.W. Malone, and D.C. Nielsen. 2012. Calibrating RZWQM2 model for maize responses to deficit irrigation. Agric. Water Manage. 103:140–149.

Magdoff, F., and H. Van Es. 2010. Building soils for better crops. Sustainable Agric. Res. and Educ., Brentwood, MD.

Matthews, H.D., and K. Caldeira. 2008. Stabilizing climate requires near-zero emissions. Geophys. Res. Lett. 35(4):L04705. doi:10.1029/2007GL032388

Meki, M.N., J.P. Marcos, J.D. Atwood, L.M. Norfleet, E.M. Steglich, J.R. Williams, and T.J. Gerik. 2011. Effects of site-specific factors on corn stover removal thresholds and subsequent environmental impacts in the Upper Mississippi River Basin. J. Soil Water Conserv. 66(6):386–399.

Merrill, S.D., D.L. Tanaka, and J.D. Hanson. 2002. Root length growth of eight crop species in Haplustoll soils. Soil Sci. Soc. Am. J. 66:913–923. doi:10.2136/sssaj2002.9130

Metivier, K.A., E. Pattey, and R.F. Grant. 2009. Using the ecosys mathematical model to simulate temporal variability of nitrous oxide emissions from a fertilized agricultural soil. Soil Biol. Biochem. 41(12):2370–2386.

Mikutta, R., M. Kleber, M.S. Torn, and R. Jahn. 2006. Stabilization of soil organic matter: Association with minerals or chemical recalcitrance? Biogeochemistry 77:25–56. doi:10.1007/s10533-005-0712-6

Millar, N., G.P. Robertson, P.R. Grace, R.J. Gehl, and J.P. Hoben. 2010. Nitrogen fertilizer management for nitrous oxide ($N_2O$) mitigation in intensive corn (Maize) production: An emissions reduction protocol for US Midwest agriculture. Mitig. Adapt. Strategies Global Change 15:185–204. doi:10.1007/s11027-010-9212-7

Miner, G.L., N.C. Hansen, D. Inman, L.A. Sherrod, and G.A. Peterson. 2013. Constraints of no-till dryland agroecosystems as bioenergy production systems. Agron. J. 105(2):364–376.

Miner, R., and J. Perez-Garcia. 2007. The greenhouse gas and carbon profile of the global forest products industry. For. Prod. J. 57:80–90.

Mosier, A.R., A.D. Halvorson, C.A. Reule, and X.J.J. Liu. 2006. Net global warming potential and greenhouse gas intensity in irrigated cropping systems in northeastern Colorado. J. Environ. Qual. 35:1584–1598. doi:10.2134/jeq2005.0232

Mudgal, A., C. Baffaut, S.H. Anderson, E.J. Sadler, N.R. Kitchen, K.A. Sudduth, and R.N. Lerch. 2012. Using the Agricultural Policy/Environmental eXtender to develop and validate physically based indices for the delineation of critical management areas. J. Soil Water Conserv. 67(4):284–299.

Mudgal, A., C. Baffaut, S.H. Anderson, E.J. Sadler, and A.L. Thompson. 2010. APEX model assessment of variable landscapes on runoff and dissolved herbicides. Trans. ASABE 53(4):1047–1058.

Murray, B.C., B.A. McCarl, and H.-C. Lee. 2002. Estimating leakage from forest carbon sequestration programs. Work. Pap. 02_06. Research Triangle Inst., Research Triangle Park, NC.

Nakagawa, Y., Y. Chin, T. Shiono, T. Miyamoto, K. Kameyama, and Y. Shinogi. 2008. Evaluating the validity and sensitivity of the DNDC model for shimajiri dark red soil. Japan Agric. Res. Q. 42(3):163–172.

Neufeldt, H., M. Schäfer, E. Angenendt, C. Li, M. Kaltschmitt, and J. Zeddies. 2006. Disaggregated greenhouse gas emission inventories from agriculture via a coupled economic-ecosystem model. Agric. Ecosyst. Environ. 112(2–3):233–240.

Nielsen, D.C., S.A. Saseendran, L. Ma, and L.R. Ahuja. 2012. Simulating the production potential of dryland spring canola in the Central Great Plains. Agron. J. 104(4):1182–1188.

Nolan, B.T., L.J. Puckett, L. Ma, C.T. Green, E.R. Bayless, and R.W. Malone. 2010. Predicting unsaturated zone nitrogen mass balances in agricultural settings of the United States. J. Environ. Qual. 39(3):1051–1065.

Norby, R.J., M.F. Cotrufo, P. Ineson, E.G. O'Neill, and J.G. Canadell. 2001. Elevated $CO_2$, litter chemistry, and decomposition: A synthesis. Oecologia 127:153–165. doi:10.1007/s004420000615

Norby, R.J., and Y. Luo. 2004. Evaluating ecosystem responses to rising atmospheric $CO_2$ and global warming in a multi-factor world. New Phytol. 162:281–293. doi:10.1111/j.1469-8137.2004.01047.x

Olander, L.P., K. Haugen-Kozyra, S. Del Grosso, C. Izaurralde, D. Malin, K. Paustian, and W. Salas. 2011. Using biogeochemical process models to quantify greenhouse gas mitigation from agricultural management projects. Technical Working Group on Agricultural Greenhouse Gasses (T-AGG) supplemental report. Rep. NI R 11-03. Duke Univ. Nicholas Inst. for Environ. Policy Solutions, Durham, NC.

Osei, E., B. Du, A. Bekele, L. Hauck, A. Saleh, and A. Tanter. 2008. Impacts of alternative manure application rates on Texas animal feeding operations: A macro level analysis. J. Am. Water Resour. Assoc. 44(3):562–576.

Ouyang, W., S. Qi, F. Hao, X. Wang, Y. Shan, and S. Chen. 2013. Impact of crop patterns and cultivation on carbon sequestration and global warming potential in an agricultural freeze zone. Ecol. Modell. 252:228–237.

Pan, Y.D., R.A. Birdsey, J.Y. Fang, R. Houghton, P.E. Kauppi, W.A. Kurz, O.L. Phillips, A. Shvidenko, S.L. Lewis, J.G. Canadell, P. Ciais, R.B. Jackson, S.W. Pacala, A.D. McGuire, S.L. Piao, A. Rautiainen, S. Sitch, and D. Hayes. 2011. A large and persistent carbon sink in the world's forests. Science 333:988–993.

Parkin, T.B. 1987. Soil microsites as a source of denitrification variability. Soil Sci. Soc. Am. J. 51:1194–1199.

Parkin, T.B. 2008. Effect of sampling frequency on estimates of cumulative nitrous oxide emissions. J. Environ. Qual. 37:1390–1395. doi:10.2134/jeq2007.0333

Parsinejad, M., and Y. Feng. 2003. Field evaluation and comparison of two models for simulation of soil-water dynamics. Irrig. Drain. 52(2):163–175.

Parton, W.J., M. Hartman, D. Ojima, and D. Schimel. 1998. DAYCENT and its land surface submodel: Description and testing. Global Planet. Change 19:35–48. doi:10.1016/S0921-8181(98)00040-X

Parton, W.J., E.A. Holland, S.J. Del Grosso, M.D. Hartman, R.E. Martin, A.R. Mosier, D.S. Ojima, and D.S. Schimel. 2001. Generalized model for $NO_x$ and $N_2O$ emissions from soils. J. Geophys. Res. 106:17403–17419. doi:10.1029/2001JD900101

Parton, W.J., A.R. Mosier, D.S. Ojima, D.W. Valentine, D.S. Schimel, K. Weier, and A.E. Kulmala. 1996. Generalized model for $N_2$ and $N_2O$ production from nitrification and denitrification. Global Biogeochem. Cycles 10:401–412. doi:10.1029/96GB01455

Parton, W.J., D.S. Schimel, C.V. Cole, and D.S. Ojima. 1987. Analysis of factors controlling soil organic matter levels in Great Plains grasslands. Soil Sci. Soc. Am. J. 51:1173–1179. doi:10.2136/sssaj1987.03615995005100050015x

Pathak, H., C. Li, and R. Wassmann. 2005. Greenhouse gas emissions from Indian rice fields: Calibration and upscaling using the DNDC model. Biogeosciences 2:113–123.

Pathak, H., C. Li, R. Wassmann, and J.K. Ladha. 2006. Simulation of nitrogen balance in rice–wheat systems of the Indo-Gangetic Plains. Soil Sci. Soc. Am. J. 70(5):1612–1622.

Paul, E.A., and F.E. Clark. 1996. Soil microbiology and biochemistry. 2nd ed. Acad. Press, San Diego, CA.

Paul, E.A., S.J. Morris, R.T. Conant, and A.F. Plante. 2006. Does the acid hydrolysis–incubation method measure meaningful soil organic carbon pools? Soil Sci. Soc. Am. J. 70(3):1023–1035.

Paul, E.A., K. Paustian, E.T. Elliot, and C.V. Cole, editors. 1997. Soil organic matter in temperate agroecosystems: Long-term experiments in North America. CRC Press, Boca Raton, FL.

Paustian, K., H.P. Collins, and E.A. Paul. 1997. Management controls on soil carbon. In: E.A. Paul, K. Paustian, E.T. Elliot, and C.V. Cole, editors, Soil organic matter in temperate agroecosystems: Long-term experiments in North America. CRC Press, Boca Raton, FL. p. 15–49.

Peltzer, D.A., R.B. Allen, G.M. Lovett, D. Whitehead, and D.A. Wardle. 2010. Effects of biological invasions on forest carbon sequestration. Global Change Biol. 16:732–746. doi:10.1111/j.1365-2486.2009.02038.x

Pielke, R.A., G. Marland, R.A. Betts, T.N. Chase, J.L. Eastman, J.O. Niles, D.D.S. Niyogi, and S.W. Running. 2002. The influence of land-use change and landscape dynamics on the climate system: Relevance to climate-change policy beyond the radiative effect of greenhouse gases. Philos. Trans. R. Soc. London, Ser. A 360:1705–1719.

Powers, S.E., J.C. Ascough II, R.G. Nelson, and G.R. Larocque. 2011. Modeling water and soil quality environmental impacts associated with bioenergy crop production and biomass removal in the Midwest USA. Ecol. Modell. 222(14):2430–2447.

Proctor, P., L.S. Heath, P. Van Deusen, J.H. Gove, and J.E. Smith. 2005. COLE: A web-based tool for interfacing with forest inventory data. In: R.E. McRoberts, G.A. Reams, P.C. Van Deusen, W.H. McWilliams, C.J. Cieszewski, eds. Proceedings of the fourth annual forest inventory and analysis symposium. Gen. Tech. Rep. NC-252. USDA Forest Service. North Central Research Station, St. Paul, MN. p. 167–172.

Puget, P., and L.E. Drinkwater. 2001. Short-term dynamics of root- and shoot-derived carbon from a leguminous green manure. Soil Sci. Soc. Am. J. 65:771–779. doi:10.2136/sssaj2001.653771x

Qi, Z., P.N.S. Bartling, J.D. Jabro, A.W. Lenssen, W.M. Iversen, L.R. Ahuja, L. Ma, B.L. Allen, and R.G. Evans. 2013. Simulating dryland water availability and spring wheat production in the Northern Great Plains. Agron. J. 105(1):37–50.

Qi, Z., M.J. Helmers, R.W. Malone, and K.R. Thorp. 2011. Simulating long-term impacts of winter rye cover crop on hydrologic cycling and nitrogen dynamics for a corn–soybean crop system. Trans. ASABE 54(5):1575–1588.

Qi, Z., L. Ma, M.J. Helmers, L.R. Ahuja, and R.W. Malone. 2012. Simulating nitrate-nitrogen concentration from a subsurface drainage system in response to nitrogen application rates using RZWQM2. J. Environ. Qual. 41(1):289–295.

Qin, X., H. Wang, Y. Li, Y. Li, B. McConkey, R. Lemke, C. Li, K. Brandt, Q. Gao, Y. Wan, S. Liu, Y. Liu, and C. Xu. 2013. A long-term sensitivity analysis of the denitrification and decomposition model. Environ. Modell. Software. 43:26–36

Qiu, J.N. 2013. China gets tough on carbon. Nature 498:145–146. doi:10.1038/498145a

Qiu, J., C. Li, L. Wang, H. Tang, H. Li, and E. Van Ranst. 2009a. Modeling impacts of carbon sequestration on net greenhouse gas emissions from agricultural soils in China. Global Biogeochem. Cycles 23(1):GB1007.

Qiu, J., H. Li, L. Wang, H. Tang, C. Li, and E. Van Ranst. 2011. GIS-model based estimation of nitrogen leaching from croplands of China. Nutr. Cycl. Agroecosyst. 90(2):243–252.

Qiu, J., L. Wang, H. Li, H. Tang, C. Li, and E. Van Ranst. 2009b. Modeling the impacts of soil organic carbon content of croplands on crop yields in China. Agric. Sci. China. 8(4):464–471.

Rafique, R., M. Peichl, D. Hennessy, and G. Kiely. 2011. Evaluating management effects on nitrous oxide emissions from grasslands using the process-based DeNitrification–DeComposition (DNDC) model. Atmos. Environ. 45(33):6029–6039.

Ramankutty, N., H.K. Gibbs, F. Achard, R. Defries, J.A. Foley, and R.A. Houghton. 2007. Challenges to estimating carbon emissions from tropical deforestation. Global Change Biol. 13:51–66. doi:10.1111/j.1365-2486.2006.01272.x

Rasmussen, P.E., and H.P. Collins. 1991. Long-term impacts of tillage, fertilizer, and crop residue on soil organic matter in temperate semiarid regions. Adv. Agron. 45:93–134. doi:10.1016/S0065-2113(08)60039-5

Rasmussen, P.E., K.W.T. Goulding, J.R. Brown, P.R. Grace, H.H. Janzen, and M. Körschens. 1998. Long-term agroecosystem experiments: Assessing agricultural sustainability and global change. Science 282:893–896. doi:10.1126/science.282.5390.893

Reeling, C.J., and B.M. Gramig. 2012. A novel framework for analysis of cross-media environmental effects from agricultural conservation practices. Agric. Ecosyst. Environ. 146:44–51.

Reganold, J.P., A.S. Palmer, J.C. Lockhart, and A.N. Macgregor. 1993. Soil quality and financial performance of biodynamic and conventional farms in New Zealand. Science 260:344–349. doi:10.1126/science.260.5106.344

Regional Greenhouse Gas Initiative. 2013a. $CO_2$ auctions, tracking & offsets. Regional Greenhouse Gas Initiative, New York. http://www.rggi.org/market (accessed 31 July 2015).

Regional Greenhouse Gas Initiative. 2013b. Regional greenhouse gas initiative offset protocol US forest projects. Regional Greenhouse Gas Initiative, New York. http://www.rggi.org/market/offsets/categories/forestry-afforestation (accessed 31 July 2015).

Reich, P.B., B.A. Hungate, and Y. Luo. 2006. Carbon-nitrogen interactions in terrestrial ecosystems in response to rising atmospheric carbon dioxide. Annu. Rev. Ecol. Syst. 37:611–636. doi:10.1146/annurev.ecolsys.37.091305.110039

Reichstein, M., J.A. Subke, A.C. Angeli, and J.D. Tenhunen. 2005. Does the temperature sensitivity of decomposition of soil organic matter depend upon water content, soil horizon, or incubation time? Global Change Biol. 11:1754–1767. doi:10.1111/j.1365-2486.2005.001010.x

Richter, D.D., M. Hofmockel, M.A. Callaham, D.S. Powlson, and P. Smith. 2007. Long-term soil experiments: Keys to managing Earth's rapidly changing ecosystems. Soil Sci. Soc. Am. J. 71:266–279. doi:10.2136/sssaj2006.0181

Rustad, L.E. 2006. From transient to steady-state response of ecosystems to atmospheric $CO_2$–enrichment and global climate change: Conceptual challenges and need for an integrated approach. Plant Ecol. 182:43–62.

Saleh, A., O. Gallego, E. Osei, H. Lal, C. Gross, S. McKinney, and H. Cover. 2011. Nutrient Tracking Tool—A user-friendly tool for calculating nutrient reductions for water quality trading. J. Soil Water Conserv. 66(6):400–410.

Saseendran, S.A., D.C. Nielsen, L.R. Ahuja, L. Ma, and D.J. Lyon. 2013. Simulated yield and profitability of five potential crops for intensifying the dryland wheat–fallow production system. Agric. Water Manage. 116:175–192.

Saseendran, S.A., D.C. Nielsen, D.J. Lyon, L. Ma, D.G. Felter, D.D. Baltensperger, G. Hoogenboom, and L.R. Ahuja. 2009. Modeling responses of dryland spring triticale, proso millet and foxtail millet to initial soil water in the High Plains. Field Crops Res. 113(1):48–63.

Saseendran, S.A., D.C. Nielsen, L. Ma, L.R. Ahuja, and M.F. Vigil. 2010. Simulating alternative dryland rotational cropping systems in the Central Great Plains with RZWQM2. Agron. J. 102(5):1521–1534.

Schaefer, K., C.R. Schwalm, C. Williams, M.A. Arain, A. Barr, J.M. Chen, K.J. Davis, D. Dimitrov, T.W. Hilton, D.Y. Hollinger, E. Humphreys, B. Poulter, B.M. Raczka, A.D. Richardson, A. Sahoo, P. Thornton, R. Vargas, H. Verbeeck, R. Anderson, I. Baker, T.A. Black, P. Bolstad, J.Q. Chen, P.S. Curtis, A.R. Desai, M. Dietze, D. Dragoni, C. Gough, R.F. Grant, L.H. Gu, A. Jain, C. Kucharik, B. Law, S.G. Liu, E. Lokipitiya, H.A. Margolis, R. Matamala, J.H. McCaughey, R. Monson, J.W. Munger, W. Oechel, C.H. Peng, D.T. Price, D. Ricciuto, W.J. Riley, N. Roulet, H.Q. Tian, C. Tonitto, M. Torn, E.S. Weng, and X.L. Zhou. 2012. A model-data comparison of gross primary productivity: Results from the North American Carbon Program site synthesis. J. Geophys. Res. 117:G03010. doi:10.1029/2012JG001960

Schwalm, C.R., C.A. Williams, K. Schaefer, R. Anderson, M.A. Arain, I. Baker, A. Barr, T.A. Black, G.S. Chen, J.M. Chen, P. Ciais, K.J. Davis, A. Desai, M. Dietze, D. Dragoni, M.L. Fischer, L.B. Flanagan, R. Grant, L.H. Gu, D. Hollinger, R.C. Izaurralde, C. Kucharik, P. Lafleur, B.E. Law, L.H. Li, Z.P. Li, S.G. Liu, E. Lokupitiya, Y.Q. Luo, S.Y. Ma, H. Margolis, R. Matamala, H. McCaughey, R.K. Monson, W.C. Oechel, C.H. Peng, B. Poulter, D.T. Price, D.M. Riciutto, W. Riley, A.K. Sahoo, M. Sprintsin, J.F. Sun, H.Q. Tian, C. Tonitto, H. Verbeeck, and S.B. Verma. 2010. A model-data intercomparison of $CO_2$ exchange across North America: Results from the North American Carbon Program site synthesis. J. Geophys. Res. 115:G00H05. doi:10.1029/2009JG001229

Sendich, E.D., B.E. Dale, and S. Kim. 2008. Comparison of crop and animal simulation options for integration with the biorefinery. Biomass Bioenergy. 32(12):1162–1174.

Shepherd, A., L. Wu, D. Chadwick, and R. Bol. 2011. A review of quantitative tools for assessing the diffuse pollution response to farmer adaptations and mitigation methods under climate change. Adv. Agron. 112:1–54. doi:10.1016/B978-0-12-385538-1.00001-9

Shirato, Y. 2005. Testing the suitability of the DNDC model for simulating long-term soil organic carbon dynamics in Japanese paddy soils. Soil Sci. Plant Nutr. 51(2):183–192.

Sistani, K.R., J.G. Warren, N. Lovanh, S. Higgins, and S. Shearer. 2010. Greenhouse gas emissions from swine effluent applied to soil by different methods. Soil Sci. Soc. Am. J. 74:429–435. doi:10.2136/sssaj2009.0076

Six, J., S.M. Ogle, F.J. Breidt, R.T. Conant, A.R. Mosier, and K. Paustian. 2004. The potential to mitigate global warming with no-tillage management is only realized when practised in the long term. Global Change Biol. 10:155–160. doi:10.1111/j.1529-8817.2003.00730.x

Skog, K.E. 2008. Sequestration of carbon in harvested wood products for the United States. For. Prod. J. 58:56–72.

Skog, K.E., and G.A. Nicholson. 1998. Carbon cycling through wood products: The role of wood and paper products in carbon sequestration. For. Prod. J. 48:75–83.

Skog, K.E., K. Pingoud, and J.E. Smith. 2004. A method countries can use to estimate changes in carbon stored in harvested wood products and the uncertainty of such estimates. Environ. Manage. 33(Supplement 1):S65–S73. doi:10.1007/s00267-003-9118-1

Sleutel, S., S. De Neve, D. Beheydt, C. Li, and G. Hofman. 2006. Regional simulation of long-term organic carbon stock changes in cropland soils using the DNDC model: 1. Large-scale model validation against a spatially explicit data set. Soil Use Manage. 22(4):342–351.

Smakgahn, K., T. Fumoto, and K. Yagi. 2009. Validation of revised DNDC model for methane emissions from irrigated rice fields in Thailand and sensitivity analysis of key factors. J. Geophys. Res. 114:G02017. doi:10.1029/2008JG000775

Smith, J.E., and L.S. Heath. 2002. A model of forest floor carbon mass for United States forest types. Res. Pap. NE-722. USDA Forest Service, Northeastern Research Station, Newtown Square, PA.

Smith, J.E., L.S. Heath, and M.C. Nichols. 2007. US Forest Carbon Calculation Tool: Forest-land carbon stocks and net annual stock change. Gen. Tech. Rep. NRS-13. USDA Forest Service, Northern Research Station, Newtown Square, PA.

Smith, J.E., L.S. Heath, and M.C. Nichols. 2010. US Forest Carbon Calculation Tool: Forest-land carbon stocks and net annual stock change. Revised for FIADB 4.0. Gen. Tech. Rep. NRS-13 (revised). USDA Forest Service, Northern Research Station, Newtown Square, PA.

Smith, K.A., and K.E. Dobbie. 2001. The impact of sampling frequency and sampling times on chamber-based measurements of $N_2O$ emissions from fertilized soils. Global Change Biol. 7:933–945. doi:10.1046/j.1354-1013.2001.00450.x

Smith, P., D. Martino, Z. Cai, D. Gwary, H. Janzen, P. Kumar, B. McCarl, S. Ogle, F. O'Mara, C. Rice, B. Scholes, O. Sirotenko, M. Howden, T. McAllister, G. Pan, V. Romanenkov, U. Schneider, S. Towprayoon, M. Wattenbach, and J. Smith. 2008b. Greenhouse gas mitigation in agriculture. Philos. Trans. R. Soc. Lond. B Biol. Sci. 363:789–813. doi:10.1098/rstb.2007.2184

Smith, P., J.U. Smith, D.S. Powlson, W.B. McGill, J.R.M. Arah, O.G. Chertov, K. Coleman, U. Franko, S. Frolking, D.S. Jenkinson, L.S. Jensen, R.H. Kelly, H. Klein-Gunnewiek, A.S. Komarov, C. Li, J.A.E. Molina, T. Mueller, W.J. Parton, J.H.M. Thornley, and A.P. Whitmore. 1997. A comparison of the performance of nine soil organic matter models using datasets from seven long-term experiments. Geoderma 81:153–225. doi:10.1016/S0016-7061(97)00087-6

Smith, W.N., B.B. Grant, C.A. Campbell, B.G. McConkey, R.L. Desjardins, R. Kröbel, and S.S. Malhi. 2012. Crop residue removal effects on soil carbon: Measured and inter-model comparisons. Agric. Ecosyst. Environ. 161:27–38. doi:10.1016/j.agee.2012.07.024

Smith, W.N. B.B. Grant, R.L. Desjardins, P. Rochette, C.F. Drury, and C. Li. 2008a. Evaluation of two process-based models to estimate soil $N_2O$ emissions in Eastern Canada. Can. J. Soil Sci. 88(2):251–260.

Sollins, P., C. Swanston, M. Kleber, T. Filley, M. Kramer, S. Crow, B.A. Caldwell, K. Lajtha, and R. Bowden. 2006. Organic C and N stabilization in a forest soil: Evidence from sequential density fractionation. Soil Biol. Biochem. 38:3313–3324. doi:10.1016/j.soilbio.2006.04.014

Spencer, S., S.M. Ogle, F.J. Breidt, J.J. Goebel, and K. Paustian. 2011. Designing a national soil carbon monitoring network to support climate change policy: A case example for US agricultural lands. Greenhouse Gas Meas. Manage. 1:167–178. doi:10.1080/20430779.2011.637696

Stehfest, E., and L. Bouwman. 2006. $N_2O$ and NO emission from agricultural fields and soils under natural vegetation: Summarizing available measurement data and modeling of global annual emissions. Nutr. Cycl. Agroecosyst. 74:207–228. doi:10.1007/s10705-006-9000-7

Stehfest, E., M. Heistermann, J.A. Priess, D.S. Ojima, and J. Alcamo. 2007. Simulation of global crop production with the ecosystem model DayCent. Ecol. Modell. 209(2–4):203–219.

Stevenson, S., B. Morris, N. Martin, and M. Grady. 2012. Compliance offset supply forecast for California's Cap-and-Trade Program (2013–2020). WinRock International, American Carbon Registry, Arlington, VA.

Stockmann, U., M.A. Adams, J.W. Crawford, D.J. Field, N. Henakaarchchi, M. Jenkins, B. Minasny, A.B. McBratney, V.D. de Courcelles, K. Singh, I. Wheeler, L. Abbott, D.A. Angers, J. Baldock, M. Bird, P.C. Brookes, C. Chenu, J.D. Jastrowh, R. Lal, J. Lehmann, A.G. O'Donnell, W.J. Parton, D. Whitehead, and M. Zimmermann. 2013. The knowns, known unknowns and unknowns of sequestration of soil organic carbon. Agric. Ecosyst. Environ. 164:80–99. doi:10.1016/j.agee.2012.10.001

Sun, J., L. Yang, Y. Wang, and D. Ort. 2009. FACE-ing the global change: Opportunities for improvement in photosynthetic radiation use efficiency and crop yield. Plant Sci. 177:511–522. doi:10.1016/j.plantsci.2009.08.003

Sutherst, R.W. 2014. Pest species distribution modelling: Origins and lessons from history. Biol. Invasions 16:239–256. doi:10.1007/s10530-013-0523-y

Svenning, J.C., and B. Sandel. 2013. Disequilibrium vegetation dynamics under future climate change. Am. J. Bot. 100:1266–1286. doi:10.3732/ajb.1200469

Syp, A., A. Faber, J. Kozyra, R. Borek, R. Pudelko, M. Borzęcka-Walker, and Z. Jarosz. 2011. Modeling impact of climate change and management practices on greenhouse gas emissions from arable soils. Pol. J. Environ. Stud. 20(6):1593–1602.

Tang, H., J. Qiu, E. Van Ranst, and L. Changsheng. 2006. Estimations of soil organic carbon storage in cropland of China based on DNDC model. Geoderma. 134(1–2):200–206.

Terhoeven-Urselmans, T., E. Scheller, M. Raubuch, B. Ludwig, and R. Georg Joergensen. 2009. $CO_2$ evolution and N mineralization after biogas slurry application in the field and its yield effects on spring barley. Appl. Soil Ecol. 42(3):297–302.

Thornley, J.H.M., and M.G.R. Cannell. 2001. Soil carbon storage response to temperature: An hypothesis. Ann. Bot. (Lond.) 87:591–598. doi:10.1006/anbo.2001.1372

Thuiller, W., C. Albert, M.B. Araújo, P.M. Berry, M. Cabeza, A. Guisan, T. Hickler, G.F. Midgley, J. Paterson, F.M. Schurr, M.T. Sykes, and N.E. Zimmermann. 2008. Predicting global change impacts on plant species' distributions: Future challenges. Perspect. Plant Ecol. Evol. Syst. 9:137–152. doi:10.1016/j.ppees.2007.09.004

Tonitto, C., M.B. David, and L.E. Drinkwater. 2006. Replacing bare fallows with cover crops in fertilizer-intensive cropping systems: A meta-analysis of crop yield and N dynamics. Agric. Ecosyst. Environ. 112(1):58–72. doi:10.1016/j.agee.2005.07.003

Tonitto, C., M.B. David, and L.E. Drinkwater. 2009. Modeling $N_2O$ flux from an Illinois agroecosystem using Monte Carlo sampling of field observations. Biogeochemistry 93:31–48. doi:10.1007/s10533-008-9271-y

Tonitto, C., M.B. David, L.E. Drinkwater, and C. Li. 2007a. Application of the DNDC model to tile-drained Illinois agroecosystems: Model calibration, validation, and uncertainty analysis. Nutr. Cycl. Agroecosyst. 78(1):51–63.

Tonitto, C., M.B. David, C. Li, and L.E. Drinkwater. 2007b. Application of the DNDC model to tile-drained Illinois agroecosystems: Model comparison of conventional and diversified rotations. Nutr. Cycl. Agroecosyst. 78(1):65–81.

Tonitto, C., C.L. Goodale, M.S. Weiss, S.D. Frey, and S.V. Ollinger. 2014. The effect of nitrogen addition on soil organic matter dynamics: A model analysis of the Harvard Forest Chronic Nitrogen Amendment Study and soil carbon response to anthropogenic nitrogen deposition. Biogeochemistry 117:431–454.

Torbert, H.A., S.A. Prior, H.H. Rogers, and C.W. Wood. 2000. Review of elevated atmospheric $CO_2$ effects on agro-ecosystems: Residue decomposition processes and soil C storage. Plant Soil 224:59–73. doi:10.1023/A:1004797123881

Torn, M.S., S.E. Trumbore, O.A. Chadwick, P.M. Vitousek, and D.M. Hendricks. 1997. Mineral control of soil organic carbon storage and turnover. Nature 389:170–173. doi:10.1038/38260

Torn, M.S., P.M. Vitousek, and S.E. Trumbore. 2005. The influence of nutrient availability on soil organic matter turnover estimated by incubations and radiocarbon modeling. Ecosystems 8:352–372. doi:10.1007/s10021-004-0259-8

Tuppad, P., C. Santhi, X. Wang, J.R. Williams, R. Srinivasan, and P.H. Gowda. 2010. Simulation of conservation practices using the APEX model. Appl. Eng. Agric. 26(5):779–794.

USDA. 2011. US Agriculture and Forestry Greenhouse Gas Inventory: 1990–2008. Tech. Bull. 1930. Climate Change Program Office, Office of the Chief Economist, USDA, Washington, DC.

USDA-NRCS. 2011. Conservation Innovation Grants—Greenhouse gas awardees: Fiscal year 2011. U.S. Gov. Print. Office, Washington, DC.

USDOE. 2007. Technical Guidelines Voluntary Reporting of Greenhouse Gases (1605(b)) Program. USDOE, Office of Policy and International Affairs, Washington, DC.

USEPA. 2012. Global anthropogenic non-$CO_2$ greenhouse gas emissions: 1990–2030. Revised December 2012. EPA 430-R-12-006, Office of Atmospheric Programs, Climate Change Division, US Environmental Protection Agency, Washington, DC.

USEPA. 2013. Inventory of US greenhouse gas emissions and sinks: 1990–2011. US Environmental Protection Agency, Office of Atmospheric Programs, Washington, DC.

Van Deusen, P., and L.S. Heath. 2010. COLE web applications suite. USDA, Forest Service, North Central Research Station, St. Paul, MN.

VandenBygaart, A.J., E. Bremer, B.G. McConkey, B. Ellert, H. Janzen, D. Angers, M. Carter, C. Drury, G. Lafond, and R. McKenzie. 2011. Impact of sampling depth on differences in soil carbon stocks in long-term agroecosystem experiments. Soil Sci. Soc. Am. J. 75:226–234.

VandenBygaart, A.J., E.G. Gregorich, and D.A. Angers. 2003. Influence of agricultural management on soil organic carbon: A compendium and assessment of Canadian studies. Can. J. Soil Sci. 83:363–380. doi:10.4141/S03-009

VandenBygaart, A.J., B.G. McConkey, D.A. Angers, W. Smith, H.D. Gooijer, M. Bentham, and T. Martin. 2008. Soil carbon change factors for the canadian agriculture national greenhouse gas inventory. Can. J. Soil Sci. 88:671–680. doi:10.4141/CJSS07015

Van Groenigen, J.W., G.L. Velthof, O. Oenema, K.J. Van Groenigen, and C. Van Kessel. 2010. Towards an agronomic assessment of $N_2O$ emissions: A case study for arable crops. Eur. J. Soil Sci. 61:903–913. doi:10.1111/j.1365-2389.2009.01217.x

Van Oost, K., T.A. Quine, G. Govers, S. De Gryze, J. Six, J.W. Harden, J.C. Ritchie, G.W. McCarty, G. Heckrath, C. Kosmas, J.V. Giraldez, J.R.M. da Silva, and R. Merckx. 2007. The impact of agricultural soil erosion on the global carbon cycle. Science 318:626–629. doi:10.1126/science.1145724

Vidon, P., C. Allan, D. Burns, T.P. Duval, N. Gurwick, S. Inamdar, R. Lowrance, J. Okay, D.S. Scott, and S. Sebestyen. 2010. Hot spots and hot moments in riparian zones: Potential for improved water quality management. J. Am. Water Resour. Assoc. 46:278–298. doi:10.1111/j.1752-1688.2010.00420.x

Wagner-Riddle, C., A. Furon, N.L. McLaughlin, I. Lee, J. Barbeau, S. Jayasundara, G. Parkin, P. Von Bertoldi, and J. Warland. 2007. Intensive measurement of nitrous oxide emissions from a corn-soybean-wheat rotation under two contrasting management systems over 5 years. Global Change Biol. 13:1722–1736. doi:10.1111/j.1365-2486.2007.01388.x

Wang, J., L.M. Cardenas, T.H. Misselbrook, S. Cuttle, R.E. Thorman, and C. Li. 2012a. Modelling nitrous oxide emissions from grazed grassland systems. Environ. Pollut. 162:223–233.

Wang, J., X. Zhang, Y. Liu, X. Pan, P. Liu, Z. Chen, T. Huang, and Z. Xiong. 2012b. Modeling impacts of alternative practices on net global warming potential and greenhouse gas intensity from rice–wheat annual rotation in china. PLoS One 7(9):e45668.

Wang, L., J. Qiu, H. Tang, H. Li, C. Li, and E. Van Ranst. 2008a. Modelling soil organic carbon dynamics in the major agricultural regions of China. Geoderma 147(1–2):47–55.

Wang, X., P.W. Gassman, J.R. Williams, S. Potter, and A.R. Kemanian. 2008b. Modeling the impacts of soil management practices on runoff, sediment yield, maize productivity, and soil organic carbon using APEX. Soil Tillage Res. 101:78–88. doi:10.1016/j.still.2008.07.014

Wang, X., D.W. Hoffman, J.E. Wolfe, J.R. Williams, and W.E. Fox. 2009. Modeling the effectiveness of conservation practices at Shoal Creek Watershed, Texas, using APEX. Trans. ASABE 52:1181–1192. doi:10.13031/2013.27794

Wang, X., N. Kannan, C. Santhi, S.R. Potter, J.R. Williams, and J.G. Arnold. 2011a. Integrating APEX output for cultivated cropland with SWAT simulation for regional modeling. Trans. ASABE 54(4):1281–1298.

Wang, X., J.R. Williams, P.W. Gassman, C. Baffaut, R.C. Izaurralde, J. Jeong, and J.R. Kiniry. 2012c. EPIC and APEX: Model use, calibration, and validation. Trans. ASABE 55(4):1447–1462.

Wang, Y., G.J. Sun, F. Zhang, J. Qi, and C.Y. Zhao. 2011b. Modeling impacts of farming management practices on greenhouse gas emissions in the oasis region of China. Biogeosciences 8:2377–2390.

Weiss, M., R. Schaldach, J. Alcamo, and M. Flörke. 2009. Quantifying the human appropriation of fresh water by African agriculture. Ecol. Soc. 14(2):25.

West, J.B., S.E. Hobbie, and P.B. Reich. 2006. Effects of plant species diversity, atmospheric [$CO_2$], and N addition on gross rates of inorganic N release from soil organic matter. Global Change Biol. 12:1400–1408. doi:10.1111/j.1365-2486.2006.01177.x

West, J.S., J.A. Townsend, M. Stevens, and B.D.L. Fitt. 2012. Comparative biology of different plant pathogens to estimate effects of climate change on crop disease in Europe. Eur. J. Plant Pathol. 133:315–331. doi:10.1007/s10658-011-9932-x

Whitehead, D. 2011. Forests as carbon sinks-benefits and consequences. Tree Physiol. 31:893–902. doi:10.1093/treephys/tpr063

Whitney, K.D., and C.A. Gabler. 2008. Rapid evolution in introduced species, 'invasive traits' and recipient communities: Challenges for predicting invasive potential. Divers. Distrib. 14:569–580. doi:10.1111/j.1472-4642.2008.00473.x

Williams, J.R. 1995. The EPIC model. In: V. P. Singh, editor, Computer models of watershed hydrology. Water Resour. Publ., Highlands Ranch, CO. p. 909–1000.

Williams, J.W., R.C. Izaurralde, and E.M. Steglich. 2008. Agricultural Policy/Environmental Extender Model theoretical documentation. Version 0604. BREC Rep. 2008-17. Texas A&M AgriLife Blackland Research and Extension Center, Temple, TX.

Woldendorp, J.W. 1962. The quantitative influence of the rhizosphere on denitrification. Plant Soil 17:267–270. doi:10.1007/BF01376229

Wolf, B., R. Kiese, W. Chen, R. Grote, X. Zheng, and K. Butterbach-Bahl. 2012. Modeling $N_2O$ emissions from steppe in Inner Mongolia, China, with consideration of spring thaw and grazing intensity. Plant Soil. 350:297–310.

Wolf, B., X.H. Zheng, N. Brueggemann, W.W. Chen, M. Dannenmann, X.G. Han, M.A. Sutton, H.H. Wu, Z.S. Yao, and K. Butterbach-Bahl. 2010. Grazing-induced reduction of natural nitrous oxide release from continental steppe. Nature 464:881–884. doi:10.1038/nature08931

Woodall, C.W., L.S. Heath, and J.E. Smith. 2008. National inventories of down and dead woody material forest carbon stocks in the United States: Challenges and opportunities. For. Ecol. Manage. 256:221–228. doi:10.1016/j.foreco.2008.04.003

Woodall, C.W., C.H. Perry, and J.A. Westfall. 2012. An empirical assessment of forest floor carbon stock components across the United States. For. Ecol. Manage. 269:1–9. doi:10.1016/j.foreco.2011.12.041

Woodbury, P.B., L.S. Heath, and J.E. Smith. 2007a. Effects of land use change on soil carbon cycling in the conterminous United States from 1900 to 2050. Global Biogeochem. Cycles 21:GB3006. doi:10.1029/2007GB002950

Woodbury, P.B., J.E. Smith, and L.S. Heath. 2007b. Carbon sequestration in the US forest sector from 1990 to 2010. For. Ecol. Manage. 241:14–27. doi:10.1016/j.foreco.2006.12.008

Xing, H., E. Wang, C.J. Smith, D. Rolston, and Q. Yu. 2011. Modelling nitrous oxide and carbon dioxide emission from soil in an incubation experiment. Geoderma 167–168:328–339.

Xu, S., X. Shi, Y. Zhao, D. Yu, C. Li, S. Wang, M. Tan, and W. Sun. 2011a. Carbon sequestration potential of recommended management practices for paddy soils of China, 1980–2050. Geoderma 166:206–213.

Xu, S., X. Shi, Y. Zhao, D. Yu, S. Wang, M. Tan, W. Sun, and C. Li. 2012. Spatially explicit simulation of soil organic carbon dynamics in China's paddy soils. Catena. 92:113–121.

Xu, S.-X., X.-Z. Shi, Y.-C. Zhao, D.-S. Yu, S.-H. Wang, L.-M. Zhang, C.S. Li, and M.-Z. Tan. 2011b. Modeling carbon dynamics in paddy soils in Jiangsu Province of China with soil databases differing in spatial resolution. Pedosphere 21(6):696–705.

Xu, S., Y. Zhao, X. Shi, D. Yu, C. Li, S. Wang, M. Tan, and W. Sun. 2013. Map scale effects of soil databases on modeling organic carbon dynamics for paddy soils of China. Catena 104:67–76.

Xu, X., J. Tang, Z. Li, C. Liu, and W. Han. 2011c. Global warming potential of emissions from rice paddies in Northeastern China. Mitigation Adaptation Strategies Global Change. 16(6):721–731.

Yu, D.S., H. Yang, X.Z. Shi, E.D. Warner, L.M. Zhang, and Q.G. Zhao. 2011. Effects of soil spatial resolution on quantifying $CH_4$ and $N_2O$ emissions from rice fields in the Tai Lake region of China by DNDC model. Global Biogeochem. Cycles 25(2):GB2004.

Zhang, F., C. Li, Z. Wang, and H. Wu. 2006. Modeling impacts of management alternatives on soil carbon storage of farmland in Northwest China. Biogeosciences 3:451–466.

Zhang, L., D. Yu, X. Shi, D. Weindorf, L. Zhao, W. Ding, H. Wang, J. Pan, and C. Li. 2009. Quantifying methane emissions from rice fields in the Taihu Lake region, China by coupling a detailed soil database with biogeochemical model. Biogeosciences 6:739–749.

Zhang, L.M., D.S. Yu, X.Z. Shi, S.X. Xu, S.H. Wang, S.H. Xing, and Y.C. Zhao. 2012a. Simulation soil organic carbon change in China's Tai-Lake paddy soils. Soil Tillage Res. 121:1–9.

Zhang, W., X. Wang, and S. Wang. 2013. Addition of external organic carbon and native soil organic carbon decomposition: A meta-analysis. PLoS ONE 8:e54779. doi:10.1371/journal.pone.0054779

Zhang, X.Y., H. Jiang, Y.Q. Wang, Y. Han, M. Buchwitz, O. Schneising, and J.P. Burrows. 2011a. Spatial variations of atmospheric methane concentrations in China. Int. J. Remote Sens. 32:833–847. doi:10.1080/01431161.2010.517804

Zhang, Y., C.S. Li, X.J. Zhou, and B. Moore. 2002. A simulation model linking crop growthand soil biogeochemistry for sustainable agriculture. Ecol. Modell. 151:75–108. doi:10.1016/S0304-3800(01)00527-0

Zhang, Y., S. Su, F. Zhang, R. Shi, and W. Gao. 2012b. Characterizing spatiotemporal dynamics of methane emissions from rice paddies in Northeast China from 1990 to 2010. PLoS One 7(1):e29156.

Zhang, Y., Y.Y. Wang, S.L. Su, and C.S. Li. 2011b. Quantifying methane emissions from rice paddies in Northeast China by integrating remote sensing mapping with a biogeochemical model. Biogeosciences 8:1225–1235.

Ziska, L.H., D.M. Blumenthal, G.B. Runion, E.R. Hunt, Jr., and H. Diaz-Soltero. 2011. Invasive species and climate change: An agronomic perspective. Clim. Change 105:13–42. doi:10.1007/s10584-010-9879-5

Ziska, L.H., and J.A. Bunce. 2007. Predicting the impact of changing $CO_2$ on crop yields: Some thoughts on food. New Phytol. 175:607–618. doi:10.1111/j.1469-8137.2007.02180.x

# Appendix—Publications Reviewed for Simulation Model Comparison

Publication†

APEX

- Anomaa Senavirante et al. (2013)
- Cavero et al. (2012)
- Gassman et al. (2010)
- Kumar et al. (2011)
- Meki et al. (2011)
- Mudgal et al. (2010)
- Mugdal et al. (2012)
- Osei et al. (2008)
- Powers et al. (2011)
- Saleh et al. (2011)
- Tuppad et al. (2010)
- Wang et al. (2008b)
- Wang et al. (2009)
- Wang et al. (2011a)
- Wang et al. (2012c)

DAYCENT

- Adler et al. (2007)
- Chamberlain et al. (2011)
- Chianese et al. (2009)
- David et al. (2009)
- Davis et al. (2009)
- Davis et al. (2012)
- De Gryze et al. (2010)
- De Gryze et al. (2011)
- Del Grosso et al. (2005)
- Del Grosso et al. (2006)
- Del Grosso et al. (2008)
- Del Grosso et al. (2009a)
- Del Grosso et al. (2009b)
- Delgado et al. (2009)
- Foereid et al. (2012)
- Gao et al. (2013)
- Gramig et al. (2013)
- Hartman et al. (2011)
- Kim et al. (2009)
- Lee et al. (2011)
- Lee et al. (2012)
- Li et al. (2005)
- Miner et al. (2013)
- Paul et al. (2006)
- Reeling and Gramig (2012)
- Sendich et al. (2008)
- Smith et al. (2008a)
- Smith et al. (2012)
- Stehfest et al. (2007)
- Weiss et al. (2009)
- Xing et al. (2011)

DNDC

- Abdalla et al. (2011)
- Abdalla et al. (2012)
- Beheydt et al. (2007)
- Britz et al. (2009)
- Chirinda et al. (2011)
- David et al. (2009)
- Deng et al. (2011)
- Deng et al. (2011)
- Desjardins (2005)
- Desjardins et al. (2010)
- Dietiker et al. (2010)
- Farahbakhshazad et al. (2008)
- Follador et al. (2011)
- Fumoto et al. (2008)
- Giltrap et al. (2010)
- Giltrap et al. (2011)
- Gopalakrishnan et al. (2012)
- Guo et al. (2012)
- Hastings et al. (2010)
- Jagadeesh Babu et al. (2006)
- Kang et al. (2011)
- Kariyapperuma et al. (2011)
- Katayanagi et al. (2012)
- Kröbel et al. (2011)
- Leip et al. (2011)
- Li et al. (2005)
- Li et al. (2006)
- Li et al. (2010)
- Li et al. (2011)
- Li et al. (2012)
- Li et al. (2012)
- Liu et al. (2006)
- Ludwig et al. (2011a)
- Ludwig et al. (2011b)
- Nakawaga et al. (2008)
- Neufeldt et al. (2006)
- Ouyang et al. (2013)
- Pathak et al. (2005)
- Pathak et al. (2006)
- Qin et al. (2013)
- Qiu et al. (2009a)
- Qiu et al. (2009b)
- Qiu et al. (2011)
- Rafique et al. (2011)
- Shirato (2005)
- Sleutel et al. (2006)
- Smakgahn et al. (2009)
- Smith et al. (2008a)
- Smith et al. (2012)

**Appendix continued.**

DNDC continued.
- Syp et al. (2011)
- Tang et al. (2006)
- Terhoeven-Ureselmans et al. (2009)
- Tonitto et al. (2007a)
- Tonitto et al. (2007b)
- Wang et al. (2008a)
- Wang et al. (2011b)
- Wang et al. (2012a)
- Wang et al. (2012b)
- Wolf et al. (2012)
- Xu et al. (2011a)
- Xu et al. (2011b)
- Xu et al. (2011c)
- Xu et al. (2012)
- Xu et al. (2013)
- Yu et al. (2011)
- Zhang et al. (2006)
- Zhang et al. (2009)
- Zhang et al. (2011b)
- Zhang et al. (2012a)
- Zhang et al. (2012b)

ECOSYS
- Grant (1994a)
- Grant (1994b)
- Grant (1995)
- Grant (1997)
- Grant (1998)
- Grant (1999)
- Grant and Flanagan (2007)
- Grant and Heaney (1997)
- Grant and Pattey (1999)
- Grant and Pattey (2003)
- Grant and Pattey (2008)
- Grant and Rochette (1994)
- Grant et al. (1995)
- Grant et al. (1997)
- Grant et al. (2001a)
- Grant et al. (2001b)
- Grant et al. (2004a)
- Grant et al. (2004b)
- Grant et al. (2006)
- Grant et al. (2007)
- Grant et al. (2011)
- Grant et al. (2012a)
- Grant et al. (2012b)
- Li et al. (2004)
- Metivier et al. (2009)
- Parsinejad and Feng (2003)

RZWQM2
- Ahuja et al. (2010)
- Deb et al. (2012)
- Fang et al. (2010)
- Fang et al. (2012)
- Islam et al. (2012)
- Ko et al. (2010)
- Ko et al. (2012)
- Ma et al. (2009)
- Ma et al. (2012)
- Nielsen et al. (2012)
- Nolan et al. (2010)
- Qi et al. (2011)
- Qi et al. (2012)
- Qi et al. (2013)
- Saseendran et al. (2009)
- Saseendran et al. (2010)
- Saseendran et al. (2013)

† APEX, Agricultural Policy/Environmental eXtender; DAYCENT, Daily CENTURY; DNDC, Denitrification–Decomposition; EPIC, Environmental Policy Integrated Climate; RZWQM2, Root Zone Water Quality Model 2.

# Modeling GHG Emissions and Carbon Changes in Agricultural and Forest Systems to Guide Mitigation and Adaptation: Synthesis and Future Needs

**Stephen J. Del Grosso,* Lajpat R. Ahuja, and William J. Parton**

## Abstract

Agricultural production systems and land use change for agriculture and forestry are important sources of anthropogenic greenhouse gas (GHG) emissions. Recent commitments by the European Union, the United States, and China to reduce GHG emissions highlight the need to improve estimates of current emissions and mitigation potentials for the agricultural and forestry sectors of the economy. Because it is not feasible to measure GHG emissions and carbon stock changes for these systems beyond the plot scale, models of varying complexity have been applied at regional and larger scales to estimate current emissions and mitigation potentials for different management change scenarios. The chapters in this volume demonstrate that both simple, empirically based methods and mechanistic, process-based models have strengths and weaknesses depending on the stakeholder interest, scale of application, and other factors. Model outputs often agree closely with observed GHG fluxes and carbon stock changes associated with management of agricultural soils, forests, and livestock production systems, but uncertainty in model results remains high when they are applied to new or novel management–landscape combinations due to limitations in model driver data, parametrizations, and representation of key processes. These limitations can be addressed by devoting resources to increase the quality and quantity of model driver data, particularly in developing countries, establishing protocols for modeling methods and making algorithms more transparent, and integrating observations using different approaches (e.g., bottom-up chamber-based vs. top-down micrometeorological-based gas flux measurements) into standardized databases to facilitate rigorous model testing.

Abbreviations: ALU, Agriculture and Land Use National GHG Inventory; GHG, greenhouse gas; LULUCF, Land Use, Land Use Change, and Forestry; SOC, soil organic carbon.

Stephen J. Del Grosso and Lajpat R. Ahuja (Laj.Ahuja@ars.usda.gov), USDA-ARS, 2150 Centre Ave., Bldg. D, Suite 100, Fort Collins, CO 80526. William J. Parton, Natural Resource Ecology Lab, Colorado State University, Campus Delivery 1499, Fort Collins, CO 80523 (William.parton@colostate.edu). *Corresponding author (steve.delgrosso@ars.usda.gov).

doi:10.2134/advagricsystmodel6.2013.0008


Synthesis and Modeling of Greenhouse Gas Emissions and Carbon Storage in Agricultural and Forest Systems to Guide Mitigation and Adaptation
S. Del Grosso, L. Ahuja, and W. Parton, Editors
Advances in Agricultural Systems Modeling, Volume 6. Lajpat R. Ahuja, Series Editor

The agricultural sector is a key source of GHG emissions. Agricultural production systems directly contribute 10 to 12% of global anthropogenic GHG emissions, primarily nitrous oxide and methane (Flynn and Smith, 2010). If carbon dioxide emissions from land use change for agriculture (Friedlingstein et al., 2010), production and transport of farm inputs, and operation of farm equipment are included, agriculture's contribution to total human induced emissions increases to ~22% (Del Grosso and Cavigelli, 2012). Agricultural systems (mainly cropped and grazed soils) are the major anthropogenic source of $N_2O$ emissions, accounting for about 60 to 80% of the total due to human activity (Davidson and Kanter, 2014; Syakila and Kroeze, 2011). Agriculture is also an important source of $CH_4$ emissions. In the United States, enteric fermentation contributed about 25% of anthropogenic $CH_4$ emissions, and manure management systems were responsible for an additional 9% of emissions for this gas (USEPA, 2015). At the global scale, enteric $CH_4$ is responsible for about one-third of direct agricultural emissions (IPCC, 2014). Flooded soils used for rice (*Oryza sativa* L.) production are another important $CH_4$ source at the global scale (particularly southern and southeastern Asia), although they are responsible for a relatively small portion of US and European emissions because the amount of land used for rice production in these countries is small.

Land use change can have substantial impacts on C stocks in biomass and soil, and Land Use, Land Use Change, and Forestry (LULUCF) are key GHG source categories identified by the Intergovernmental Panel on Climate Change (IPCC, 2008). At the global scale, deforestation for livestock grazing and growing crops is an important GHG source (Del Grosso and Cavigelli, 2012). Most of the carbon loss from deforestation occurs in developing countries. In contrast, LULUCF in developed countries is close to C neutral, or, is a C sink, as is the case in the United States, where C storage from forest regrowth, tree plantations, and harvested wood products offset about 13% of national emissions (USEPA, 2015). Similarly, the C sink from forests and grasslands in Europe fully compensates the agricultural $N_2O$ and feedstock methane emissions (Schulze et al., 2009). In addition to C storage (or loss) in forest biomass, soil organic carbon (SOC) changes can also be an important C flux vector in some agricultural systems. For example, SOC typically declines when perennial vegetation is converted to annual cropping but often increases when management practices, such as reduced tillage intensity, crop residue retention and cover cropping are introduced, or when marginal land is taken out of production.

Because it is not feasible to measure agricultural emissions and carbon sequestration in biomass and soil beyond the research plot scale, models of varying complexity were developed. Initially, several empirical models were used to estimate GHG emissions and carbon stock changes at field and regional scales (e.g., IPCC, 2006). During the last half century, several process-based models have been developed for soil emissions, initially for research purposes, but then extended for practical applications at field to larger scales. Some of the most commonly used models belong to the DNDC (Li et al., 1992), Century-DayCent (Parton et al., 1998), and APSIM (Probert et al., 1998) families of models. Many other models have been used for different applications and often incorporate concepts from the above families, such as CQESTER (Gollany et al., 2013), Ecosys (Grant and Pattey, 1999), ExpertN (Engel and Priesack, 1993), FASSET (Olesen et al., 2002), NASA-Ames version of the CASA (Carnegie-Ames-Stanford approach) model (Potter et al., 1997), NOE (Hénault et al., 2005), RZWQM (Ahuja et al., 2000), and WNMM (Li et al., 2007). In addition to models for soil emissions, empirical (Hillier et al., 2011) and process-based models have recently been developed for emissions from livestock manure management (e.g., Li et al., 2012) and enteric fermentation (e.g., Bannink et al., 2011).

To formalize the distinctions between simple and more complex models, the Intergovernmental Panel on Climate Change (IPCC, 2006) defines three tier approaches to estimate GHG emissions from agricultural systems. Tier 1 methods that use empirical equations based on emission factors are the simplest, Tier 3 methods employing process-based models that represent the mechanisms responsible for emissions are the most complex, and Tier 2 methods are of intermediate complexity. Lower and higher tier methods have been used to estimate livestock emissions from enteric fermentation and manure management systems (e.g., Caro et al., 2014; Herrero et al., 2013), as well as soil emissions and forest C stock changes (e.g., USEPA, 2015; Wang et al., 2011; Herold and Skutsch, 2011). In addition, both empirical (Tiers 1 and 2) and process-based models (Tier 3) have been used to quantify the effects of alternative agricultural management practices on GHG emissions (e.g., Deng et al., 2013; Cheng et al., 2014; Millar et al., 2013; Li et al., 2012; Hillier et al., 2011), but both approaches have weaknesses. Tier 1 empirical methods are easy to apply, but produce estimates with very large uncertainty ranges, and the mitigation options that can be represented are highly limited. For example, the only soil $N_2O$ mitigation practice that can be addressed with the Tier 1 method is reduction in N inputs. Similarly, the only enteric $CH_4$ mitigation practice that can be addressed with the Tier 1 method is reduction in animal numbers. Process-based models (Tier 3) in theory are more accurate and can represent a wider suite of mitigation options. Consequently, they have recently been applied in different regions and climates.

The DNDC family of models has been used to a limited extent globally for GHG emission simulation and identifying management strategies for mitigation. In the United States, soil tillage, manure management, and fertilizer use were found to be the most important farming practices contributing to $N_2O$ emission at the national scale. However, their effects were multi-faceted and varied depending on soil and climate conditions. For example, while manure applications to substitute for N fertilizer enhance carbon storage in soils, they may also increase $N_2O$ emission (Li et al., 1996). On the other hand, manure application and retaining residues (assumedly with no tillage) in northern China decreased emissions (Li et al., 2010); this effect could vary with climate. Similarly, while no-till cultivation reduces nitrate N leaching, it may also increase $N_2O$ emissions in the United States (Farahbakhshazad et al., 2008). In Ireland, no- tillage decreased $N_2O$ emissions (Lugato et al., 2010). Deeper fertilizer application and the use of a nitrification inhibitor can reduce $N_2O$ emissions (Parkin and Hatfield, 2010). The Century–DayCent family of models has also been used widely for GHG emission and mitigation research. In the US Great Plains, Parton et al. (2015) used DayCent and other methods to estimate that regional emissions could be reduced by about 68% if recommended practices (use of no-till cultivation, enhanced efficiency fertilizers, and improved livestock feed) were adopted at rates of approximately 50%. At the national scale, no-till could mitigate 20% of US agricultural GHG emissions (Del Grosso et al., 2005). At the global scale, no-tillage combined with nitrification inhibitors resulted in 50% reduction in GHG emissions (7% with inhibitors alone) and 7% increase in crop yield (Del Grosso et al., 2009). Replacing fertilizers with manures had great potential for soil GHG flux reduction, followed by cover cropping and conservation tillage (De Gryze et al., 2010). Combining these practices led to the highest reductions in emissions (De Gryze et al., 2011). Recently, the APSIM model was modified for $N_2O$ emissions (Thorburn et al., 2010) and is being used to evaluate GHG emissions in cropping systems in Australia.

As these examples show, models are being applied to help quantify the mitigation potentials of different management practices. However, most, if not all of these model applications have a major weakness: uncertainty ranges for model results were not rigorously quantified. Model results are often highly uncertain, but uncertainty can be difficult to quantify. One reason for this is that the models have not been extensively tested for all practices and all the conditions in the regions for which they have been applied. Because of this lack of evidence-based confidence in process-based models to accurately represent the full suite of management options, results from these models have recently been combined with empirically derived equations to estimate mitigation potentials (Ogle et al., 2014). But, this hybrid approach also embodies substantial uncertainty, and interactions among different management practices and environmental conditions are not realistically represented.

To help reduce uncertainty in model predictions for GHG fluxes, much further research involving model development and testing is needed to accurately quantify the extent to which emissions from the agricultural sector can be reduced and help accomplish GHG mitigation goals. One reason it is difficult to quantify GHG emissions from agricultural systems is that the observations of these gas fluxes exhibit high spatial and temporal variability. Agricultural management practices (fertilizer and irrigation application, tillage, cover crop, and cropping system) interact with climate and soil conditions (Halvorson et al., 2010, 2011; Halvorson and Del Grosso, 2008, 2012) to control soil emissions in complex ways that are very difficult to characterize and simulate for accurate field-scale assessments. Similarly, forage chemistry and digestibility, intake rate, rumen microbial community, and other factors interact to control enteric $CH_4$ emissions from livestock production systems (Kebreab et al., 2016, this volume). Process-based models have shown great potential for quantifying these complex effects of alternative management on GHG emissions, at least when calibrated for particular location and conditions. However, the accuracy of model results when applied to new management–location combinations remains questionable. Improved understanding of the underlying factors that cause these variations across soils, climates, forage quality, ruminant processes, and management is needed to identify strategies for mitigation under specific conditions. For example, there is a particular need to quantify soil $N_2O$ emissions due to nitrification and denitrification processes separately.

The goal of this volume Advances in Agricultural Systems Modeling was to present the current state of science of modeling GHG emissions and carbon changes in agricultural and forest systems and to identify further improvements needed to more accurately predict the emissions and guide strategies for mitigation and adaptation. We have presented eight recent case studies and two methodological comparisons describing the application of models for assessing soil and enteric GHG emissions and forest C stock changes, which reflect the current state of modeling. The articles include studies in the United States, Canada, China, South Africa, Germany, Philippines, and the UK, conducted at different spatial and temporal scales. Note that modeling of emissions from manure management systems is not extensively covered in this volume.

Qin et al. (2016, this volume) presented measurement data on $N_2O$ emissions in the major crops at several different locations in the North China Plain, which show a large variation in the $N_2O$ emission factor depending on crop, weather, and N fertilizer rate. Nitrogen fertilization increased $N_2O$ emissions, but low yield-scaled emissions ($N_2O$ per unit of grain harvest) were obtained at intermediate fertilization rates. Grant et al. (2016, this volume) evaluated the capability of DNDC and DayCent to capture the long-term inter-annual variability in spring wheat grain yield, soil C change and trace gas emissions in continuous wheat and fallow–wheat cropping systems with different fertilizer options on three different soils in the Canadian prairie. Both models were effective in estimating yields under adequate N in dry to normal years. They also did well in estimating crop residue C inputs, soil C changes, and $N_2O$ emissions. The authors suggest model improvements and noted that a much higher portion of total $N_2O$ emissions were from denitrification in DNDC compared to DayCent. Gollany (2016, this volume) verified the ability of the CQESTR model to simulate SOC stocks, then applied the model to investigate the effects of tillage, cropping intensity and N fertilization on SOC changes in dryland wheat rotations, compared to grass pasture, over 30 yr of different climate change scenarios. Conventional tillage generally resulted in loss of soil C while no-tillage and increased cropping intensity (with residue cover) resulted in increase in soil C. Climate change tended to accelerate SOC losses and inhibit SOC gains. Del Grosso et al. (2016, this volume) compared the abilities of two process-based models (CQESTR and DayCent) and one empirical model (IPCC Tier 1 methodology) to represent observed SOC changes in a dryland wheat (*Triticum aestivum* L.) system in Oregon and an irrigated maize (*Zea mays* L.) system in Colorado. As expected, the process-based models matched observations more closely than the empirical model, but tended to underestimate SOC losses associated with conventional tillage and limited N amendments.

Smith and Heath (2016, this volume) evaluated a collection of models that convert the USDA Forest Inventory and Analysis program survey data to estimate all forest C components, and thence calculate the total C stock and change between surveys. The C stock changes show a large inter-annual variability, related to the uncertainty of the underlying data. Agriculture and Land Use National GHG Inventory (ALU) software is an advanced system that employs the IPCC inventory methods and has the functionality to assess GHG mitigation potentials. Ogle et al. (2016, this volume) provided two ALU case studies, for rice management in Philippines and livestock and manure management in South Africa. Burning rice straw can reduce GHG emissions, but returning straw to the soil is essential for maintaining long-term soil productivity. Enteric methane emission can be reduced by feed additives, and GHG emissions from manure can be reduced by more use of digesters that recover $CH_4$ and produce energy.

Klatt et al. (2016, this volume) quantified regional-scale parameter-induced uncertainty on $N_2O$ and nitrate leaching in agricultural soils of Saxony, Germany, using LandscapeDNDC (a combination of agricultural-DNDC and Forest-DNDC) and a joint parameter distribution, obtained by a Bayesian approach. The mean $N_2O$ emission rate calculated by this approach was slightly lower than the IPCC Tier 1 emission factor estimates, but the latter was well within the large uncertainty around the mean, CV of 87%. Similar results were obtained for nitrate leaching.

Kebreab et al. (2016, this volume) addressed the utility and limitations of the mathematical models used to estimate enteric methane emissions from livestock production. Whereas the driving variables of both the empirical models and mechanistic models are similar, empirical models cannot be extrapolated to explore mitigation options, and the mechanistic models are subject to errors of the input data and parameter estimates. They concluded that global collaboration is required to reach a consensus on methodological protocols. Hillier et al. (2016, this volume) surmised that different models should not be judged by their precision alone. To achieve desired goals, the ability of model results to allow communication and achieve understanding with multiple stakeholders is important as well. The authors illustrate these points with application of a range of models with varying complexities and describe the ELUM simplified meta-modeling approach.

Tonitto et al. (2016, this volume) presented a comprehensive review of current models and approaches to simulate GHG emissions in agriculture and forestry. The authors describe methods used to estimate $N_2O$ emissions and carbon storage changes, their strengths and weaknesses, and the observational data available to parameterize them, in relation to using the results for mitigation policy development. They show that limited observations of GHG fluxes and management effects, particularly for $N_2O$, limit our understanding of the process controls that are essential for modeling. Nonlinear responses to driving variables (e.g., soil water content), spatial and temporal variability, scale effects, and impacts of extreme events also contribute to modeling challenges. Soil C changes in both agriculture and forest systems are difficult to quantify compared to biomass C changes. Forest inventory data needed to accurately estimate forest biomass C changes are available mainly in advanced countries. These limitations can greatly influence the application of simulation results for GHG management. Policymakers and users should be informed of these limitations, and the researchers and modelers should address these limitations by, for example, quantifying uncertainty in model outputs. Modelers also need to improve documentation of model development, coding, and testing.

We propose the following list of future research and modeling needs:

1. Coordinate global efforts for more complete spatial and temporal measurements of GHG fluxes at field and greater scales by some simpler means, along with more data for related soil and plant growth processes, soil and plant parameters, and climate effects.
2. Organize GHG measurements along with model driver information into databases that are easily accessible to facilitate model building and testing (e.g., the GRACEnet/REAP data management system, Del Grosso et al., 2013).
3. Coordinate global efforts for more thorough evaluation of the models against the above data sets to fill knowledge gaps in the models and to improve existing concepts. Seek better understanding of the biophysical processes and factors that drive or influence the results, for example, nonlinear soil $N_2O$ emission responses to N amendments.
4. Document the model components for easy international sharing and further enhancements
5. Increase rigor in characterizing the basic and the soil-, climate-, and scale-dependent model parameters from the above analyses.
6. Improve understanding and quantification of the spatial and temporal changes and variability of processes and parameters, as well as interfield variability due to soil and topography differences.
7. Conduct comprehensive further research on how management influences key processes and investigate the impacts of extreme events to improve model algorithms and parameterizations.
8. Improve presentation of the model results to various stakeholders and policymakers. There is a strong need for bilateral exchange between modelers, experimentalists, and stakeholders on the knowledge, meta-data and its implementation in models, limitations and uncertainties in knowledge and models, and needs for further research and improvements.

# References

Ahuja, L.R., K.W. Rojas, J.D. Hanson, M.J. Shaffer, and L. Ma. 2000. Root Zone Water Quality Model—Modeling management effects on water quality and crop production. Water Resources Publ., Highlands Ranch, CO.

Bannink, A., M.W. Van Schijndel, and J. Dijkstra. 2011. A model of enteric fermentation in dairy cows to estimate methane emission for the Dutch National Inventory Report using the IPCC Tier 3 approach. Anim. Feed Sci. Technol. 166–167:603–618. doi:10.1016/j.anifeedsci.2011.04.043

Caro, D., S.J. Davis, S. Bastianoni, and K. Caldeira. 2014. Global and regional trends in greenhouse gas emissions from livestock. Clim. Change 126:203–216. doi:10.1007/s10584-014-1197-x

Cheng, K., S.M. Ogle, W.J. Parton, and G. Pan. 2014. Simulating greenhouse gas mitigation potentials for Chinese Croplands using the DAYCENT ecosystem model. Glob. Change Biol. 20(3):948–962. doi:10.1111/gcb.12368

Davidson, E.A., and D. Kanter. 2014. Inventories and scenarios of nitrous oxide emissions. Environ. Res. Lett. 9(10):105012. doi:10.1088/1748-9326/9/10/105012

De Gryze, S., J. Lee, S. Ogle, K. Paustian, and J. Six. 2011. Assessing the potential for greenhouse gas mitigation in intensively managed annual cropping systems at the regional scale. Agric. Econ. Environ. 144:150–158.

De Gryze, S., A. Wolf, S.R. Kaffka, J. Mitchell, D.E. Rolston, S.R. Temple, J. Lee, J. Six. 2010. Simulating greenhouse gas budgets of four California cropping systems under conventional and alternative management. Ecol. Appl. 20:1805–1819. doi:10.1890/09-0772.1

Del Grosso, S.J., and M.A. Cavigelli. 2012. Climate stabilization wedges revisited: Can agricultural production and greenhouse gas reduction goals be accomplished? Front. Ecol. Environ 10:571–578. doi:10.1890/120058

Del Grosso, S.J., H.T. Gollany, and M. Reyes-Fox. 2016. Simulating soil organic carbon stock changes in agroecosystems using CQESTR, DayCent, and IPCC Tier 1 methods. In: S.J. Del Grosso, L.R. Ahuja, and W.J. Parton, editors, Synthesis and modeling of greenhouse gas emissions and carbon storage in agricultural and forest systems to guide mitigation and adaptation. Advances in Agricultural Systems Modeling 6. ASA, CSSA, SSSA, Madison, WI. doi:10.2134/advagricsystmodel6.2013.0001.5

Del Grosso, S.J., A.R. Mosier, W.J. Parton, and D.S. Ojima. 2005. DAYCENT model analysis of past and contemporary soil N2O and net greenhouse gas flux for major crops in the USA. Soil Tillage Res. 83:9–24. doi:10.1016/j.still.2005.02.007

Del Grosso, S.J., D.S. Ojima, W.J. Parton, E. Stehfest, M. Heistemann, B. DeAngelo, and S. Rose. 2009. Global scale DAYCENT model analysis of greenhouse gas emissions and mitigation strategies for cropped soils. Global Planet. Change. 67:44–50. doi:10.1016/j.gloplacha.2008.12.006

Del Grosso, S.J., J.W. White, G. Wilson, B. Vandenberg, D.L. Karlen, R.F. Follett, J.M.F. Johnson, A.J. Franzluebbers, D.W. Archer, H. Gollany, M. Liebig, J. Ascough, M. Reyes-Fox, L. Pellack, J. Starr, N. Barbour, M. Gutwein, D. James. 2013. Introducing the GRACEnet/REAP Data Contribution, Discovery and Retrieval System. J. Environ. Qual. 42:1274–1280. doi:10.2134/jeq2013.03.0097

Deng, J., Z. Zhou, X. Zheng, and C. Li. 2013. Modeling impacts of fertilization alternatives on nitrous oxide and nitric oxide emissions from conventional vegetable fields in southeastern China. Atmos. Environ. 81:642–650. doi:10.1016/j.atmosenv.2013.09.046

Engel, T., and E. Priesack. 1993. Expert-N, a building block system of nitrogen models as resource for advice, research, water management and policy. In: H.J.P. Eijsackers and T. Hamers, editors, Integrated soil and sediment research: A basis for proper protection. Kluwer Academic Publishers, Dordrecht, The Netherlands. p. 503–507.

Farahbakhshazad, N., D. Dinnes, C. Li, D. Jaynes, and W. Salas. 2008. Modeling biogeochemical impacts of alternative management practices for a row-crop field in Iowa. Agric. Ecosyst. Environ. 123:30–48. doi:10.1016/j.agee.2007.04.004

Flynn, H., and P. Smith. 2010. Greenhouse gas budgets of crop production—Current and likely future trends. International Fertilizer Industry Assoc., Paris.

Friedlingstein, P.H.R., R.A. Houghton, G. Marland, et al. 2010. Update on $CO_2$ emissions. Nat. Geosci. 3:811–812. doi:10.1038/ngeo1022

Gollany, H. 2016. CQESTR simulation of dryland agroecosystem soil organic carbon changes under climate change scenarios. In: S.J. Del Grosso, L.R. Ahuja, and W.J. Parton, editors, Synthesis and modeling of greenhouse gas emissions and carbon storage in agricultural and forest systems to guide mitigation and adaptation. Advances in Agricultural Systems Modeling 6. ASA, CSSA, SSSA, Madison, WI. doi:10.2134/advagricsystmodel6.2013.0004

Gollany, H.T., A. Fortuna, M. Samuel, F.L. Young, W. Pan, and M. Pecharko. 2013. Estimating soil organic carbon accretion vs. sequestration using physicochemical fractionation and CQESTR simulation. Soil Sci. Soc. Am. J. 77:618–629. doi:10.2136/sssaj2012.0303

Grant, B.G., W.N. Smith, C.A. Campbell, R.L. Desjardins, R.L. Lemke, R. Kröbel, B.G. McConkey, E.G. Smith, and G.P. Lafond. 2016. Comparison of DayCent and DNDC models: Case studies using data from long-term experiments on the Canadian prairies. In: S.J. Del Grosso, L.R. Ahuja, and W.J. Parton, editors, Synthesis and modeling of greenhouse gas emissions and carbon storage in agricultural and forest systems to guide mitigation and adaptation. Advances in Agricultural Systems Modeling 6. ASA, CSSA, SSSA, Madison, WI. doi:10.2134/advagricsystmodel6.2013.0035

Grant, R.F., and E. Pattey. 1999. Mathematical modelling of nitrous oxide emissions from an agricultural field during spring thaw. Global Biogeochem. Cycles 13:679–694. doi:10.1029/1998GB900018

Halvorson, A.D., and S.J. Del Grosso. 2008. Nitrogen, tillage, and crop rotation effects on nitrous oxide emissions from irrigated cropping systems. J. Environ. Qual. 37:1337–1344. doi:10.2134/jeq2007.0268

Halvorson, A.D., and S.J. Del Grosso. 2012. Nitrogen source and placement effects on soil nitrous oxide emissions from no-till corn. J. Environ. Qual. 41:1349–1360. doi:10.2134/jeq2012.0129

Halvorson, A.D., S.J. Del Grosso, and F. Alluvione. 2010. Tillage and inorganic nitrogen source effects on nitrous oxide emissions from irrigated cropping systems. Soil Sci. Soc. Am. J. 74:436–445. doi:10.2136/sssaj2009.0072

Halvorson, A.D., S.J. Del Grosso, and C.P. Jantalia. 2011. Nitrogen source effects on soil nitrous oxide emissions from strip-till corn. J. Environ. Qual. 40:1775–1786. doi:10.2134/jeq2011.0194

Hénault, C., F. Bizouard, P. Laville, B. Gabrielle, B. Nicoullaud, J.C. Germon, and P. Cellier. 2005. Predicting in situ soil $N_2O$ emission using NOE algorithm and soil database. Global Change Biol. 11:115–127. doi:10.1111/j.1365-2486.2004.00879.x

Herold, M., and M. Skutsch. 2011. Monitoring, reporting and verification for national REDD+ programmes: Two proposals. Environ. Res. Lett. 6:014002. doi:10.1088/1748-9326/6/1/014002

Herrero, M., P. Havlík, H. Valin, A. Notenbaert, M.C. Rufino, P.K. Thornton, M. Blümmel, F. Weiss, D. Grace, and M. Obersteiner. 2013. Biomass use, production, feed efficiencies, and greenhouse gas emissions from global livestock systems. Proc. Natl. Acad. Sci. 110: 20888–20893. doi:10.1073/pnas.1308149110

Hillier, J., M. Abdalla, J. Bellarby, F. Albanito, A. Datta, M. Dondini, N. Fitton, P. Hallett, A. Hastings, E. Jones, M. Kuhnert, D. Nayak, M. Pogson, M. Richards, J. Smith, S. Vetter, J. Yeluripati, and P. Smith. 2016. Mathematical modeling of greenhouse gas emissions from agriculture for different end users. In: S.J. Del Grosso, L.R. Ahuja, and W.J. Parton, editors, Synthesis and modeling of greenhouse gas emissions and carbon storage in agricultural and forest systems to guide mitigation and adaptation. Advances in Agricultural Systems Modeling 6. ASA, CSSA, SSSA, Madison, WI. doi:10.2134/advagricsystmodel6.2013.0038

Hillier, J., C. Walter, D. Malin, T. Garcia-Suarez, L. Mila-i-Canals, and P. Smith. 2011. A farm-focused calculator for emissions from crop and livestock production. Environ. Model. Softw. 26:1070–1078. doi:10.1016/j.envsoft.2011.03.014

IPCC. 2006. $N_2O$ Emissions from managed soils, and $CO_2$ emissions from lime and urea application. In: Intergovernmental Panel on Climate Change Guidelines for National Greenhouse Gas Inventories. Volume 4: Agriculture, forestry and other land use. http://www.ipcc-nggip.iges.or.jp/public/2006gl/pdf/4_Volume4/V4_11_Ch11_N2O&CO2.pdf (accessed 6 Oct. 2015).

IPCC. 2008. 2006 IPCC Guidelines for National Greenhouse Gas Inventories—A primer. Prepared by the National Greenhouse Gas Inventories Programme, H.S. Eggleston, K. Miwa, N. Srivastava, and K. Tanabe, editors. IGES, Japan. http://www.ipcc-nggip.iges.or.jp/support/Primer_2006GLs.pdf (accessed 6 Oct. 2015).

IPCC. 2014. Climate Change 2014: Synthesis report. Contribution of Working Groups I, II and III to the Fifth Assessment Report of the Intergovernmental Panel on Climate Change. R.K. Pachauri and L.A. Meyer, editors. IPCC, Geneva, Switzerland. https://www.ipcc.ch/pdf/assessment-report/ar5/syr/AR5_SYR_FINAL_Front_matters.pdf (accessed 6 Oct. 2015).

Kebreab, E., L. Tedeschi, J. Dijkstra, J.L. Ellis, A. Bannink, and J. France. 2016. Modeling greenhouse gas emissions from enteric fermentation. In: S.J. Del Grosso, L.R. Ahuja, and W.J. Parton, editors, Synthesis and modeling of greenhouse gas emissions and carbon storage in agricultural and forest systems to guide mitigation and adaptation. Advances in Agricultural Systems Modeling 6. ASA, CSSA, SSSA, Madison, WI. doi:10.2134/advagricsystmodel6.2013.0006

Klatt, S., D. Kraus, K.-H. Rahn, C. Werner, R. Kiese, K. Butterbach-Bahl, and E. Haas. 2016. Parameter-induced uncertainty quantification of regional $N_2O$ emissions and $NO_3$ leaching using the biogeochemical model LandscapeDNDC. In: S.J. Del Grosso, L.R. Ahuja, and W.J. Parton, editors, Synthesis and modeling of greenhouse gas emissions and carbon storage in agricultural and forest systems to guide mitigation and adaptation. Advances in Agricultural Systems Modeling 6. ASA, CSSA, SSSA, Madison, WI. doi:10.2134/advagricsystmodel6.2013.0001

Li, C., S. Frolking, and T.A. Frolking. 1992. A model of nitrous oxide evolution from soil driven by rainfall events: I. Model structure and sensitivity. J. Geophys. Res. 97:9759–9776. doi:10.1029/92JD00509

Li, C., V. Narayanan, and R.C. Harriss. 1996. Model estimates of nitrous oxide emissions from agricultural lands in the United States. Global Biogeochem. Cycles 10:297–306. doi:10.1029/96GB00470

Li, C., W. Salas, R. Zhang, C. Krauter, A. Rotz, and F. Mitloehner. 2012. Manure-DNDC: A biogeochemical process model for quantifying greenhouse gas and ammonia emissions from livestock manure systems. Nutr. Cycl. Agroecosyst. 93:163–200. doi:10.1007/s10705-012-9507-z

Li, H., J.J. Qiu, L.G. Wang, H.J. Tang, C.S. Li, and E. Van Ranst. 2010. Modelling impacts of alternative farming management practices on greenhouse gas emissions from a winter wheat–maize rotation system in China. Agric. Ecosyst. Environ. 135:24–33. doi:10.1016/j.agee.2009.08.003

Li, Y., D.L. Chen, R.E. White, J.B. Zhang, B.G. Li, Y.M. Zhang, Y.F. Huang, and R. Edis. 2007. A spatially referenced Water and Nitrogen Management Model (WNMM) for (irrigated) intensive cropping systems in the North China Plain. Ecol. Modell. 203:395–423. doi:10.1016/j.ecolmodel.2006.12.011

Lugato, E., M. Zuliani, G. Alberti, G. DelleVedove, B. Gioli, F. Miglietta, and A. Peressotti. 2010. Application of DNDC biogeochemistry model to estimate greenhouse gas emissions from Italian agricultural areas at high spatial resolution. Agric. Ecosyst. Environ. 139(4):546–556. doi:10.1016/j.agee.2010.09.015

Millar, N., G.P. Robertson, A. Diamant, R.J. Gehl, P.R. Grace, and J.P. Hoben. 2013. Quantifying $N_2O$ emissions reductions in US agricultural crops through N fertilizer rate reduction. Verified Carbon Standard, Washington, DC.

Ogle, S.M., P.R. Adler, F.J. Breidt, S.J. Del Grosso, J. Derner, A. Franzluebbers, M. Liebig, B. Linquist, G.P. Robertson, M. Schoeneberger, J. Six, C. Van Kessel, R. Venterea, and T. West. 2014. Quantifying Greenhouse Gas Sources and Sinks in Cropland and Grazing Land Systems. In: M. Eve et al., editors, Quantifying greenhouse gas fluxes in agriculture and forestry: Methods for entity-scale inventory. Tech. Bull. 1939. USDA, Office of the Chief Economist, Washington, DC.

Ogle, S.M., S. Spencer, M. Hartman, L. Buendia, L. Stevens, L. du Toit, and J. Witi. 2016. Developing national baseline GHG emissions and analyzing mitigation potentials for agriculture and forestry using an advanced national GHG inventory software system. In: S.J. Del Grosso, L.R. Ahuja, and W.J. Parton, editors, Synthesis and modeling of greenhouse gas emissions and carbon storage in agricultural and forest systems to guide mitigation and adaptation. Advances in Agricultural Systems Modeling 6. ASA, CSSA, SSSA, Madison, WI. doi:10.2134/advagricsystmodel6.2013.0009

Olesen, J.E., B.M. Petersen, J. Berntsen, S. Hansen, P.D. Jamieson, and A.G. Thomsen. 2002. Comparison of methods for simulating effects of nitrogen on green area index and dry matter growth in winter wheat. Field Crops Res. 74:131–149. doi:10.1016/S0378-4290(01)00204-0

Parkin, T.B., and J.L. Hatfield. 2010. Influence of nitrapyrin on $N_2O$ losses from soil receiving fall-applied anhydrous ammonia. Agric. Ecosyst. Environ. 136:81–86. doi:10.1016/j.agee.2009.11.014

Parton, W.J., M.P. Gutmann, E.R. Merchant, M.D. Hartman, P.R. Adler, F.M. McNeal, and S.M. Lutz. 2015. Measuring and mitigating agricultural greenhouse gas production in the US Great Plains, 1870–2000. Proc. Natl. Acad. Sci. 112(34):E4681-8. doi:10.1073/pnas.1416499112.

Parton, W.J., M. Hartman, D.S. Ojima, and D.W. Schimel. 1998. DAYCENT and its land surface submodel: Description and testing. Global Planet. Change 19:35–48. doi:10.1016/S0921-8181(98)00040-X

Potter, C.S., R.H. Riley, and S.A. Klooster. 1997. Simulation modeling of nitrogen trace gas emissions along an age gradient of tropical forest soils. Ecol. Modell. 97:179–196. doi:10.1016/S0304-3800(96)01903-5

Probert, M.E., J.P. Dimes, B.A. Keating, R.C. Dalal, and W.M. Strong. 1998. APSIM's water and nitrogen modules and simulation of the dynamics of water and nitrogen in fallow systems. Agric. Syst. 56(1):1–28. doi:10.1016/S0308-521X(97)00028-0

Qin, S., C. Hu, L.R. Ahuja, Y. Zhang, and Y. Wang. 2016. Nitrous oxide emission from uplands in northern China. In: S.J. Del Grosso, L.R. Ahuja, and W.J. Parton, editors, Synthesis and modeling of greenhouse gas emissions and carbon storage in agricultural and forest systems to guide mitigation and adaptation. Advances in Agricultural Systems Modeling 6. ASA, CSSA, SSSA, Madison, WI. doi:10.2134/advagricsystmodel6.2013.0006.5

Schulze, E.D., S. Luyssart, P. Ciais, A. Freibauer, I.A. Janssens, et al. 2009. Importance of methane and nitrous oxide for Europe's terrestrial greenhouse-gas balance. Nat. Geosci. 2:842–850. doi:10.1038/ngeo686

Smith, J.E., and Linda S. Heath. 2016. Measuring and modeling carbon stock change estimates for us forests and uncertainties from apparent inter-annual variability. In: S.J. Del Grosso, L.R. Ahuja, and W.J. Parton, editors, Synthesis and modeling of greenhouse gas emissions and carbon storage in agricultural and forest systems to guide mitigation and adaptation. Advances in Agricultural Systems Modeling 6. ASA, CSSA, SSSA, Madison, WI. doi:10.2134/advagricsystmodel6.2013.0005

Syakila, A., and C. Kroeze. 2011. The global nitrous oxide budget revisited. Greenhouse Gas Meas. Manage. 1:17–26. doi:10.3763/ghgmm.2010.0007

Thorburn, P.J., J.S. Biggs, K. Collins, and M.E. Probert. 2010. Using the APSIM model to estimate nitrous oxide emissions from diverse Australian sugarcane production systems. Agric. Ecosyst. Environ. 136:343–350. doi:10.1016/j.agee.2009.12.014

Tonitto, C., N.P. Gurwick, and P.B. Woodbury. 2016. Quantifying greenhouse gas emissions from agricultural and forest landscapes for policy development and verification. In: S.J. Del Grosso, L.R. Ahuja, and W.J. Parton, editors, Synthesis and modeling of greenhouse gas emissions and carbon storage in agricultural and forest systems to guide mitigation and adaptation. Advances in Agricultural Systems Modeling 6. ASA, CSSA, SSSA, Madison, WI. doi:10.2134/advagricsystmodel6.2013.0007

USEPA. 2015. Inventory of US greenhouse gas emissions and sinks: 1990–2013. USEPA, Office of Atmospheric Programs, Washington, DC.

Wang, J., L.M. Cardenas, T.H. Misselbrook, and S. Gilhespy. 2011. Development and application of a detailed inventory framework for estimating nitrous oxide and methane emissions from agriculture. Atmos. Environ. 45:1454–1463. doi:10.1016/j.atmosenv.2010.12.014

Printed and bound by CPI Group (UK) Ltd, Croydon, CR0 4YY
16/09/2026
14963421-0002